Joseph Meites

Monographs on Endocrinology

Volume 3

Edited by

F. Gross, Heidelberg · A. Labhart, Zürich

T. Mann, Cambridge · L. T. Samuels, Salt Lake City

J. Zander, München

F. G. Sulman

Hypothalamic Control of Lactation

In Collaboration with

M. Ben-David, A. Danon, S. Dikstein, Y. Givant
K. Khazen, J. Mishkinsky-Shani, I. Nir, C. P. Weller

With 58 Figures

Springer-Verlag New York · Heidelberg · Berlin 1970

Department of Applied Pharmacology
The Hebrew University
School of Pharmacy
Jerusalem, Israel

This work is subject to copyright. All rights are reserved, whether the whole or part of the material is concerned, specifically those of translation, reprinting, re-use of illustrations, broadcasting, reproduction by photocopying machine or similar means, and storage in data banks.

Under § 54 of the German Copyright Law where copies are made for other than private use, a fee is payable to the publisher, the amount of the fee to be determined by agreement with the publisher.

The use of general descriptive names, trade names, trade marks etc. in this publication, even if the former are not especially identified, is not to be taken as a sign that such names, as understood by the Trade Marks and Merchandise Marks Act, may accordingly by used freely by anyone.

© by Springer-Verlag Berlin · Heidelberg 1970. Library of Congress Catalog Card Number 70—125281. Printed in Germany.
Title No. 3973

Preface

The contents of this monograph "Hypothalamic Control of Lactation" are the result of studies by a nine-man team which has worked together for nine years. The topic is so fascinating that new research students are attracted to it every year, and the circle is growing steadily. This is, in fact, part of a much more extensive research field to which our Department is applying itself at present, *viz.* "Hypothalamic Pharmacology".

Our work in this sphere was made possible by a generous grant from the United States Department of Agriculture, which helped us during the first 5 years of research with Grant No. FG-IS-147, Project No. A10-AH-3, and recently with another grant for a further 5-year period, No. FG-IS-259, Project No. A10-AH-33. Our thanks are due to Dr. JOEL BITMAN, Hormone Physiology Leader of the USDA Agricultural Research Service at Beltsville, Md., the sponsor of this research.

As this monograph contains mainly references connected with our own work, the reader who wishes to make a survey of lactation should study the references cited in the bibliography, such as COWIE (1961, 1966), COWIE and FOLLEY (1957, 1961), FOLLEY (1952, 1956), JACOBSOHN (1961), LYONS and DIXON (1966), MEITES (1959 a, 1961, 1966), MEITES and NICOLL (1966).

In this monograph the individual team members have each reported on their own special field of research. Prof. H. Z. WINNIK, Head of our Psychiatric Department who has instigated, and Prof. J. ZANDER, Head of the Universitäts-Frauenklinik in Munich, who has encouraged our studies, deserve our special gratitude. Our work would not have been possible without the untiring devotion of Mrs. SHEILA WOOLFSON.

Special thanks are due to Miss YOCHEVED SUSSMANN, who has been editing and proof-reading our papers for many years past and has now devoted herself to putting this monograph into proper shape. Finally, we wish to thank Dr. H. GÖTZE and Miss E. PFISTERER of the Springer-Verlag Heidelberg, for their invaluable advice and guidance.

Jerusalem, August 1970   F. G. SULMAN

Contents

Part I
Physiological Regulation of Lactation

Chapter 1.	Mechanism of Lactation	1
Chapter 2.	Hypothalamus-Pituitary Axis	4
	a) Hypothalamic Control of Prolactin Secretion .	4
	b) Prolactin-Inhibiting Factor (PIF)	7
	c) Prolactin-Releasing Factor (PRF)	13
	d) Hypothalamic Feedback of Prolactin . . .	17
	e) Effect of Steroids	23
	f) Interrelation between Pituitary Tropins . . .	24
	g) Suckling	25
	h) Conclusions	28
Chapter 3.	Prolactin	29
	a) Introduction	29
	b) Storage and Release	30
	c) Properties	30
	d) Prolactin in Body Fluids	32
	e) Prolactin in Pituitary Tumors	32
	f) Effect of Prolactin on Mammary Growth . .	33
	g) Levels of Pituitary Prolactin in Various Stages of Lactation	34
	h) Theories on the Mechanism of Onset of Lactation	36
	i) Conclusion	37
Chapter 4.	Growth Hormone	37
Chapter 5.	Gonadotropins and Sex Steroids	42
Chapter 6.	Adrenocorticotropin and Corticosteroids . . .	44
Chapter 7.	Thyrotropin and Thyroid Hormones	48
Chapter 8.	Oxytocin and Vasopressin	49
Chapter 9.	Parathormone and Calcitonin	52

Chapter 10. Insulin and Glucagon 53
Chapter 11. Pineal Gland 54
Chapter 12. Placental and Amniotic Fluid 55

Part II

Pharmacological Regulation of Lactation

Chapter 13. Hypothalamic Lactation Produced by Psychopharmaca 59
    a) History 59
    b) Dissociation of Mammotropic and Sedative Effects 61
    c) Biological Dissociation through Combined Treatment with Tranquilizers and Psychostimulants 61
    d) Chemical Dissociation by Molecular Changes of Tranquilizers 63
    e) Studies in Hypophysectomized Rats 64
    f) Conclusions 67

Chapter 14. Mammotropic Effect of Phenothiazine Derivatives 67
    a) History 67
    b) Mammotropic Index (MTI) 68
    c) Body Temperature Drop (BTD) 69
    d) Screening of Phenothiazine Derivatives . . . 69
    e) Screening of Phenothiazine Sulfoxides and Sulfodioxides 69
    f) Stimulation of Synthesis and Release of Prolactin by Perphenazine 88
    g) Intracerebral Implantation of Phenothiazine Derivatives 94
    h) Mammotropic Effect of Fluphenazine Enanthate in the Rat 97
    i) Comparative Study of Phenothiazine Effect on Hypothalamic Prolactin-Inhibiting Factor and MSH-Releasing Factor 100
    j) Conclusions 101

Chapter 15.	Mammotropic Effect of Phenothiazine-like Compounds	102
Chapter 16.	Mammotropic Effect of Reserpine Derivatives	113
Chapter 17.	Mammotropic Effect of Butyrophenones	120
Chapter 18.	Mammotropic Effect of Miscellaneous Psychotropic Drugs	131
Chapter 19.	Structure-Activity Relationship of Mammotropic Drugs and Receptor Theory	133
Chapter 20.	Pituitary-Ovary Axis in Hypothalamic Lactation	141
Chapter 21.	Pituitary-Adrenal Axis in Hypothalamic Lactation	146
Chapter 22.	Pituitary-Thyroid Axis in Hypothalamic Lactation	150
	a) Role of Thyroid in vivo and in vitro	150
	b) Thyroid and Hypothalamic Mammary Growth	151
	c) Thyroid and post-partum Galactopoiesis	152
	d) Conclusion	152
Chapter 23.	Prolactin Release by Hypothalamus of Nursing Rat Mothers	153
Chapter 24.	Prolactin as Luteotropic Hormone	156

Part III
Prolactin Assay

Chapter 25.	In vitro Methods	162
	a) Effect of Hypothalamus	162
	b) Effect of Pituitary	164
	c) Effect of Steroids	164
	d) Effect of Neurohormones	166
	e) Effect of Phenothiazines	166
	f) Effect of Reserpine	167
	g) Effect of Thyroid	167
	h) Effect of Oxytocin	167
	i) Conclusion	168
Chapter 26.	In vivo Methods	169
	a) Rabbit Tests	169
	α) Rabbit Intraductal Method	169
	β) Rabbit Intracerebral Test	170

	b)	Rat Tests	170
		α) Mammotropic Index in Female Rats	170
		β) Milk Yield Test in post-partum Rats	170
	c)	Pigeon Tests	173
		α) Systemic Pigeon Crop Gland Weight Method	173
		β) Intradermal Pigeon Crop Gland Method	174
	d)	Concentration Methods	175
Chapter 27.	Radioimmunoassay		179

Part IV
Problems of Lactation

Chapter 28.	Lactation in Humans	182
	a) Lactation and post-partum Amenorrhea	182
	b) Inhibition of Lactation and Breast Engorgement	182
	c) Persistence of Lactation and Menstrual Cycle	185
	d) Sterility during Lactation	186
	e) Galactorrhea as a General Syndrome	187
	f) Forbes-Albright-Castillo Syndrome	188
	g) Chiari-Frommel Syndrome	189
	h) Sheehan's Syndrome	191
	i) Hypothalamic Lactation	191
	j) Promotion of post-partum Lactation	191
	k) Gynecomastia	192
	α) Morphology	192
	β) Physiologic States	192
	γ) Pathological Conditions	193
	l) Conclusion	194
Chapter 29.	Conclusions	195
Chapter 30.	Research Problems	196
References		199
Subject Index		231

Abbreviations

The following abbreviations were used in this book:

ACTH	=	Adrenocorticotropic Hormone
AP	=	Anterior Pituitary
BIT	=	Barbitone Sleep Induction Time
BTD	=	Body Temperature Drop
CPE	=	Cataleptic Effect
CRF	=	Corticotropin (ACTH)-Releasing Factor
FRF	=	FSH-Releasing Factor
FSH	=	Follicle Stimulating Hormone
HCG	=	Human Chorionic Gonadotropin (LH)
HGH	=	Human Growth Hormone (STH)
HPL	=	Human Placental Lactogen
IU	=	International Unit
LH	=	Luteinizing Hormone
LRF	=	LH-Releasing Factor
LTH	=	Prolactin, Lactotropic Hormone, Luteotropic Hormone
ME	=	Median Eminence: The center of the portal vessel-chemo-transmitters (releasing or inhibiting factors). It consists of a neural component (the infundibulum of the hypothalamus), a vascular component (the hypophysial-portal capillaries and veins) and an epithelial component (the pars tuberalis of the adenohypophysis).
MIF	=	MSH-Inhibiting Factor
MRF	=	MSH-Releasing Factor
MSH	=	Melanophore Stimulating Hormone
MTI	=	Mammotropic Index (cf. Chapter 26)
MU	=	Mouse Unit
PIF	=	Prolactin-Inhibiting Factor
PMS	=	Pregnant Mare Serum Gonadotropin (FSH)
PRF	=	Prolactin-Releasing Factor
PTS	=	Eyelid Ptosis
PTU	=	Propylthiouracil

RPL	=	Rat Placental Lactogen
RU	=	Rat Unit
SAR	=	Structure-Activity Relationship
SRF	=	STH-Releasing Factor
STH	=	Somatotropic Hormone, Growth Hormone
T-3	=	Triiodothyronine
TRF	=	TSH-Releasing Factor
TSH	=	Thyroid Stimulating Hormone

Part I

Physiological Regulation of Lactation

Chapter 1

Mechanism of Lactation

The mechanism of lactation is poorly understood. For this reason no hormonal preparation for stimulation of lactation in man or animals has as yet been accepted. The following working hypothesis and nomenclature is proposed by us:

Stages of Lactation	↓ *Active Hormones*
Preparation of milk gland Mammogenesis	Pituitary and later placental FSH and LH (in rodents LTH) Estradiol and progesterone
Initiation of lactation Colostrogenesis	Prolactin-inhibiting factor depressed Prolactin-releasing factor stimulated Removal of placental hormones
Secretion of milk Lactogenesis	Pituitary LTH (prolactin) Pituitary ACTH and corticosteroids
Maintenance of lactation Galactopoiesis	Pituitary STH: intrinsic lactogenic effect and augmentation of prolactin Pituitary TSH, tri-iodo-thyronine and thyroxine
Promotion of excretion Ejection: "Let-Down"	Oxytocin: contraction of mammary myoepithelium and slight stimulation of prolactin Neurohormones inhibit "let-down"

The main problem derives from the fact that, while the common hormone reactions, e. g. ovulation etc., are brought about by hyperemia of the target organ and enzymatical changes involving low-energy processes which can be induced by one hormone alone, lactation additionally requires the action of many metabolic hormones which have to mobilize casein, butter fat, lactose, calcium etc. for the excessive requirements of milk production. It is obvious that this goal cannot be achieved by one hormone alone, i. e. prolactin, without the aid of STH, ACTH, TSH, parathyroid hormone, etc. A harmonious concurrence of all these hormones is apparently only possible through appropriate stimulation by the "conductor of such a symphony"—the hypothalamus.

With this approach in mind we studied the different hormone mechanisms involved. The three main objectives of that study can be summarized as follows:

a) contribution to the basic knowledge of the mechanism of lactation by interplay of all body hormones;

b) developing exact methods for assay of all hormones involved in the different species;

c) stimulation of lactation in animals and women postpartum.

Experimental data strongly suggest that the mode of action of the ovarian hormones on mammary tissue may be direct as well as indirect. Today we know that high doses of LTH and STH can substitute for the effect of estrogen + progesterone (cf. Chapters 5 and 20).

The action of the steroids on lactation is only comprehensible if we assume the existence of an acceptor or receptor for them, both in the brain (hypothalamus and hypophysis) and in the mammary gland. As we shall show in Chapter 19, these receptors have a striking similarity with those required for the modern psychopharmaca which elicit lactation, with one difference: Lactation elicited in women by psychopharmaca—apparently induced by impairment of a prolactin-inhibiting factor (PIF) in the posterior hypothalamus—cannot be suppressed by steroids, which, as it now seems, cannot enter a receptor occupied by specific psychopharmaca. The same holds for the Chiari-Frommel syndrome and Ahumada-del Castillo syndrome. These are called forth by a disturbance of the PIF center and do not react to steroids (Chapter 28). On the other hand, puerperal lactation is apparently induced by a prolactin-releasing factor (PRF)—its area in the anterior hypothalamus is extremely sensitive to estrogen; it can be stimulated by it and it can also be suppressed by an overdose of estrogens. Thus, it seems that a double-center theory would explain all the manifold problems of hypothalamic lactation described in this monograph. Release or inhibition of additional hypophysiotropins can further contribute to lactation (DIKSTEIN and SULMAN, 1966).

A new aspect of hypothalamic lactation is its dependence upon the temporal cortex of the brain. Bilateral lesions produced by suction in the basal portion of the temporal lobes in ovariectomized estrogen-primed rabbits resulted in milk secretion (MENA and BEYER, 1968). This effect was apparent within one week after the operation. The common area destroyed by these lesions included the entorhinal cortex and the ventral part of the amygdala. Control lesions in the neocortex and olfactory bulbs did not induce this effect. Using stereotaxic approaches, we have not been able to confirm these results which—if correct—would suggest that the temporal lobes influence the hypothalamic pituitary systems.

Summarizing, it seems that the difficulty of producing lactation by prolactin alone stems from the necessity of a harmonious interplay of all hypophysiotropins. This can only be obtained by suitable stimulation (or suppression) originating in the hypothalamus. The importance of the temporal lobe of the cortex and its ability to evoke feelings of motherhood in animals and humans (cf. Solomon's judgement: I Kings: 3: 16—28) needs further study.

Peripheral hormonal factors are no less important for regulation of mammary gland growth and function. Yet, the most vital for mammary proliferation are the ovarian hormones—estrogen and progesterone. Normal cyclic changes in these two hormonal components govern the normal cyclic changes in mammary gland development associated with estrus or menstrual cycle. The same changes can be reproduced experimentally by exogenous hormonal treatment. In this respect we can distinguish between two different types of reaction: Animals such as the mouse, rat, rabbit or cat will respond to physiological amounts of estrogen by proliferation of the duct system, while for lobulo-alveolar growth progesterone is also required. Detailed studies on the optimal amounts and proportions of the two hormones have been carried out by many researchers (for details cf. D. JACOBSOHN, 1961), and are obviously different according to the species and specific strain studied. A different kind of reaction is observed in the guinea pig, goat and cow. In these species estrogen alone has been shown to cause both ductal and lobulo-alveolar development. Closer observation reveals, however, that even here treatment with estrogen alone results in anomalous gland structures (COWIE, FOLLEY, MALPRESS and RICHARDSON, 1952). For optimal mammary growth a combination of estrogen and progesterone is essential.

In still other species, such as the bitch and the ferret, estrogen alone does not even produce ductal growth.

A further point to be remembered, apart from the species differences, are the differences in the individual response of animals belonging to the same species and strain. This applies both to natural and to experimental mammary development. Lactation studies should, therefore, preferably be carried out on estrogen-primed animals. The dose given varies between 8 and 10 µg daily per rat for 10 days. We prefer the lower dosage since we found that priming with 10 µg per day may give unspecific reactions. Thus, NICOLL, TALWALKER and MEITES (1960) obtained lactation after non-specific stresses, such as injection of formalin, exposure to extreme cold or heat, restraint, etc., using 10 µg for daily priming. They also reported lactation following injection of meprobamate after identical priming, whereas this substance proved negative in our set-up when only 8 µg/day was used for priming (BEN-DAVID, DIKSTEIN and SULMAN, 1965).

In short: mammary physiology is similar in all mammals, but hormone requirements for mammary development vary according to species. For experimental mammogenesis estrogen priming at subthreshold doses is of utmost importance.

Chapter 2

Hypothalamus-Pituitary Axis

a) Hypothalamic Control of Prolactin Secretion

The theory of the anterior pituitary gland being controlled by neurohumors (Fig. 1) carried through the hypophysial portal circulation, advanced mainly by HARRIS (1955), and the concept of neurosecretion, pioneered by SCHARRER and his group (1963), have led to the present, widely held view that the median eminence of the hypothalamus and the adjacent pituitary stalk make up a specialized neurohumoral organ whose function it is to control the anterior lobe (HARRIS and DONOVAN, 1966; BROWN-GRANT and CROSS, 1966; MARTINI and GANONG, 1966; REICHLIN, 1966). Within the median

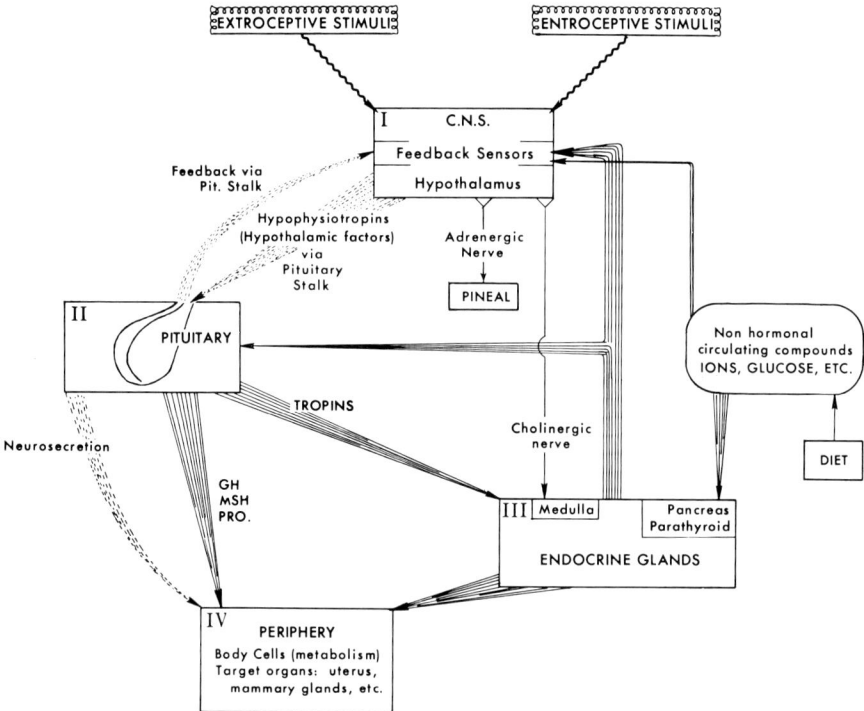

Fig. 1. Diagram of mammary stimulation by different hormones, drawn by M. BEN-DAVID (1968). Effect of I CNS (hypothalamus) and II pituitary, on III endocrine glands and IV peripheral target organs (mammary gland, etc.). Note absence of feedback from the mammary gland

eminence there are nerve endings containing neurosecretory substances which have the property of regulating the secretion of the cells of the anterior pituitary. These "hypophysiotropic hormones" or "releasers" diffuse into the capillary plexus of the median eminence and are carried by the veins of the pituitary stalk to the sinusoids of the anterior pituitary gland. This vascular system has been designated the hypophysial-portal circulation, by analogy with the hepatic-

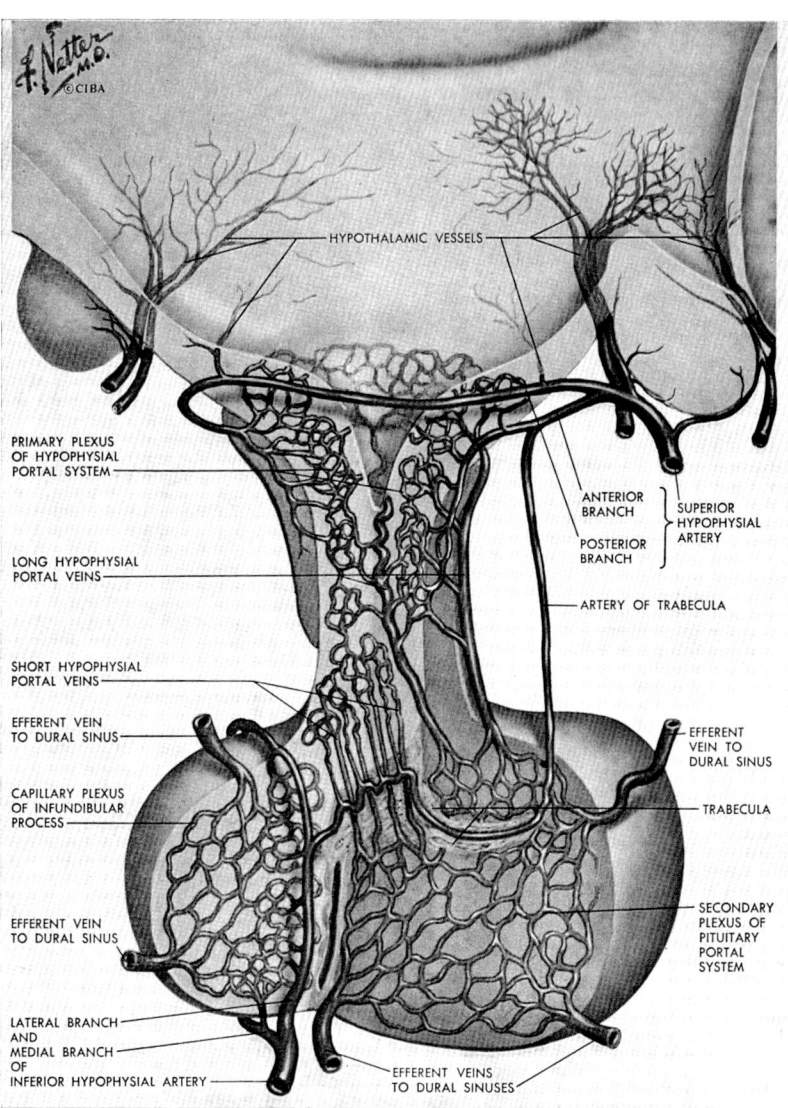

Fig. 2. Scheme of hypophysial-portal circulation. The sinusoids of the anterior lobe receive their blood supply from the hypophysial portal vessels, which arise from the capillary beds within the median eminence (above) and enter the infundibular stem (below). In the anterior lobe (r) they form a secondary plexus of the pituitary portal system, leading to the venous dural sinuses, which surround the pituitary and connect it to the general peripheral circulation. The releasing or inhibiting factors of the hypothalamus or the median eminence enter the pituitary circulation at the primary plexus of the hypophysial portal system which runs from the median eminence via the infundibular stem to the pars distalis (anterior pituitary lobe). (Copyright: The Ciba Collection of Medical Illustrations by FRANK H. NETTER, M. D.)

portal system (Fig. 2). The portal vessel chemotransmitter hypothesis has been supported by the demonstration that compounds extracted from the stalk and median eminence region of the hypothalamus possess the property of "releasing" or "inhibiting" anterior pituitary tropic hormones.

In mammals, prolactin is a unique anterior pituitary hormone because the hypothalamus exerts a tonic and chronic inhibition on its pituitary secretion by a prolactin-inhibiting factor (PIF). This fact does not exclude the possibility of the existence of a prolactin-releasing factor (PRF) as will be shown in paragraph c. In the mammalian animal the inhibitory hypothalamic regulation of prolactin secretion is the predominant one; it is, however, reversed by lactation following delivery. With regard to MSH, it seems that it, too, is subject to the regulation by a MSH-releasing factor (MRF) in the median eminence and a MSH-inhibiting factor (MIF) in the supra-optic nucleus area (TALEISNIK and TOMATIS, 1967). Research in this field will be reported in Chapter 14.

b) Prolactin-Inhibiting Factor (PIF)

A considerable body of evidence, direct and indirect, in both experimental animals and humans, has been accumulated to indicate the presence of the prolactin-inhibiting factor (PIF). Its location near the median eminence is shown in Fig. 3.

PIF activity has been found in extracts of hypothalamic fragments which inhibit release of prolactin from pituitaries cultured *in vitro* (PASTEELS, 1961 b; DANON, DIKSTEIN and SULMAN, 1963; TALWALKER, RATNER and MEITES, 1963; GALA and REECE, 1964 a; SCHALLY, STEELMAN and BOWERS, 1965). Moreover, changes in the hypothalamic PIF content during the estrus cycle of rats were recently reported by SAR and MEITES (1967). They found that pro-estrous and estrous rats had significantly less PIF in the hypothalamus than di-estrous rats. It was postulated that ovarian estrogen secreted during pro-estrus and estrus would increase prolactin release by depressing hypothalamic PIF production and by directly stimulating the pituitary. On the other hand, GALA and REECE (1964 b) reported that the concentration of PIF activity in hypothalamic extracts from rats killed in mid-lactation did not differ greatly from that in extracts from non-lactating rats. Similar findings were described by GROSvenor (1965 b, c) and by Grosvenor, McCann and Nallar (1965) using suckling rats for assay *in vivo*. They showed that hypothalamic, but not cerebral cortical, extracts could block the nursing-induced decline in pituitary prolactin. Kuroshima, Arimura, Bowers and

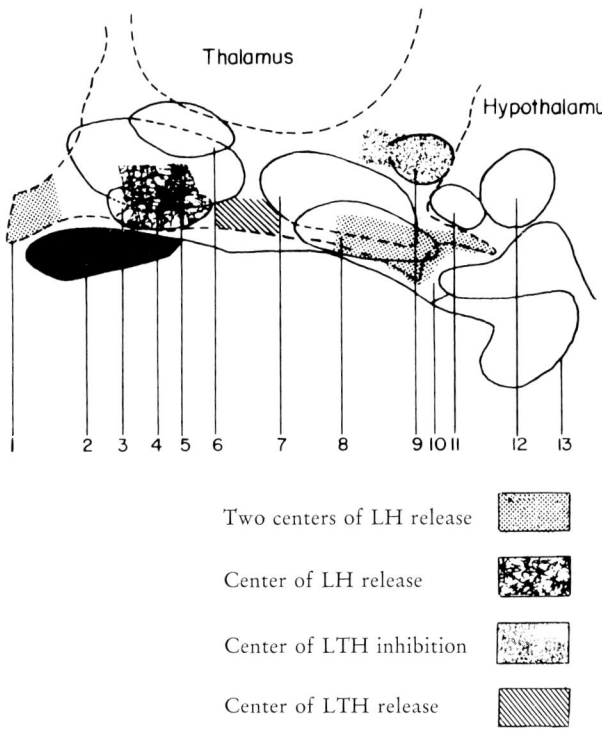

Fig. 3. Hypothalamic control of pituitary FSH, LH, LTH, indicating the mooted localization of the hypothalamic prolactin-inhibiting factor (PIF) and prolactin-releasing factor (PRF). — *1* Preoptic area; *2* Optic chiasma; *3* Suprachiasmatic nucleus; *4* Supraoptic nucleus; *5* Anterior hypothalamic nucleus; *6* Paraventricular nucleus; *7* Ventromedial nucleus; *8* Arcuate nucleus; *9* Dorso-medial nucleus; *10* Median eminence; *11* Premamillary nucleus; *12* Mamillary nucleus; *13* Hypophysis

Schally (1966) demonstrated that pig hypothalamic extracts could block the depletion of pituitary prolactin which follows cervical stimulation, according to the method of Herlyn, Geeler, I. Berswordt-Wallrabe and R. Berswordt-Wallrabe (1965).

Psychopharmaca which depress the PIF center can induce prolactin release *in vivo* and initiate mammary secretion in rats, rabbits, or humans. RATNER and MEITES (1964) and MITTLER and MEITES (1967), therefore, investigated whether different lactogenic agents acted by depressing production of PIF in the hypothalamus. The effects of suckling, estradiol, reserpine, adrenaline, acetylcholine, and electrical stimulation of the uterine cervix on the content of PIF in the hypothalamus of rats were tested. After each *in vivo* treatment period, the hypothalami were removed, extracted, and tested for PIF by a 2-hour incubation with rat pituitary and subsequent assay of the medium for prolactin. Each of the agents used produced a significant decrease in hypothalamic content of PIF. DANON, DIKSTEIN and SULMAN (1963) reported that perphenazine—a chlorpromazine derivative—depressed hypothalamic inhibition of prolactin release. Both reserpine (KANEMATSU and SAWYER, 1963a) and suckling (REECE and TURNER, 1937; GROSVENOR, 1965a) were found to decrease pituitary prolactin content in rats, indicating release of this hormone. Some agents may elicit an increase in hypothalamic release of PIF and thereby decrease pituitary prolactin release. Thus the odor of strange males interrupts early pregnancy in mice (BRUCE and PARKES, 1963), and this phenomenon is presumably mediated through inhibition of prolactin release. If such be the case, PIF production by the hypothalamus may be increased.

The prolactin-inhibiting activity of hypothalamic extracts is not obtainable by oxytocin, vasopressin, epinephrine, acetylcholine, serotonin, histamine, substance "P", or bradykinin, suggesting that the PIF is different from any of these substances (DHARIWAL et al., 1968). The chemical nature of PIF has not been identified, although it has been observed that it passes through a semipermeable membrane, which suggests that it is a small molecule (MEITES, 1966). Prolactin-inhibiting factor activity was not found in purified extracts of the luteinizing hormone-releasing factor (LRF), indicating that these are separate factors (SCHALLY, MEITES, BOWERS and RATNER, 1964). Assays of several samples of corticotropin-releasing factor (CRF) did not show any PIF activity either (MEITES, 1966).

According to the PIF theory, all stimulating effects can only be indirectly explained by "initiation of inhibition". If the connection between the hypothalamus and the pituitary is interrupted, prolactin secretion will be invariably increased. Such an interruption has been produced experimentally by hypothalamic lesions, pituitary stalk section, pituitary transplantation, *in vitro* incubation of the pituitary together with the hypothalamus (Chapter 25) and treatment with hypothalamic depressant drugs (Chapters 13—18). Interruption of the normal connection between the hypothalamus and the pituitary may induce one of the following effects: changes in pituitary prolactin content, prevention of mammary involution or induction of mammary growth, induction of spontaneous milk secretion, maintenance of luteal function, induction of persistent pseudopregnant state with formation of deciduomata in the traumatized uterus, and, finally, induction of continuous diestrus pattern of the vaginal smear. All these phenomena indicate hypersecretion of pituitary prolactin. Therefore it is believed that accelerated synthesis and/or release of LTH is brought about by elimination of the prolactin-inhibiting factor (PIF) of the hypothalamus.

Little is yet known about the chemical nature of PIF, nor has this factor definitely been shown to exist as a separate entity in the hypothalamus. The question of whether the "luteinizing hormone-releasing factor" (LRF) and PIF are similar has been of interest ever since EVERETT (1956) suggested that the same hypothalamic region might control release of both LH and LTH. This is based in part on the many observations that under some conditions LH release may be inhibited when LTH secretion is high. Though purified preparations of LH-releasing factor were found to be free PIF by SCHALLY et al. (1964), activity of the latter factor was detected in preparations of the former by DHARIWAL, GROSVENOR, ANTUNES-RODRIGUES and McCANN (1965). Different assay procedures for PIF were employed by the two groups of workers, and further purification of these factors will be necessary before this question can be resolved.

Attempts to demonstrate a direct prolactin releasing activity in intact rats *in vivo* have so far failed (RATNER, TALWALKER and MEITES, 1965). GROSVENOR (1965 a) carried out ingenious experiments in which the injection of rat hypothalamic extracts inhibited the fall in pituitary prolactin which followed 30 min nursing or stress. The specificity and reliability of this method for the measurement of PIF have, however, been questioned (MINAGUCHI and MEITES, 1967), since the failure of intact rats to respond to exogenous PIF could be attributed to the constant endogenous hypothalamic inhibition of prolactin release from the pituitary. Depression of this inhibitory hypothalamic control by a phenothiazine derivative would eliminate this difficulty and furnish a suitable set-up for the detection of exogenous prolactin-inhibiting activity.

This reasoning led DANON, DIKSTEIN and SULMAN (1970) to devise a new assay method for the *in vivo* detection of PIF: Male rats, weighing 200 ± 20 g, were injected s. c. with 10 mg/kg perphenazine, followed 30 min later by an i. p. injection of 30 mg/kg sodium pentobarbitone to induce anesthesia. One hour after the injection of perphenazine, a common carotid artery was cannulated. Then 1 ml of neutralized acid extract from $^1/_2$ or 2 hypothalami, respectively, was infused into the artery at a rate of 0.1 ml/min. 20 min later the rats were killed by decapitation, their pituitaries were removed and homogenized in a glass homogenizer in ice-cooled 0.9% NaCl solution. The homogenate was centrifuged and the supernatant frozen for the prolactin assay. Control rats were infused with extracts of cerebral cortex prepared in the same manner.

Perphenazine injected 1 hour prior to intracarotid infusion of cerebral cortical extract significantly reduced the prolactin content of the anterior pituitary ($P < 0.0025$) as compared with intact rats infused with similar extracts of cerebral cortex. The crop gland test revealed a 50% reduction in the pituitary prolactin content. On the other hand, intracarotid infusion of hypothalamic extracts into perphenazine-treated rats elevated the pituitary hormone content as compared to that of animals infused with cerebral cortical extract. As the increase in pituitary prolactin content induced by hypothalamic extracts was dose-dependent in the range tested, it can be used for measurement of PIF levels in the hypothalamus extracts. The response produced by $^1/_2$ hypothalamic equivalent is significant at the level of $P < 0.05$, while 2 hypothalamic equivalents evoke a highly significant increase ($P < 0.0005$).

Storage of PIF in the hypothalamus of perphenazine-treated rats has been described by DANON and SULMAN (1970). Diagram 1 (p. 12) explains the set-up. It represents the results of 3 identical experiments, the results of which were pooled and the combined data evaluated as to their significance. Analysis of these results shows that the hypothalami taken from rats 1 hour after injection of 10 mg/kg perphenazine were more effective in increasing the pituitary prolactin content of assay rats pre-treated with perphenazine and pentobarbital sodium than were hypothalami taken from vehicle-

Diagram. 1. *Effect of acute treatment with perphenazine on hypothalamic PIF content in rats*

Procedure:

Stage I: Hypothalamic blockage: s. c. injection of perphenazine 10 mg/kg.
Stage II: Anesthesia: i. p. injection of pentobarbital sodium 30 mg/kg, to permit cannulation of carotid artery.
Stage III: Assay of PIF: intracarotid infusion of hypothalamic extract, or cerebral cortex extract as control.
Stage IV: Assay of pituitary prolactin: rats killed and pituitaries removed for prolactin assay.

Results expressed as average differences between mucosal reactions of two hemicrops (15 pigeons for one assay point) injected with pituitary extracts from control and experimental rats (10 rats for one assay point).

No.	Experimental set-up	Left hemicrop: control		Right hemicrop: experimental		Average difference mg	P
		Treatment of assay rats	Average mucosal dry weight, mg	Treatment of assay rats	Average mucosal dry weight, mg		
1	Effect of hypothalamus versus cerebral cortex	Cerebral cortex extract	5.25	Hypothalamic extract	6.00	0.75 ± 0.27	<0.01
2	Effect of hypothalamus from perphenazine rats versus hypothalamus from control rats	Hypothalamus from control rats	6.47	Hypothalamus from perphenazine-treated rats	7.43	0.96 ± 0.33	<0.005

treated rats (Exp. 2). The difference between the two hemicrop reactions was highly significant ($p < 0.005$). A control series, in which the response of assay rats to hypothalamic extracts from intact rats was compared with the response of other rats to cerebral cortical extracts, demonstrated the sensitivity of the assay procedure (Exp. 1). Hypothalamic PIF activity resulted in elevation ($p < 0.01$) of the pituitary prolactin content of assay rats, previously depleted by perphenazine.

The increase in PIF content of the hypothalamus, while pituitary prolactin content decreases and blood-plasma prolactin rises (BEN-DAVID and CHRAMBACH, 1970), can best be interpreted as being due to blockage of PIF release from the hypothalamus. This results in accumulation of PIF in the hypothalamic neurons. The apparent contradiction of the present results with an earlier report showing depletion of PIF after long-term treatment (DANON et al., 1963) is easily resolved on the basis of the different treatment schedules used in the two cases. It becomes obvious that perphenazine blocks the release of PIF from the hypothalamus in the acute phase, whereas synthesis is inhibited only later, as evidenced after 10 days' treatment. Whether the impaired synthesis of PIF is directly affected by perphenazine, or is secondary to accumulation of this factor in the hypothalamic neurons, is now being studied by us. Still, the first-mentioned possibility appears more acceptable, otherwise one would expect hypothalamic PIF content not to fall beyond normal values after long-term treatment.

c) Prolactin-Releasing Factor (PRF)

The equilibrium between PIF and a tentative PRF has been a matter of controversy for a long time. As originally shown by KRAGT and MEITES (1965), control of prolactin release in the bird seems to differ fundamentally from the pattern observed in mammals. Pituitary autotransplants in the pigeon fail to stimulate the crop sac (BAYLE and ASSENMACHER, 1965), and the pigeon pituitary does not release appreciable amounts of prolactin when cultured *in vitro* (GALA and REECE, 1965 a, b). In the bird, a stimulatory hypothalamic influence on prolactin secretion seems to operate which is mediated by a prolactin-stimulating factor (KRAGT and MEITES, 1965). PRF has been found in the hypothalamus of pigeons, chickens, quails, blackbirds, turkeys and ducks (NICOLL, 1965; MEITES, 1966).

Release of prolactin by the pigeon pituitary appears to be controlled differently than in mammals. Injections of reserpine, chlorpromazine, epinephrine, acetylcholine, estrogen, progesterone, or testosterone did not, as in rats, have any effect on pituitary prolactin content or release in pigeons (MEITES and TURNER, 1947; KRAGT and MEITES, 1965); nor did cultures of pigeon pituitary show increased release of prolactin as is the case with the mammalian pituitary (NICOLL and MEITES, 1962c). Therefore, KRAGT and MEITES (1965) studied the effect of pigeon hypothalamic extract on the release of prolactin by the pigeon pituitary *in vitro*. Prolactin release into the medium by the pituitary incubated with hypothalamic extract from parent pigeons was 2—4 times greater than that released by the pituitary incubated with cerebral cortical extract or with no extract. Slight stimulation of prolactin release was also found in the hypothalamic extract from 4—6-week-old pigeons, although this was not significant at the dose level tested. Hypothalamic extract from rats or pigeons had no effect on release of prolactin by the incubated pituitary of the other species, indicating that the hypothalamic factors regulating prolactin release in two species were not the same. These observations suggest that the pigeon hypothalamus contains a prolactin-stimulating factor in contrast to the presence of a prolactin-inhibiting factor in the mammalian hypothalamus.

Earlier studies have shown that injections of crude acid extracts of rat hypothalamus initiated lactation in rats (MEITES, TALWALKER and NICOLL, 1960a), indicating that such extracts stimulate rather than inhibit pituitary prolactin secretion *in vivo*. The use of highly purified PIF extracts, free of other anterior pituitary regulating neurohumors, would obviate any criticism of *in vivo* methods for testing PIF activity, but such extracts are not yet available. The *in vitro* procedure has the advantage of measuring only the direct effect of PIF in hypothalamic extracts on pituitary prolactin release, since it is not complicated by the presence of target organs and is uninfluenced by other hypothalamic agents.

There is as yet no direct evidence for the existence of a PRF in mammals, though the result of several studies tend to favor it: one of these deals with the inability of suckling to influence the PIF content in the rat hypothalamus (GROSVENOR, 1965c); another study—on the prolactin-releasing capacity of hypothalami from suckling rat mothers—carried out by MISHKINSKY, KHAZEN and

SULMAN (1968) (Chapter 23), demonstrated enhanced mammary development in rats injected with pituitaries of suckling rat mothers. DESCLIN and FLAMENT-DURAND (1969) and NICOLL (1970) also claim evidence for the existence of a hypothalamic PRF.

Another indication for the existence of a PRF was supplied by the discovery of an internal feedback mechanism for prolactin (CLEMENS and MEITES, 1968; MISHKINSKY et al., 1968), when it was shown that there exists a certain parallelism between the mechanisms of internal feedback of all tropins of the anterior pituitary. The tropins of the anterior pituitary, e.g. FSH, LH, ACTH, TSH and STH are known to exert a short feedback on the production and/or release of the endogenous tropins, by depressing their corresponding hypothalamic releasing centers. If this is the case for the above five tropins, it may well be the case for prolactin as well and may point to the existence of a PRF in addition to the PIF. PIF would then be active in normal individuals and PRF only post partum. The possibility of enhancing the lactogenic effect of a perphenazine-suppressed PIF by injecting hypothalamic extracts from suckling rat mothers—which may contain PRF—has been demonstrated by MISHKINSKY and SULMAN (unpublished results—Fig. 4).

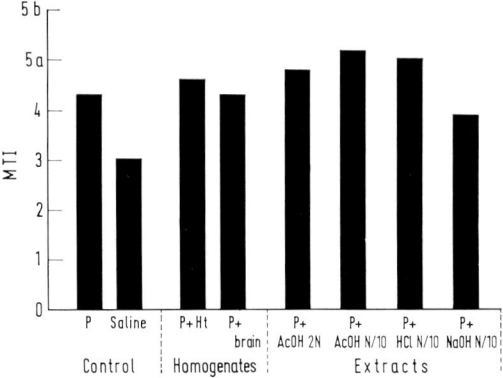

Fig. 4. Enhancing the effect of PIF suppression by perphenazine (P) followed by PRF stimulation: Average mammotropic index (MTI) of rats, after 10 days priming with estradiol (8 µg/d), followed by 5 days' injections of perphenazine (P: 5 mg/kg) and homogenates from hypothalami (Ht) taken from rat mothers 2—21 days post partum. Hypothalamus homogenate increases MTI up to 4.6; hypothalamus extracts ($N/_{10}$ acetic acid) increase MTI up to 5 a (for MTI cf. Fig. 52)

It has previously been shown that implantation of estrogen into the basal tuberal hypothalamus will evoke prolactin secretion in both rats (RAMIREZ and McCANN, 1964) and rabbits (KANEMATSU and SAWYER, 1963 b). TINDAL, KNAGGS and TURVEY (1967) later reported that estrogen implanted in the amygdala or stria terminalis will evoke prolactin release in the rabbit. These results also point to the existence of a PRF center in the hypothalamus.

Another approach to research on the releasing mechanism for prolactin arose from papers by McCANN, MACK and GALE (1959) and AVERILL (1965) in rats and by BAN, SHIMIZU and KUROTSU (1958) in rabbits, reporting cessation of lactation by specific hypothalamic lesions. Both McCANN and AVERILL postulated that bilateral destruction of hypothalamic tissues in the lateral pre-optic area and the lateral hypothalamic area, immediately posterior to the optic tracks, interferes with the production or release of prolactin from the anterior lobe of the pituitary. In 1962 TURNER and GRIFFITH advanced a tentative theory for the hormonal release of LTH, assuming that the hypothalamus contains a "lactogen-releasing factor". TURNER postulated that under normal conditions this factor builds up in the hypothalamus, producing LTH storage in the anterior pituitary, and through the suckling stimulus "this hypothalamic factor is discharged into the portal blood system to stimulate discharge of LTH from the anterior pituitary".

Similar observations were made by ANDERSON and McCANN (1955) in goats during attempts to locate the hypothalamic center of milk ejection. They confirmed ANDERSON's previous data on the relation between milk ejection and the supra-optic nuclei, and established that stimulation of these nuclei evoked milk ejection (ZAKS, 1962). Using ANDERSON's methods, POPOVICH (1958) showed that, when the supra-optic nucleus is stimulated in goats, the magnitude of milk ejection is dependent on the strength and duration of the stimulus: if the stimulus is weak—milk ejection can easily be obtained several times in the course of the same experiment. POPOVICH reasons that this inhibitory effect is not associated with the depletion of adrenaline. After destruction of one of the supra-optic nuclei by electrocoagulation, milk ejection declines on the side of the destruction, and is maintained on the intact side. After some time the reflex is partially restored on the side of the destruction (ZAKS, 1962). Thus, it seems that the failure of rats and ewes to eject milk after supra-optic lesions is due to some hormonal control of the anterior hypothalamus, since (1) oxytocin injections are not followed by milk "let-down"; (2) the histology of the mammary glands of the lesioned animals reveals a deficiency of alveolar milk constituents; and (3) lesions which do not damage the median eminence may affect a prolactin-releasing center.

EVERETT and QUINN (1966) demonstrated by electrical stimulation of the rat hypothalamus that the areas which produce ovulation indicative of luteinizing hormone release, and those which produce pseudopregnancy indicative of prolactin release, are separable. Consequently, it would appear that different hypothalamic regions are involved in the regulation of each of these tropins. QUINN and EVERETT (1967) were able to induce delayed pseudopregnancy by suitable stimulation of the area which controls prolactin release.

These studies prove that lesions in the anterior hypothalamus may interfere with the active process of lactogenesis. Since such lesions seem to inhibit prolactin release, it was of interest to determine the effect of electrical stimulation on this anterior hypothalamic area of prolactin discharge. In sexually mature, primed rats, bipolar concentric electrodes were implanted chronically in various parts of the hypothalamus. Electrical stimulation at a frequency of 25 square waves/sec, of 0.01 msec duration and at a peak current of 10 μA (10″ on—20″ off for 60 min), caused slight mammotropic development when applied to the anterior hypothalamus (Fig. 5), and none when given given to other sites of the brain (MISHKINSKY and SULMAN, unpublished results). It seems that this site of the hypothalamus is identical with that suggested by RAJKOVITZ (1962), FLERKO and SZENTAGOTHAI (1957) and AVERILL (1965), and it may be a center of prolactin discharge due to the presence of PRF. More exact experiments to localize such a center in the anterior hypothalamus are now being carried out (cf. Fig. 3).

d) Hypothalamic Feedback of Prolactin

The functions of the pituitary and hypothalamus are interrelated through two feedback mechanisms—"internal" or "short" (Fig. 6) and "external" or "peripheral" (Fig. 7). The internal feedback operates direct between the hypophysiotropins and the releasing or inhibiting factors of the hypothalamus, or *vice versa*. The peripheral feedback is indirect, being brough about via peripheral target organs which secrete their hormones into the blood and thus affect the pituitary tropins (SULMAN, 1956 b). When the hormone output from the target organ is increased, it has a negative (inhibitory) effect on the pituitary, while its decrease has the opposite, a positive (stimulatory), effect on tropin production. In certain cases different doses of a hormone may elicit a positive or a negative effect. Most peri-

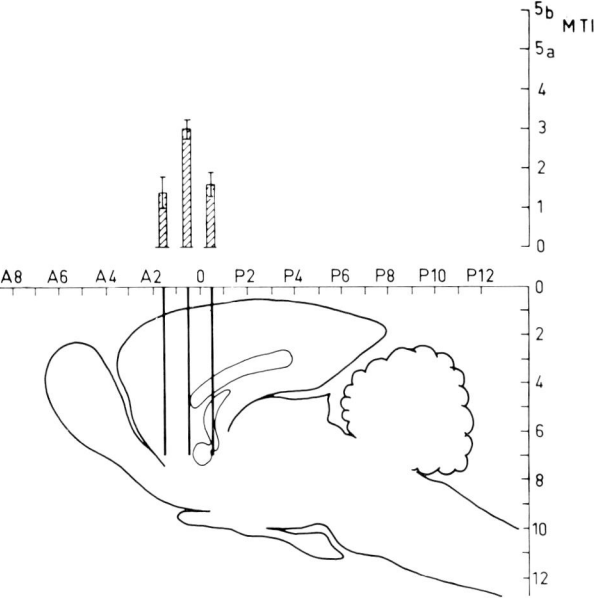

Fig. 5. Mooted center of PRF: Prolactin release following electrical stimulation of the rat brain, after estradiol priming (8 µg/d/rat) for 10 days. Prolactin release (MTI 3) is obtained in a special region of the anterior hypothalamus A-1, at a depth coordinate H-7 bilateral at L-0.5

pheral hormones, when given in small doses, are known to enhance, but in large quantities to block, the secretion of their pituitary tropins. Studies on implantation of most of the known tropins into the median eminence of the rat brain have been reported (FLERKO, 1966; MESS and MARTINI, 1968). The mechanism of a peripheral prolactin feedback has been discussud on many occasions (SULMAN, 1956 b), while CLEMENS and MEITES (1967, 1968) were the first to find that LTH, when implanted into the median eminence of female rats, creates a "feedback loop" by enhancing hypothalamic PIF secretion and ultimately causing a decrease in the pituitary prolactin level and mammary involution.

The work of Török (1964), showing the existence of a two-directional blood flow in the hypophysial portal system from and to the hypothalamus, has exposed an anatomical route by which a pituitary tropin may pass from the pituitary to the hypothalamus (Fig. 6). This

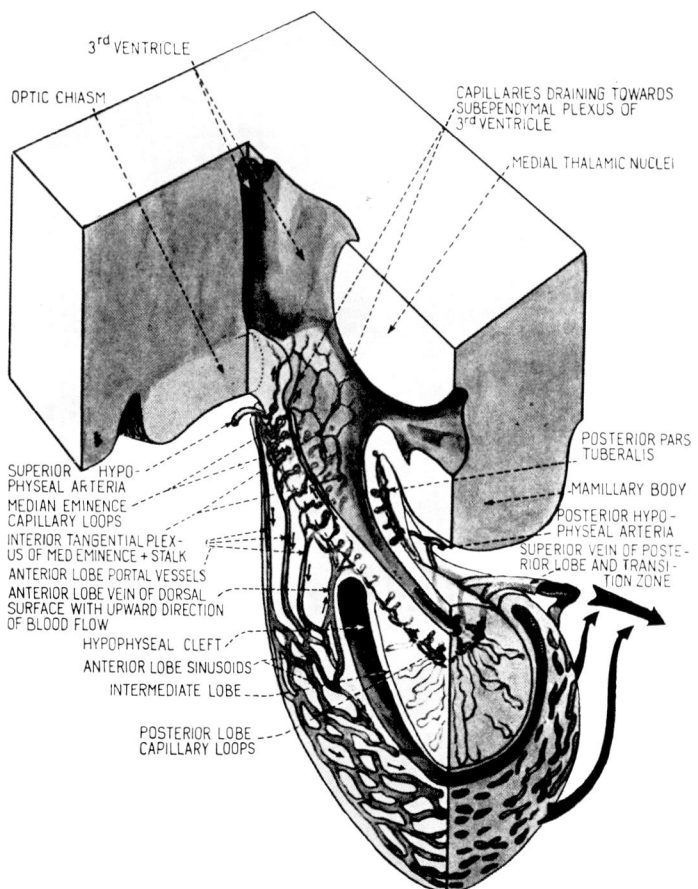

Fig. 6. Diagram showing the circulatory pathways of hypothalamic hypophysiotropin release and short or internal feedback: small arrows indicate direction of blood flow (by courtesy of Dr. Török, 1964, Publ. House of the Hungarian Academy of Sciences, Budapest)

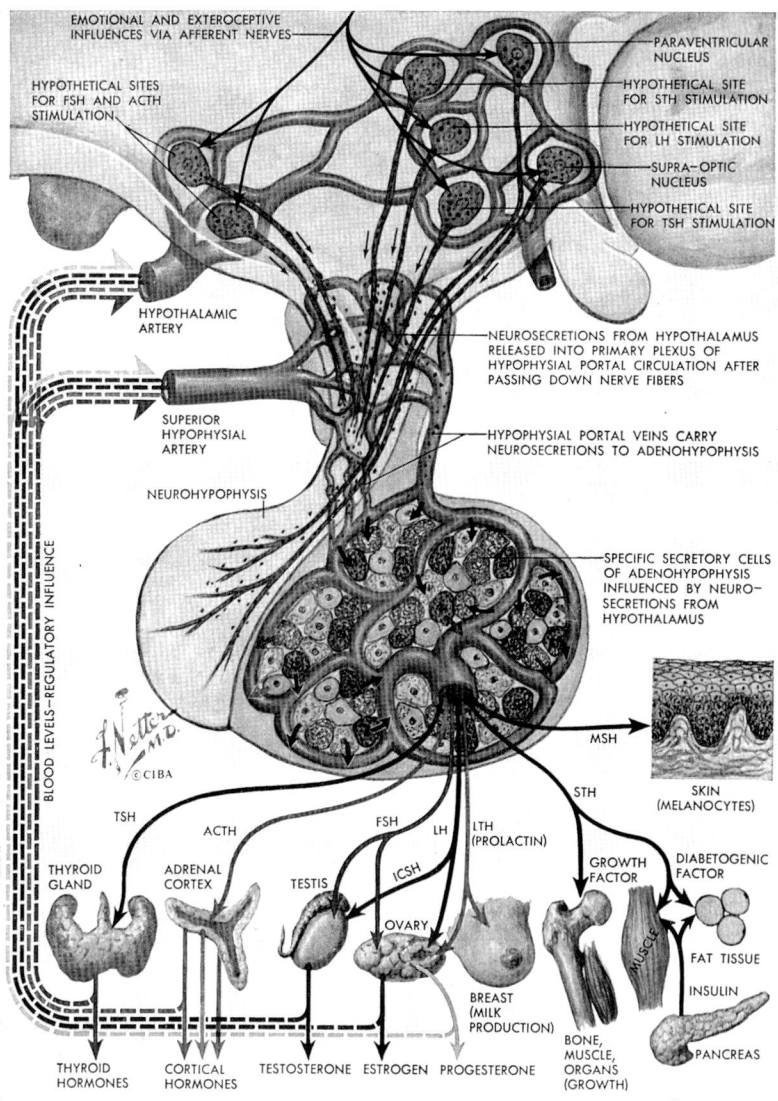

Fig. 7. Diagram showing the circulatory pathways of hypothalamus-pituitary-peripheral endocrine glands relationship: long or external feedback. (Copyright: The Ciba Collection of Medical Illustrations by FRANK H. NETTER, M. D.)

morphological observation provides grounds for assuming that anterior pituitary hormones participate in their own feedback control (internal, short, direct or auto feedback) by acting directly on the hypothalamus. Such an internal feedback was reported for ACTH by LEGORI, MOTTA, ZANISL and MARTINI (1965), for FSH by CORBIN and STORY (1967), for LH by CORBIN and COHEN (1966) and for STH by KATZ, MOLITCH and McCANN (1967). Recently, CLEMENS and MEITES (1968) showed that implantation of ovine prolactin into the ME of mature, intact and ovariectomized rats resulted in increased hypothalamic content of PIF, reduced pituitary prolactin concentration and pituitary weight, and marked regression of the mammary glands. This effect was enhanced by simultaneous implantation of ACTH. These results suggest that the implanted prolactin inhibits pituitary prolactin synthesis and release by directly increasing hypothalamic PIF secretion. It is not known whether the small physiological amounts of prolactin reaching the hypothalamus can really exert such negative feedback, but the findings of GUILLEMIN (1967) who found pituitary tropins in hypothalamic extracts provide evidence for the existence of such an internal feedback (Fig. 7).

MISHKINSKY, NIR and SULMAN (1969) have recently shown that the internal prolactin feedback is dose-response related; this would mean that prolactin depresses its own production by exerting an inhibitory effect on the hypothalamic mechanism which in turn governs pituitary prolactin release (Fig. 8). This study was carried out on four groups of virgin female albino rats, each weighing 200 ± 10 g. After 10 days of estradiol priming, stereotaxic implantation in the median eminence of the hypothalamus was carried out. Three groups were implanted with tubes containing prolactin in doses of 1, 4 and 16 IU/rat respectively, and the fourth group, which served as control, was implanted with empty tubes. The implants were carried out in the following manner: NIH-PS-8 ovine prolactin was tapped into a micro glass tube (Rudmond's disposable pipettes) and, after all the overflow material had been cleaned from the external surface of the tube, it was immediately implanted in the ME. A follow-up showed that this technique permits complete absorption of the implanted hormone.

From Fig. 8 it is evident that the MTI values of the prolactin-treated animals are significantly decreased when compared with those of the controls: The degree of significance is $p < 0.01$ with the implantation of 4 IU, and $p < 0.005$ with 16 IU, but only $0.1 > p > 00.5$

with 1 IU of prolactin. Comparing the responses to the various doses of prolactin, a highly significant difference (p<0.005) between 1 IU and 4 IU, but no significant difference between 4 and 16 IU is discernible.

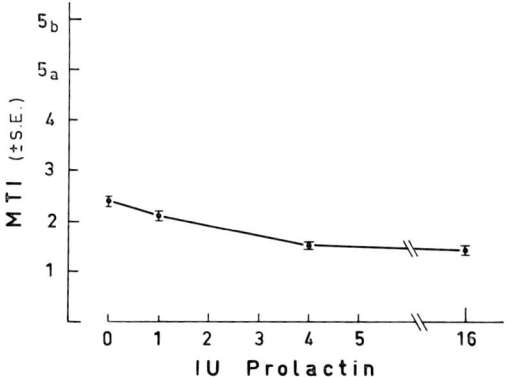

Fig. 8. Diagram of short or internal feedback after hypothalamic implantation of prolactin. Average MTI values of estradiol-primed female rats, following stereotaxic implantation of various doses of prolactin in the median eminence

For many years LTH and MSH have been considered unique among the pituitary tropins, their release being governed by a permanent inhibiting effect of the hypothalamus, in contrast to all other known tropins which were shown to be controlled by hypothalamic-releasing factors. These results have been attributed to a prolactin-inhibiting factor (PIF), and increased prolactin release has always been explained by a depression of this factor (MEITES and NICOLL, 1966; MISHKINSKY, NIR and SULMAN, 1969). Moreover, no feedback mechanism has ever been found to exist between the peripheral target organ and pituitary prolactin.

In our study, decreased MTI values were obtained following implantation in the median eminence of three different doses of prolactin. Moreover, a dose-response relationship was found to exist. These results, which show the action of prolactin to parallel that of all the other anterior-pituitary tropins, are rather surprising. The depressant effect of all the other anterior-pituitary tropins implanted in the median eminence has been attributed to the fact that they block the corresponding hypothalamic releasing factor. If in the case of prolactin there exists only an inhibiting factor (PIF) in the hypothalamus of mammals, then depression of this PIF should result in an increase in pituitary prolactin content and mammary gland stimulation. As this was not the case in our experiments, we were led to assume that, beside a prolactin-inhibiting factor (PIF), a prolactin-releasing factor (PRF) also exists in the hypothalamus. PIF and PRF may balance each other. Prolactin implanted in the median eminence may have the same depressant effect on PRF as the other pituitary tropins have on their releasing factors. When the PRF is depressed, the PIF is intensified to suppress pituitary prolactin. The existence of a prolactin-realeasing factor in the hypothalamus was previously postulated in our observations on prolactin-releasing activity of the hypothalamus in post-partum rats (MISHKINSKY, KHAZEN and SULMAN, 1968).

e) Effects of Steroids

Low doses of estradiol have been reported to stimulate prolactin secretion by direct action on the pituitary (RAMIREZ and MCCANN, 1964; BEN-DAVID, DIKSTEIN and SULMAN, 1964) or indirectly by way of the hypothalamus (SAR and MEITES, 1967). Other steroids, too, have been shown to affect the PIF and pituitary prolactin content: a recent study by SAR and MEITES (1968) showed that progesterone, testosterone propionate, and cortisol acetate injections into ovariectomized rats significantly reduced the hypothalamic prolactin-inhibiting factor content, increased pituitary prolactin concentration, and promoted mammary development; testosterone propionate and cortisol acetate also induced mammary secretion. The findings on pituitary prolactin concentration are in agreement with previous studies showing that the pituitaries of rats treated with progesterone, testosterone propionate and cortisol acetate contain more prolactin than the pituitaries of control rats. The results of investigations on the hypothalamic prolactin-inhibiting factor at least partially explain how these steroids act to increase pituitary prolactin concentration and to stimulate mammary growth. Since these steroids were previously found to have no direct effect on pituitary prolactin release *in vitro*, it may be concluded that they act through the hypothalamus *in vivo*. This does not exclude the possibility that progesterone may be converted partly into estrogens *in vivo* before acting on the hypothalamus.

The above results indicate that both protein molecules and steroids can affect the hypothalamic cells responsible for the production of the factors which regulate prolactin and other tropin secretion in the pituitary. We shall show in Chapters 13—19 that drugs resembling the steroids can act the same way.

f) Interrelation between Pituitary Tropins

Just as there is an interplay of the PIF and the PRF, a reciprocal relationship seems to exist between the secretion of the gonadotropins FSH and LH, on the one hand, and prolactin, on the other. Thus, in all the experimental conditions associated with increased prolactin secretion, such as pituitary stalk section, heterotropic pituitary transplantation or treatment with hypothalamic depressant drugs, there was impairment of gonadotropin secretion. Conversely, treatment with estrogens or androgens in medium doses, known to inhibit gonadotropin secretion by a negative feedback mechanism, resulted in enhanced prolactin secretion. Similar depression of gonadotropin secretion has been observed during lactation in rats and mice, resulting in the phenomenon of "lactation-anestrus".

VOOGT, CLEMENS and MEITES (1969) implanted prolactin into the median eminence of 21-day-old female rats and studied its effects on pituitary FSH secretion. Rats killed 5 or 8 days after implantation showed a significant reduction in pituitary FSH concentration as compared with control rats implanted with cholesterol, or with control rats given implants of other hormones in amounts stated to be present in the prolactin preparation. Examination of the ovaries and uterus indicated that prolactin implantation resulted in follicular and uterine endometrial growth and increased uterine weight. These findings indicate that an implant of prolactin in the median eminence results in stimulation of FSH release in prepuberal rats.

CLEMENS, MINAGUCHI, STOREY, VOOGT and MEITES (1969) have shown that this assumption is correct. Injections of prolactin significantly advanced the onset of puberty in immature rats by about 6 days when given at 20 days of age and by 2 days when given at 25 days of age. Injections of other anterior pituitary hormones (FSH, LH, TSH, GH) in amounts specified to be present in the NIH prolactin preparations did not advance the onset of puberty. Prolactin injections into hypophysectomized rats were also ineffective. Implants of prolactin into the median eminence hastened the onset of puberty by an average of 6.6 days, whereas implants of doses of other hormones stated to be present in the prolactin preparation had no effect on the onset of puberty. The advancement of puberty by prolactin administration is believed to be mediated through the hypothalamus and to result in decreased prolactin and increased pituitary FSH secretion.

As to the mechanism of this reciprocal relationship between gonadotropins and prolactin, it was suggested by EVERETT (1956) that the same neurogenic agent responsible for the stimulation of gonadotropin secretion may inhibit the secretion of prolactin. Studies with purified LRF, however, have shown that this hypothalamic factor is free of any prolactin-inhibiting activity (ARIMURA et al., 1967).

WELLER, DANON and SULMAN (unpublished data) found that, while acute adrenalectomy only slightly impaired perphenazine-lactation, chronic adrenalectomy resulted in a remarkable inhibition of perphenazine-lactation with a concomitant absence of corpora lutea stimulation (BEN-DAVID, 1968 b). Trying to elucidate the mechanism of this phenomenon, we explanted pituitaries and hypothalami from perphenazine-treated rats after chronic adrenalectomy and ovariectomy (perphenazine treatment started 10 days after operation). These pituitaries released considerably less prolactin than pituitaries with hypothalami from perphenazine-treated rats after acute adrenalectomy-ovariectomy (perphenazine treatment started on the day following operation). We would explain these observations on the basis of the assumption that, after an established over-shoot of one tropin resulting from extirpation of its target organ, the pituitary is unable to increase its prolactin secretion in spite of a proper stimulans. The mechanism of such a phenomenon could also be mediated through the hypothalamus. Measurement of hypothalamic releasing or inhibitory factors along with the pituitary tropins in these experiments could provide an answer.

Summarizing, one would assume that reactions ascribed to the PIF or to the PRF could also result from the interplay between different hypophysiotropins which may mimic prolactin release or suppression.

g) Suckling

Stimulation of the nipples results in rapid release of prolactin, corticotropin, and oxytocin (MEITES, 1959 a; COWIE and FOLLEY,

1961). This mechanism may also release growth hormone as has been shown by GROSVENOR, KRULICH and MCCANN (1968). Earlier investigations had already shown that appetite and water intake, too, are increased by the suckling stimulus (CROSS, 1961), suggesting a possible release of GH and TSH.

It is of interest that the suckling stimulus can prolong lactation considerably beyond the usual post-partum interval if suckling and milk removal are permitted to continue. The litters of rats and mice are normally weaned at the end of about 3 weeks, but if there is regular replacement of the young with fresh litters from other mother rats or mice, lactation may be prolonged for as long as 9 months with relatively little decrease in milk yield (BRUCE, 1958; NICOLL and MEITES, 1959). Continued application of the suckling stimulus has been observed to prolong lactation considerably also in other species, including the human. Regular applications of the suckling stimulus without removal of milk, i. e. by ligating the galactophores of mice or rats while permitting litters to suckle, retards mammary involution and maintains slight secretory activity (SELYE, COLLIP and THOMSON, 1934). In the absence of both the suckling stimulus and removal of milk in post-partum rats, injections of prolactin, adrenal glucocorticoids, or oxytocin, either individually or in combination, may also retard mammary involution and sustain secretory activity far beyond the usual post-partum period (MEITES and NICOLL, 1959). Some maintenance of mammary lobulo-alveolar tissue and secretory activity was observed in post-partum rats for as long as 75 days after litter removal, by daily injections of prolactin, cortisol acetate and oxytocin. Even in hypophysectomized post-partum rats after litter removal, administration of these three hormones was effective in retarding mammary involution (MEITES and HOPKINS, 1961).

The suckling or milking stimulus results in a reflex release of oxytocin from the neurohypophysis into the circulation, which then acts to contract the myoepithelial elements surrounding the alveoli and finer ducts (CROSS, 1961). This reflex can be conditioned, and oxytocin can be released in response to psychic stimuli (Fig. 9). Electrical stimulation of the para-ventricular nucleus (ANDERSON and MCCANN, 1955) or of the nerve tracts to the neurohypophysis (CROSS and HARRIS, 1952), results in milk ejection, suggesting that these sites are involved in oxytocin release. Epinephrine has been shown to inhibit the local action of oxytocin on the mammary gland (BRAUDE and

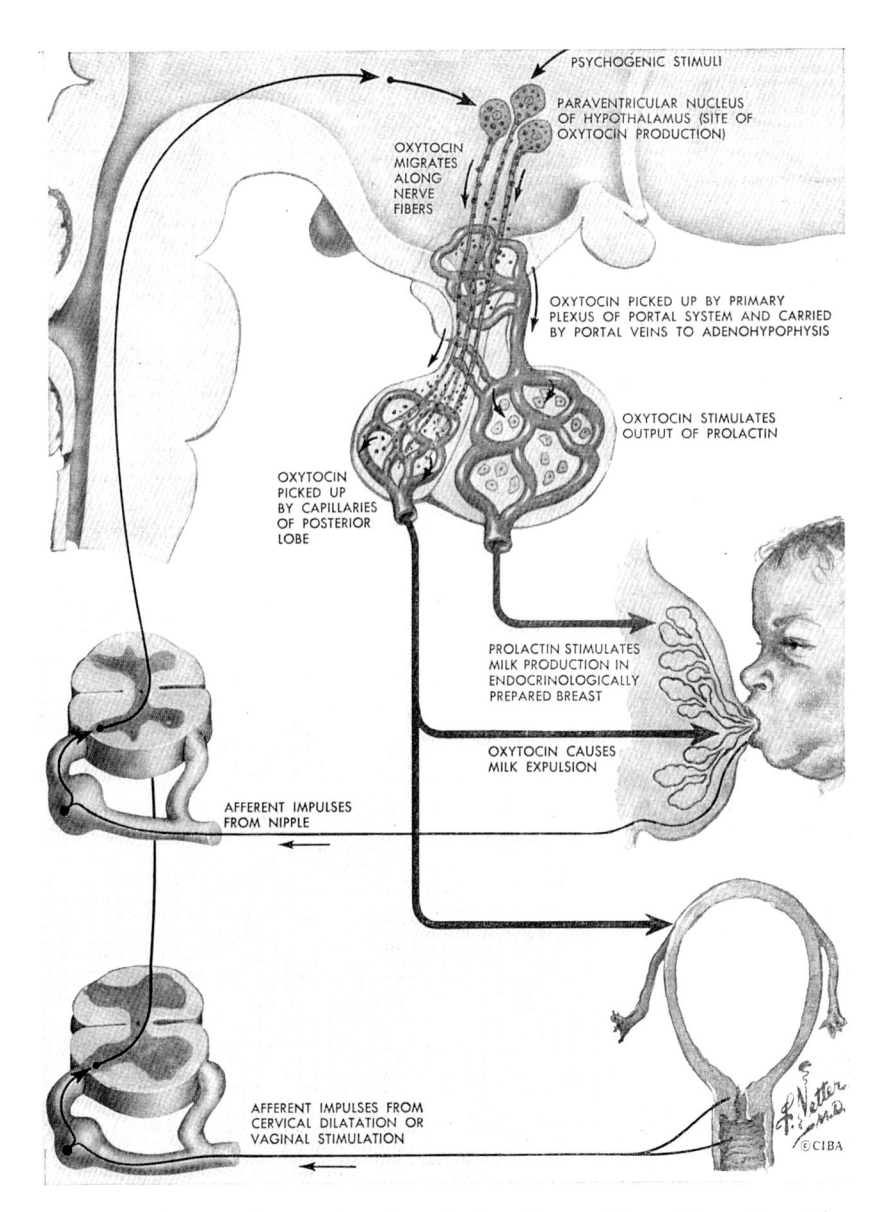

Fig. 9. Pathways of oxytocin release induced by suckling. (Copyright: The Ciba Collection of Medical Illustrations by Frank H. Netter, M. D.)

MITCHELL, 1952; CROSS, 1953), but emotional stress (cf. Fig. 13) apparently prevents milk ejection by a partial or complete block of oxytocin release from the posterior pituitary (CROSS, 1955).

Recent studies in Meites' laboratory (1966) suggest that serotonin may be involved in the reflex release of oxytocin in response to the suckling stimulus. Serotonin, its precursors (tryptophan and 5-hydroxytryptophan), and a monoamine oxidase inhibitor (iproniazid) significantly inhibited milk ejection; the inhibitory effects of these substances were overcome by injections of oxytocin or by a specific serotonin antagonist (2-bromo-D-lysergic acid diethylamide).

GROSVENOR (1965 b, c) reported that suckling for 30 min failed to alter the prolactin-inhibiting activity of rat hypothalamus, as compared to the nonsuckled controls. As the hypothalamus contains numerous pharmacologically active substances (e. g. serotonin, histamine, catecholamines, and acetylcholine) in large quantities, the possibility must be considered that these agents may influence the *in vivo* release of prolactin.

h) Conclusions

The fact that the hypophysiotropins are under the influence of hypothalamic releasers has been proved with regard to FSH, LH, ACTH, STH and TSH. Their synthesis, storage and secretion is regulated by releasing factors carried through the portal system into the anterior pituitary. All these hypophysiotropins are "balanced" tropins, i. e. they are also regulated by a feedback of the peripheral hormones created by them. MSH and LTH are unbalanced hormones, i. e. they do not create secondary hormones which may exert a feedback. It has been proved that MSH secretion is regulated both by a hypothalamic releasing factor and an inhibiting factor (cf. Chapter 14 — i).

With regard to prolactin, the existence of a hypothalamic inhibiting factor (PIF) is well established by prolactin release after hypothalamic lesions, pituitary stalk section, pituitary transplantation, treatment with hypothalamic depressants and by hypothalamus-pituitary co-culture.

The existence of a hypothalamic prolactin-releasing factor (PRF) is postulated by us on the basis of the following body of evidence:

a) inability of suckling to influence the PIF content of the rat hypothalamus;

b) steady increase of prolactin-releasing activity in the hypothalamus of suckling rat mothers during the 21 days of their nursing activity;

c) direct demonstration of such an increase of PRF activity in organ culture;

d) increase in the lactogenic effect of a hypothalamus depressant (perphenazine) by the addition of extracts from the hypothalamus of nursing rat mothers;

e) existence of an "internal" feedback mechanism for prolactin, which is subject to a dose-response relationship;

f) cessation of lactation after bilateral destruction of hypothalamic tissues in the lateral pre-optic area;

g) stimulation of lactation and deciduoma formation by electrical stimulation of the same area of the anterior hypothalamus;

h) stimulation of lactation by estrogen implanted into the amygdala or stria terminalis;

i) existence of a PRF in birds;

j) clinical observation that lactation due to the PRF can be abolished by steroids, whereas galactorrhea due to PIF suppression does not react to steroids and barely to clomiphene.

These findings do not, admittedly, constitute a direct proof, since oxytocin stimulation as well as a change in the balance of other auxiliary hypophysiotropins could also account for most of the observations cited. Obviously, unequivocal evidence for the existence of a PRF must depend on its chemical isolation. NICOLL has recently presented new evidence for the existence of a PRF in mammals (1970).

Chapter 3

Prolactin

a) Introduction

Interest in the physiology of pituitary prolactin greatly increased in recent years, as indicated by the large number of investigations on this subject since 1960. Prolactin is distinguished from other pituitary tropins by its inhibitory regulation originating in the central nervous system, i.e. the hypothalamus, by lack of a reciprocal peripheral feedback mechanism, by its diverse effects among the vertebrates, including growth-promoting actions in several species, and by its close association with growth hormone.

b) Storage and Release

Prolactin is probably present in the hypophysis of all vertebrates (NICOLL, 1968), but it is more plentiful in some species and at certain stages of the life cycle than at others. Ox and sheep glands usually yield 30 to 40 IU per gram of fresh tissue, while pork glands contain much less. Measurements of the prolactin content of human pituitaries indicate that the hormone accounts for about 2 percent of the acetone-dried residue of the gland. More prolactin is contained in the adenohypophyses of goats, sheep and cows, and less in the glands of cats and rabbits. In most species the prolactin content of the pituitary increases steadily during pregnancy, with an increase sometimes occurring soon after parturition. This is also reflected in the MTI (Fig. 10). When nursing is permitted, the prolactin content remains at this higher level for a longer period than when nursing is prevented.

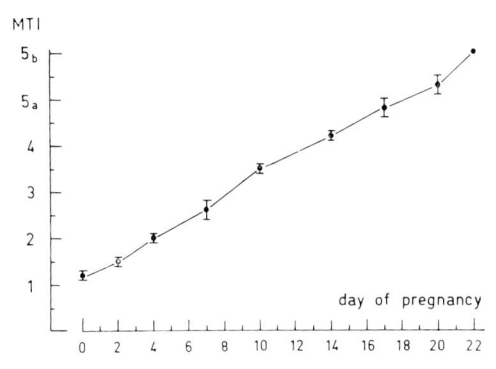

Fig. 10. Increase in mammary development during pregnancy in rats: Natural mammotropic index (cf. Fig. 52)

c) Properties

Prolactin has been prepared in a highly purified form from ox and sheep pituitaries. Isolation procedures have been described by LI (1961) in a review on the biochemistry of prolactin. The physical properties of sheep (198 aminoacids) and ox (205 aminoacids) prolactin are remarkably similar. The hormone from both species has a molecular weight of 24,000 to 26,000, an isoelectric point at pH 5.73, and a sedimentation constant of 2.19. DIXON and LI (1964) showed that the hormone molecule consists of a single peptide chain with one free amino acid group, but no free carboxy end-group. They suggested that the peptide chain has an intra-chain disulfide bridge forming a ring similar to that present in vasopressin and oxytocin. The results of their structural studies are summarized in the following schematic representation of prolactin:

$$H_2N\text{-Thr-Pro-Val-Thr-Pro-}(CyS)_4\text{-}(CyS)\text{-Tyr-Leu-Asp}(NH_2)\text{-CyS-COOH}.$$
$$\phantom{H_2N\text{-}}12345202203204205$$

In order to investigate aggregation and charge differences between prolactin components, a combination of gel filtration, ion-exchange chromatography, and zone electrophoresis was used by SLUYSER and LI (1964). As a result of these investigations, a method was developed for the isolation of monomer prolactin in highly purified form. The experiments presented in their paper show that the various peaks or bands which appear when prolactin is submitted to either ion exchange chromatography or zone electrophoresis are due to the presence of molecular aggregates in the preparations. The degree of aggregation differed from batch to batch. Some prolactin preparations contained about 20% monomer whereas others had a much lower content of unassociated material. In view of the reported finding that some aggregation of prolactin can occur when solutions are submitted to lyophilization, pervaporation, sucrose concentrates, or extremes of pH, it seems likely that the differences in degree of aggregation between various batches of lactogenic hormone arise at least partly during the preparation of the samples. In order to prevent aggregation, the hormone solution was concentrated by chromatography on small columns of DEAE-cellulose. A separation was achieved between fractions representing different states of polymerization of prolactin molecules. One of the fractions contained monomer prolactin, as revealed by exclusion chromatography and sedimentation equilibrium experiments. The molecular weight of monomer prolactin was estimated to be 23,000.

Prolactin is not soluble in any fat solvent. It is, however, soluble in water, except in the isoelectric region pH 5—6, and in strongly acid solution. It may be salted out with sodium or ammonium sulfate or sodium chloride, the required amount depending on the pH. It is also precipitated by basic salts of heavy metals at or near pH 7. It is rapdily inactivated by trypsin and pepsin. Salt-free solutions at pH 8 withstand boiling for one hour with only negligible loss; under other conditions, prolactin may be rapidly destroyed. Frozen preparations are not stable, once they have been thawed.

Despite the great similarities of bovine and ovine prolactin, certain differences have been observed. It seems that, proportionate to weight, ovine prolactin is more active than bovine prolactin. There is less tyrosine in ovine prolactin, and distribution of the two hormones in countercurrent analysis is different. However, BRYANT and GREENWOOD (1968) showed by radioimmunoassay of blood that ovine prolactin antigen can substitute for bovine prolactin. Human prolactin has yet to be characterized, but the fact that methods effective for isolating ox and sheep prolactin have completely failed when applied to human pituitaries suggests that primate prolactin may have peculiarities of structure which make it very similar to human growth hormone.

d) Prolactin in Body Fluids

Prolactin-like activity in urine has been demonstrated both by pigeon crop sac assay and by observation of the effect of urine on the hormonally-primed mammary gland of the immature hypophysectomized male rat. SEGALOFF and STEELMAN (1959) found daily up to 175 IU of prolactin in urine from menopausal women. In normal women the excretion was somewhat higher in the luteal phase of the cycle as compared with the follicular stage. Prolactin excretion also increased during the course of pregnancy. Prolactin-like activity of serum has been detected with both pigeon and mouse assays. The blood of lactating women has been found to contain prolactin activity (SIMKIN and GOODART, 1960). Somewhat smaller amounts of hormone were found in blood during the luteal phase of women with normal menstrual cycles. No prolactin activity was evident in the blood of women in the follicular phase of the menstrual cycle or in normal young men. These observations still require confirmation (cf. Chapter 26 — d).

e) Prolactin in Pituitary Tumors

Persistent lactation, with or without previous pregnancy, is encountered with considerable frequency in endocrine practice. The occurrence of this complaint in acromegalic women is well known. When lactorrhea and amenorrhea follow pregnancy (Chiari-Frommel syndrome), a functional disturbance in prolactin secretion may be present. General awareness that persistent lactation may also be a sign of pituitary tumor in the non-acromegalic patient followed the description in 1954 of fifteen such cases from Albright's Clinic (Forbes-Albright syndrome—cf. Chapter 28). The problem has been studied in rats bearing pituitary "mammotropic" tumors by MIZUNO, TALWALKER and MEITES (1964).

f) Effect of Prolactin on Mammary Growth

Present evidence indicates that mainly one anterior pituitary hormone—prolactin, and perhaps also STH, are involved in the development of the mammary gland. The most convincing and impressive work was done by LYONS (1942), who injected small amounts of purified prolactin (according to standards of 1942) into the mammary ducts of spayed rabbits and obtained not only clear-cut mammary development and lactational response, but also histological evidence of cellular hyperplasia of the alveolar tissue. This "lactational" growth was characterized by the presence of numerous mitoses in the alveoli and also by the fact that the number of cells per alveolus was on the average greater in the prolactin-injected sectors than in control sectors of the same gland.

COWIE and FOLLEY (1947) found that unfractionated ox anterior pituitary extracts will evoke mammary alevolar growth in the immature gonadectomized-adrenalectomized rat which presumably has no source of steroid-hormones.

LYONS (1951) elucidated experimentally the hormonal factors needed to evoke growth of the mammary lobulo-alveolar system in the hypophysectomized rat to a degree equal to that attained by the end of pregnancy. In the course of this work it was found that considerable, though not complete, growth of the mammary lobuloalveolar system in hypophysectomized-ovariectomized rats could be evoked by suitable treatment with estrogen, progesterone and purified prolactin, the steroids themselves being ineffective in the absence of prolactin. The mammary lobulo-alveolar growth generated by this triad of hormones was not, however, as extensive as that obtained by substituting for the purified prolactin a cruder preparation known to contain ACTH and STH. Later, LYONS, LI and JOHNSON (1952) reported that the mammary lobulo-alveolar growth-promoting action of the hormone triad—estrogen + progesterone + purified prolactin—in hypophysectomized-ovariectomized rats was enhanced by addition of purified STH. Omitting prolactin from this hormonal tetrad resulted in loss of mammary alveolar growth-promoting activity, though duct growth, as evidenced by the presence of club-shaped end buds, continued.

A most interesting development during the past few years is the growing evidence that prolactin and STH in combination can induce full mammary lobulo-alveolar growth in the apparent absence of ovarian hormones in the rat. Frequent injections of relatively large doses of prolactin and STH were found to induce mammary lobulo-alveolar growth in male and female rats that had undergone gonadectomy and adrenalectomy (MEITES, 1965) or ovariectomy, adrenalectomy, and hypophysectomy (TALWALKER and MEITES, 1961). MEITES and KRAGT (1964) showed that hypophysectomized rats, with a pituitary implanted adjacent to one of the mammary glands, showed localized lobulo-alveolar mammary growth in the vicinity of the transplant.

Thus, it becomes evident that prolactin and STH together can induce full lobulo-alveolar mammary growth in the rat, which suggests that ovarian hormones are of secondary importance. The ovarian hormones are believed to promote mammary growth under natural conditions by stimulating secretion of prolactin and possibly STH by the adenohypophysis, by synergizing with these adenohypophysial hormones, and by sensitizing the mammary gland to these hormones.

g) Levels of Pituitary Prolactin in Various Stage of Lactation

SIMKIN and ARCE (1963) found no prolactin activity in the blood of children, normal young men, or normal young women in the first half of the menstrual cycle. However, prolactin activity was present in the blood of normal women during the second half of the cycle. This evaluation of the role of prolactin in the normal human menstrual cycle suggests that the time of its peak concentration in blood differs from the peaks of the other two known pituitary gonadotropins—follicle stimulating hormone (FSH) and luteinizing hormone (LH)—which fall before the midcycle (the time of ovulation). On the other hand, ample evidence has been accumulated that FSH and LH are the gonadotropins responsible for the entire process of human ovulation, icnluding the post-ovulatory development of the corpus luteum. This has been corroborated by studies on the effect of the contraceptive agents which prevent ovulation by inhibiting pituitary release of FSH and LH (cf. Chapter 28). The role of prolactin in this set-up has not yet been studied. The presence of prolactin activity during the latter part of the cycle would, however, suggest that prolactin acts as an antagonist to LH. This may be deduced from the fact that the estrus cycle of rodents is inhibited by prolactin or by a prolactin releaser (Fig. 11). Yet, it may also be due to the luteotropic effect of prolactin.

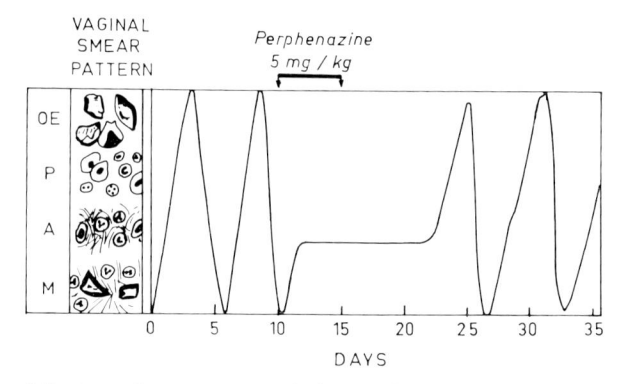

Fig. 11. Inhibition of rat estrus cycle by prolactin: Prolactin is released by perphenazine which suppresses the hypothalamic prolactin-inhibiting factor (PIF). The vaginal smear shows anestrus (A) during 14 days following injection. Thereafter, the normal cycle returns with proestrus (P), estrus (OE) and metestrus (M)

Later, in 1968, MINAGUCHI, CLEMENS and MEITES showed that pituitary content and concentration of prolactin are much lower in sexually immature than in mature rats. The levels of pituitary prolactin remain low before, and at the beginning of, puberty despite increases in body and anterior pituitary weights during this period. A marked increase in pituitary prolactin content and concentration occurs soon after onset of puberty in rats (vaginal opening). These changes in pituitary prolactin levels of rats occur before and after the onset of puberty and are strikingly different from those observed for the pituitary gonadotropins. Thus both FSH and LH reach their highest levels in the pituitary prior to puberty, and show a marked decrease shortly before and after the onset of puberty. Pituitary prolactin levels showing the opposite trend provide another example of the divergence between prolactin and gonadotropin and their functions during different physiological states. The large increase in pituitary prolactin after puberty is, therefore, believed to be due mainly to increased estrogen secretion by the ovaries.

Studies of pituitary prolactin content in the rat have shown that it is low during pregnancy but increases markedly after parturition. GROSVENOR and TURNER (1958, 1960) determined the prolactin content of the pituitary in the rat at different stages of pregnancy and lactation: Prolactin content of the pituitary is low during the first part of pregnancy, later on it exceeds that of the controls and on the 16th day it reaches its peak. Just before parturition it falls to a very low level.

Pituitary histology shows similar trends. It is well known that lactogenic hormone is secreted by acidophile cells of the anterior lobe of the pituitary gland. The number and secretory activity of acidophile cells in the rat normally increases in late pregnancy and following parturition; it gradually declines as lactation advances. It is believed that the initial increase in pituitary lactogenic hormone is due to an estrogen surge which reaches a peak in blood and urine near parturition but, as the estrogen level subsides after parturition, maintenance of lactogenic hormone secretion becomes dependent upon the regular application of suckling or nursing stimuli. This conforms with the investigations of REECE and TURNER (1937) who showed that nervous stimuli resulting from milking or nursing are transmitted to the anterior lobe, causing a release of lactogenic hormone into the blood stream. The prenursing level of lactogenic hormone increased from the 2nd to the 6th day postpartum, then declined towards the end of lactation. The fall in hormone concentration as a result of 30 minutes' nursing stimuli was steepest on days 6 and 16 (95% and 90% respectively), while on the 21st day—when milk production by the rat is at its most intensive—the prolactin content of the pituitary was equal to the control level and was not affected by the suckling stimulus. The loss in prolactin content as a result of the suckling stimulus could be proved by direct determination with starch-gel electrophoresis of the rat pituitary (KURCZ, 1967; BARANYAI, NAGY and KURCZ, 1967).

h) Theories on the Mechanism of Onset of Lactation

TURNER (1939, 1952, 1956), FOLLEY (1956) and MEITES (1956, 1959 a, b) have brought forward different theories to explain the onset of lactation. However, all investigators have stressed the importance of estrogen-progesterone combinations in inhibiting milk secretion during pregnancy, an effect which this combination appears to exert by blocking the release of lactogenic hormone from the pituitary and by rendering the alveolar cells unresponsive to the lactogenic complex.

At the end of pregnancy, the fall in the rates of progesterone and estrogen removes this inhibition which is replaced by the positive "lactogenic" effect of estrogen acting unopposed (FOLLEY, 1956); alternatively, the decline in both estrogen and progesterone leaves the mammary gland receptive to prolactin stimulation and, at the same time, estrogen is still present in sufficient quantity to increase the secretion of prolactin from the pituitary and initiate lactation (MEITES, 1959 a, b).

i) Conclusion

Prolactin is the specific hypophysiotropin which promotes lactation in the pregnant mammal. Its effect seems to be dependent on adequate priming by ovarian steroids which can, however, be dispensed with by applying very high doses of prolactin fortified by STH. This combination can induce full lobulo-alveolar growth in the absence of ovarian hormones. Whereas in rodents and ferrets prolactin is luteotropic, it is an antagonist of LH in most other mammals, including the human. Its antagonism to FSH is not absolute.

During pregnancy lactation is suppressed by ovarian and placental steroids; prolactin release begins with the expulsion of the placenta. At that stage, its activity is enhanced by corticosteroids. In the cow, prolactin plays a lesser role which will be described in Chapter 27.

In male rodents prolactin stimulates the development of the prostate, seminal vesicles and preputial glands. Its effect on the mammary gland is normally inhibited by the PIF which is only overcome under pathological conditions.

Chapter 4

Growth Hormone

Early experiments demonstrated that increases in milk production could be obtained by giving injections of extracts of the anterior pituitary to lactating cows. The best responses were obtained when cows were treated during declining lactation. Experiments by FOLLEY and YOUNG (1940) demonstrated that this response was largely due to the growth hormone present in the extracts, and they showed that preparations of highly-purified bovine growth hormone gave similar increases in yield. This yield-boosting effect of growth hormone has since been confirmed by several investigators. The experiments of BRUMBY and HANCOCK (1955), who studied the effect of daily injections of purified bovine growth hormone in identical twin cows for a 12-week period over the peak phase of lactation, and a 4-week period over the end phase of lactation, showed a very substantial increase in milk yield. We have not observed this effect in rats which received perphenazine as prolactin releaser with bovine growth hormone added for milk yield boosting (Fig. 12).

Both ACTH and STH have a prolactin-sparing effect on the restoration of lactation in hypophysectomized animals (SHAW, 1955; COWIE, 1957). Adrenocortical hormones and STH may replace prolactin completely in mice (NANDI, 1958 a, b, 1959) and are at least as

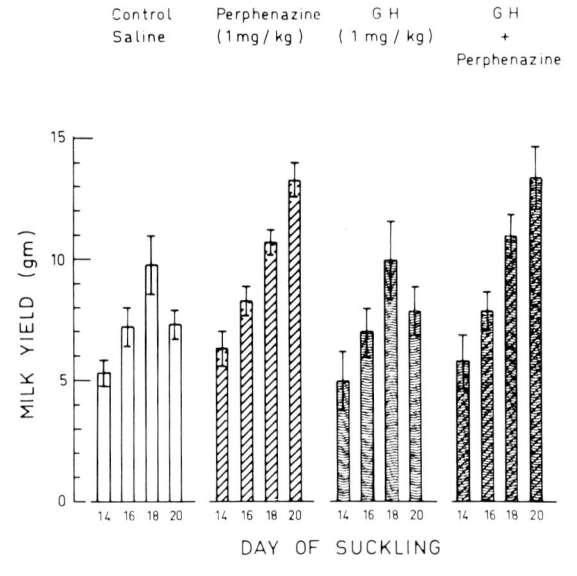

Fig. 12. Effect of growth hormone (GH) on milk yield. Perphenazine increases milk yield by prolactin release. Growth hormone does not improve milk yield

important as prolactin for maintaining lactation in ruminants. TSH and STH both appear to be necessary for maximum milk production, and they have pronounced stimulatory effects in cattle and goats (MEITES, 1959 a; MEITES, HOPKINS and TALWALKER, 1963). It has also been reported that STH can increase milk yields in rats (GROSVENOR and TURNER, 1959 a; MOON, 1965), but other workers have failed to observe this effect (MEITES, 1957 b, 1959 a; COWIE and FOLLEY, 1961).

These observations raise the question of interplay between STH and LTH in relation to the physiological phenomenon of lactation. The fact that bovine LTH injections do not raise milk yield in cows (SULMAN and TWERSKY, 1948) while STH injections do—only means that in normal conditions of lactation the amount of LTH in the blood is sufficient whereas that of STH is not. This has been proved by SCHAMS and KARG (1969) who could not detect any significant prolactin levels in the blood of cows before or during lactation. Their method of radioimmunoassay revealed, however, a significant peak one day before parturition when estrogen levels are high in the cow.

The apparent difficulty in separating prolactin from growth hormone in chemical fractionation of human pituitaries (WILHELMI, 1961; LI, 1961, 1962), and the demonstration in animals of overlapping biological activity of these two human hormones (CHADWICK, FOLLEY and GEMZELL, 1961; FERGUSON and WALLACE, 1961; LYONS, LI and JOHNSON, 1961; BARRETT, FRIESEN and ASTWOOD, 1962; BUTT, 1962) prompted an initial study by a team headed by BECK et al. (1964 a, b) on the metabolic effect in man of ovine prolactin. BERGENSTAL and LIPSETT (1958) had reported nitrogen retention in four hypophysectomized women receiving ovine prolactin, and MCCALISTER and WELBOURN (1962) had demonstrated hypercalciuria in hypophysectomized patients receiving a preparation of sheep prolactin. BECK et al. (1964 a, b) found that in pituitary dwarfs, receiving a highly purified preparation of ovine prolactin (PLR) prepared by a modification of the method of REISFELD, TONG, RICKES, BRINK and STEELMAN (1961), this preparation induced many of the metabolic changes that occur when these same individuals receive human growth hormone (MCGARRY and BECK, 1961). BECK et al. also found that their purified ovine prolactin induced a decrease in the blood urea nitrogen levels, potassium and phosphorus retention, and hypercalciuria.

MCGARRY, RUBINSTEIN and BECK (1968) extended these investigations. Their approach was based on comparison of metabolic effects, since, although immunologically STH and LTH might or might not differ, this has no bearing on their physiological task. Even the fact that, in a certain test, purified LTH fails to give a STH effect, or *vice versa*, does not attest to its task in the complicated mechanism of lactation. They feel that, chemically, STH and LTH are different materials, but their effects may overlap. A comparison of all the known biochemical and physiological effects of STH and LTH preparations shed light on the question in how far they are physiologically identical and can replace each other. They summarize the biochemical, immunological and physiological actions of STH and LTH preparations from different species on humans. The main biochemical effects of human STH were: decrease in blood sugar, decrease in free fatty acids (FFA) which later increased above the initial level, decrease in urinary and blood urea nitrogen. These effects of human STH were closely reproduced by ovine LTH, but not by other non-primate LTH or STH. In spite of the similarity of action of human STH and ovine LTH, no immunological cross-reaction was discovered either for the intact molecules or for their peptic, chymotryptic or papaic digests.

APOSTOLAKIS (1964) found that highly purified human LTH gave a very limited STH effect. The question may be raised whether, during the purification, certain physiologically reactive sites are not destroyed or polymerized. If we imagine STH and LTH as proteins with many active sites, rather than one active molecule with many possible actions, most of the contradictory findings in literature can be easily explained. If we accept this view, then LTH would appear to be a more or less anomalous STH, depending on the method of preparation and on the species. Such an anomaly—natural or artificial—might, of course, cause new effects and we should not expect all the physiological and biochemical properties of LTH to be present in STH. There is no doubt that all human STH preparations contain LTH activity, as exemplified by biological assay (KOVACIC, 1962 a) and immunoassay (LARON and ASSA, 1964; LARON, 1965). Its activity in promoting lactation in nursing women has been demonstrated by LYONS, LI, AHMAD and RICE-WRAY (1968). Its failure to promote crop sac development has been shown by FORSYTH (1967) and by COWIE, HARTMANN and TURVEY (1969).

PEAKE, MCKEEL, JARETT and DAUGHADAY (1969) described a human pituitary tumor which contained LTH only, but no STH. This observations was extended by RIMOIN et al. (1968) who described an isolated deficiency of human growth hormone in 7 female ateliotic dwarfs. They had normal pregnancies and lactated normally in the post-partum period. One such female was studied during the last trimester and the early post-partum period of a normal pregnancy. Plasma STH could not be detected during the period of post-partum lactation when measured following insulin-induced hypoglycemia, arginine infusion and suckling. Since STH was not detectable during active lactation, neither in the basal state nor following provocative stimuli, the human growth hormone molecule, or at least those parts of it which are immunologically active and involved in growth promotion, do not seem to be necessary for post-partum lactation (PASTEELS, 1967).

Although the biochemical separation of the growth-promoting and prolactin activities of human and monkey growth hormone has not been accomplished, several lines of evidence suggest that these two activities reside in different molecules in primates, as they do in other species. CHEN and WILHELMI (1964) found that the ratio of the two activities in a variety of human STH preparations differed more than was to be expected if they were associated with the same molecule. ROTH, GLICK, CUATRECASAS and HOLLANDER (1967) found that serum LTH activity does not parallel the marked rise in immunoreactive serum STH observed following insulin-induced hypoglycemia, and that it is normal in acromegalics. Moreover, prolactin activity is elevated in the plasma of normal lactating women and in women with galactorrhea not associated with acromegaly—two states in which STH levels are not elevated (CANFIELD and BATES, 1965). Evidence derived from the above study of women with an isolated deficiency in human growth hormone supports and extends these observations.

One of the most interesting findings in regard to the biologic properties of STH in experimental animals is that human STH (HGH) possesses the various characteristics previously demonstrated by ovine lactogenic hormone: 1) it promotes pigeon-crop sac growth in both local and systemic tests; 2) it stimulates functional hypertrophy of corpora lutea cells, as shown by histologic study of the ovary and by the induction of deciduoma reactions; 3) it synergizes with the ovarian hormones to induce ductal and alveolar mammary growth; and 4) it induces localized milk secretion. However, these activities of HGH reach only 10 to 20 per cent of the extent of those of the animal lactogenic hormone. Nevertheless, no matter how we modify HGH, it still exercises these lactogenic activities side by side with the growth-promoting activity as long as the latter persists. Moreover, if the growth-promoting activity is destroyed, the lactogenic activities are abolished as well. There can, therefore, be no doubt that HGH has intrinsic lactogenic properties. Still, no similarity in chemical structure is evident when gross structures of HGH and ovine lactogenic hormone are compared. It seems that both hormones have an active core in common.

In summary, the similarity between human STH and LTH on the one hand, and the LTH activity of the human STH preparations on the other, raises serious doubts as to their purity and method of preparation. The first step in resolving this problem must be the preparation of crystallized protein hormones.

Chapter 5

Gonadotropins and Sex Steroids

Estrogen and progesterone are important for mammogenesis and have been shown to play a role in colostrogenesis as well (DE JONGH, 1932). On the other hand, they are of secondary importance for the maintenance of lactation, i.e. lactogenesis and galactopoiesis. The gonadotropins are important in so far as they regulate the ovarian function. In some animals, such as rats (EVANS, SIMPSON, LYONS and TURPEINEN, 1941; EVERETT, 1944), mice (DRESEL, 1935), and probably the ferret (DONOVAN, 1963 a, b), prolactin must be included among the gonadotropins, along with FSH and LH, on account of its luteotropic action in these species.

It is now well established that the stimulus for lactogenesis is supplied by the anterior pituitary (FOLLEY, 1956; LYONS, 1958). The ovarian hormones are believed to be involved in the control of the anterior pituitary factors, or lactogenic complex, of which LTH and STH seem to be the most important components (FOLLEY, 1956). Milk is present in the mammary gland during the second half of pregnancy of animals and man, but copious milk secretion does not set in until parturition. NELSON (1936) hypothesized that the high levels of estrogen during the second half of pregnancy inhibited both anterior pituitary release of prolactin and mammary tissue response to this hormone. This theory could not be proved for all species. MEITES and TURNER (1942 a, b; 1948), whose theory is based on determinations of the prolactin content of the pituitary, found that estrogen induces the secretion of prolactin from the anterior pituitary, thereby bringing about lactogenesis, whereas progesterone inhibits the action of estrogen, so that lactation is induced by the fall in the body level of progesterone occurring at parturition and the estrogen surge at term.

As to the question whether the ratio of progesterone to estrogen, or the absolute amounts of these hormones, are the critical factors in mammogenesis, it seems clear now—on the basis of thorough experiments carried out in the guinea-pig (BENSON, COWIE, COX and GOLDZWEIG, 1957)—that the absolute amounts of estrogen and progesterone determine the growth response of the mammary gland. Recent experiments performed in the goat give further proof of the secondary importance of the ovary in mammogenesis, since full mammary development could be obtained by regular milking of ovariectomized animals (COWIE, KNAGGS, TINDAL and TURVEY, 1968). It may be argued that in these experiments adrenal steroidogenesis could have replaced the missing ovarian hormones. This would also be consistent with the fact that male goats have sometimes been found to produce milk. Then, too, experiments performed with adrenalectomized-ovariectomized-hypophysectomized rats show that the ovaries and adrenals are of secondary importance for maintenance of mammary structure in this species, as measured by DNA (HAHN and TURNER, 1967).

The integrity of the ovary is not essential for the maintenance of lactation (galactopoiesis), since it depends on STH and TSH, and ovariectomy has no noxious effect on lactation (KURAMITSU and LOEB, 1921; DE JONGH, 1932; FOLLEY and KON, 1938; FLUX, 1955). In the intact rat very low doses of estrogen inhibit milk secretion, while in the ovariectomized rat much larger doses are required (COWIE, 1966). It is believed that ovarian progesterone synergizes with exogenous estrogen in the intact animal, the combination blocking the action of the lactogenic complex on mammary tissue (COWIE, 1966). Progesterone alone has no effect on milk secretion (FOLLEY, 1952 a, b), save in the adrenalectomized animal (FLUX, 1955), in which it may compensate for lack of adrenal steroids. The inhibitory action of progesterone would, therefore, become apparent only in combination with estrogen (FAURET, 1941; WALKER and MATTHEWS, 1949; COWIE, FOLLEY, MALPRESS and RICHARDSON, 1952; MEITES and SGOURIS, 1954).

The mechanism of the inhibitory action of estrogen + progesterone must, at least partially, be attributed to a direct effect on the mammary parenchyma (DESCLIN, 1950; MEITES and SGOURIS, 1953). *In vitro* experiments with dog mammary tissue, on the other hand, showed a dual effect of the estrogen-progesterone combination: very low levels proved to be stimulatory for lactogenesis, while higher levels exerted an inhibitory influence (BARNAWELL, 1967).

To sum up, it would seem that the gonadotropins and their secondary steroids—estrogen and progesterone—are important for the development of the mammary glands in the normal female. They are, however, not required for active lactation.

Chapter 6

Adrenocorticotropin and Corticosteroids

It has long been established that adrenal cortical hormones, as well as prolactin, are essential for maintenance of lactation. In hypophysectomized rats, guinea pigs or goats, these two hormones were shown to constitute the minimal requirements for milk secretion. For many years, the adrenal cortical hormones were considered by some workers—but not by FOLLEY (1956) and COWIE and FOLLEY (1961)— to have a permissive, and hence a relatively passive, role, whereas prolactin was believed to be the primary stimulator of lactation (MEITES, 1959 a).

Early evidence on the role of the adrenal in mammogenesis is conflicting (FOLLEY, 1952 a), and the whole question has been experimentally re-examined during the last fifteen years. The 11-oxygenated corticoids exert no mammogenic effect on the ovariectomized mouse, and they block the duct-developing effects of estrogen. On the other hand, 11-desoxycorticosterone synergizes the duct-developing action of estrogen in the spayed mouse (FLUX, 1954). ACTH has no mammogenic effect in intact mice (FLUX and MUNFORD, 1957), while hydrocortisone acetate is mammogenic at low doses and without effect at higher doses (MUNFORD, 1957). The rat responds differently to the corticoids, since glucocorticoids have been reported to be both mammo- and lactogenic in virgin animals (SELYE, 1954; JOHNSON and MEITES, 1955).

As pointed out by AHRÉN and JACOBSOHN (1957), the effect of corticoids on the mammary gland is conditioned by the endocrine state of the animal. They found that, in the hypophysectomized-ovariectomized rat, cortisone was unable to induce normal growth, even after addition of ovarian hormones. In the non-hypophysectomized spayed rat cortisone was lactogenic, its mammogenic effect becoming apparent only after addition of ovarian hormones. These authors consider that corticoids are mammo- and lactogenic only in conditions of homeostatic balance, while they may inhibit or alter mammogenesis in higher doses where they induce metabolic imbalance. In hypophysectomized-ovariectomized-adrenalectomized ("triply operated") rats, full duct development can be obtained when a corticoid is added to a combination of STH + estrogen (LYONS, LI and JOHNSON, 1958). In these rats, lobulo-alveolar development can be achieved by administering estrogen + progesterone + LTH + STH, provided that saline is supplied for drinking (LYONS, LI, COLE and JOHNSON, 1953).

Research during the past few years seems to indicate that ACTH and the adrenal cortex play as active a role in initiating and maintaining lactation as does prolactin itself. Thus, in intact pregnant rats (TALWALKER, NICOLL and MEITES, 1961) and in mice (NANDI and BERN, 1961), adrenal cortical hormones elicited lactation, whereas prolactin alone was ineffective. This suggests that the availability of adrenal cortical hormones rather than prolactin is the limiting factor in inducing lactation during pregnancy in these species; presumably, sufficient prolactin is already present during gestation. BALDWIN and MARTIN (1968 a, b) showed that both prolactin and cortisol are required for casein synthesis during lactation. An adrenal glucocorticoid hormone (9-fluoro-prednisolone acetate) was also demonstrated to induce lactation in pregnant cows (TUCKER and MEITES, 1965); prolactin alone was not tested. In pregnant rabbits, either cortisol acetate or prolactin could bring about lactation, and a combination of the two hormones was more effective than either was alone

(MEITES, HOPKINS and TALWALKER, 1963). This suggests that, in the pregnant rabbit, both types of hormones are limiting factors in initiation of lactation. It remains to be determined whether the rat and mouse are more representative of the mammalian species than is the rabbit in regard to the hormonal requirements for production of milk secretion during gestation.

BEN-DAVID and SULMAN (1969) have shown that acute adrenalectomy affects only slightly the full lobulo-alveolar development induced by high doses of perphenazine. WELLER, DANON and SULMAN (unpublished data) confirmed this finding. It was found that chronic adrenalectomy (for two weeks or longer) is clearly detrimental to the mammotropic effect of perphenazine. Whether chronic adrenalectomy impairs perphenazine-induced mammogenesis only on account of chronic deficiency of corticosteroids at body and mammary tissue level, or whether tropin disbalance at pituitary level also plays a role, can only be determined by future research. Preliminary experiments point to the validity of the second hypothesis. Many drugs, non-specific agents, stress interventions which have been shown to initiate mammary secretion in rats, rabbits, or other species, apparently do so because they elicit release of both prolactin and ACTH from the adenohypophysis (NICOLL, TALWALKER and MEITES, 1960; MEITES, 1961 a, b; MEITES, NICOLL and TALWALKER, 1963). In our rat strain, however, we have been unable to confirm the mammotropic and lactogenic effects of such agents or stress (WELLER and SULMAN, unpublished data).

The primary importance of pituitary integrity for the action of adrenal steroids on the mammary gland is emphasized by the findings of HAHN and TURNER (1967), according to which adrenalectomy has negligible effect on the DNA content of the mammary glands of hypophysectomized-ovariectomized rats. In these experiments, striking differences were detected only between the hypophysectomized and non-hypophysectomized groups, irrespective of the additional gland excisions. In the intact pregnant rat an optimal level of corticosterone was found which is able to induce maximum increases in the DNA and RNA titers of the mammary tissue; lower and higher levels show a reduced effect on DNA and RNA (KUMARESAN, ANDERSON and TURNER, 1967).

The role of the adrenal in galactopoiesis is of admitted importance, but opinions are at variance as to whether it is the gluco- or the mineralcorticoids which are primarily involved. Glucocorticoids are considered critical by GAUNT, EVERSOLE and KENDALL (1942), while FOLLEY and COWIE (1944) and COWIE and FOLLEY (1947) argue that mineralocorticoids are crucial. Later experiments (COWIE, 1952; COWIE and TINDALL, 1955) suggest that both types of corticoids are necessary for optimal milk secretion. In general, the same conclusions were drawn from experiments with adrenalectomized goats, though in these studies mineralocorticoids (DOC) were found to play a more critical role than glucocorticoids, e. g. cortisone (COWIE and TINDAL, 1958).

Since the ovary may compensate for adrenal steroids by way of progesterone production (FLUX, 1955), it must be realized that differences in the degree of ovarian contribution may be due to strain differences in rats. This holds specially with regard to the relative importance of gluco- or mineralocorticoids for galactopoiesis in adrenalectomized animals (COWIE and FOLLEY, 1961). In intact cows, ACTH and cortisone depress lactation (COTES, CRICHTON, FOLLEY and YOUNG, 1949; FLUX, FOLLEY and ROWLAND, 1954; SHAW, CHUNG and BUNDING, 1955). In the rat, ACTH and cortisone have galactopoietic effects (JOHNSON and MEITES, 1958); large doses of cortisone, however, inhibit milk secretion in this animal (MERCIER-PAROT, 1955).

There is growing evidence that increased amounts of prolactin, corticotropin and adrenal cortical hormones are released at about the time of parturition (COWIE and FOLLEY, 1961; MEITES, 1959 a), and they all serve to initiate lactation. After parturition, the milking or suckling stimulus results in a realease of LTH, ACTH, and corticosteroids (COWIE and FOLLEY, 1961; MEITES, 1959 b), and these act to maintain milk secretion. Other hormones (STH, thyroid hormones, insulin, and parathormone) are necessary to elicit maximum production of milk, but are not essential for milk secretion. However, the importance of adrenal steroids in galactopoiesis may lie, at least in part, in their permissive action as concerns prolactin activity (BARNAWELL, 1967), as well as in their more general metabolic effects (AHRÉN and JACOBSOHN, 1957). Thus, for instance, they increase the formation of mammary enzymes (HEITZMAN, 1968).

In conclusion it should be said that the participation of all corticoid hormones is required for lactation. The presence of still unknown corticosteroids, e. g. Hartman's "Corticolactin" in the amorphous fraction of cortical extracts, cannot yet be excluded.

Chapter 7

Thyrotropin and Thyroid Hormones

The thyroid would seem to influence mammogenesis by its general metabolic effect (Jacobsohn, 1958, 1959). A certain degree of hypothyroidism enhances alveolar development in the rat, but it has an inhibitory effect on mammogenesis in the mouse (Folley, 1956). Treating rats with propylthiouracil (an antithyroid drug) stimulates mammary growth (Ben-David, Dikstein and Sulman, 1966). Mammogenesis can be elicited in the hypophysectomized-adrenalectomized-thyroidectomized rat by giving ovarian hormones + prolactin, STH and hydrocortisone, without the addition of thyroid hormones (Chen, Johnson, Lyons, Li and Cole, 1955).

The thyroid does not seem to be essential for lactogenesis and galactopoiesis either, though its absence may deleteriously affect the intensity and duration of milk secretion (Folley, 1952 b). Treating lactating rats with propylthiouracil depresses milk yield (Ben-David, Dikstein and Sulman, 1966). That thyroid activity is enhanced during lactation is supported by histologic (Racadot, 1957) and biochemical (Grosvenor and Turner, 1958 b) studies. Graham (1934) showed that, by feeding dried thyroid gland or by injecting thyroxine, the milk yield of cows could be increased. Some years later it was found that certain iodinated proteins possessed thyroxine-like activity when given in the feed (Cowie, 1956). It was shown that thyroxine is galactopoietic when fed to lactating cows in daily doses of about 100 mg (Bailey, Bartlett and Folley, 1949). As regards 3, 5, 3'-triiodothyronine, it was ineffective when orally administered to cows, probably because it is inactivated in the rumen. On the other hand, it is more galactopoietic than thyroxine when administered subcutaneously (Bartlett, Burt, Folley and Rowland, 1954). There is a wide variation in the degree of stimulation of milk production resulting from the administration of thyroxine or of thyroid-active substances; on the whole, a better response is obtained during the decline of lactation than at the peak or end of lactation (Herman, Graham and Turner, 1938; Ralston, Cowsert, Ragsdale, Herman and Turner, 1940; Blaxter, 1945), but it is not possible to make a good producer out of a poor producer by feeding thyroprotein (Thomas, 1953). Moreover, it has been found that the use of thyroxine or thyroid-active substances during the first to third lactations is inadvisable, for the boost in yield obtained is largely cancelled out by a shortening of the lactation period (THOMAS and MOORE, 1953; COWIE, 1956).

The best results are obtained when the thyroid-active substances are administered for relatively short periods, i. e. not exceeding two months. Short-term administration at favorable periods can result in dramatic increases (REINEKE, 1942). Unfortunately, because it takes some time for the animal's own thyroid to recover from its quiescent state, when administration ends there is often an equally spectacular decrease in yield, which is due to the sudden fall in the thyroxine level in the blood (SWANSON, 1954; COWIE, 1956).

The careful administration of thyroid-active substances to dairy cows has no deleterious effect on the health and reproductive ability of the cows (LEECH and BAILEY, 1953). Unfortunately, there is only a net gain in yield of about 3 per cent, so that the practical application of the procedure seems to be limited (COWIE and FOLLEY, 1961).

Thyrotropin, as was to be expected, increases milk yield by increasing the output of thyroxine from the thyroid (COWIE, 1956). It is unlikely that thyroxine has a specific effect on the alveolar cells, so that its effect on galactopoiesis should probably be ascribed to its general metabolic action (MEITES, 1961 a, b).

Summarizing, it seems that TSH and its secondary hormones are not required for the first stages of lactation. Similarly, thyroxine and thiouracil do not affect hypothalamic PIF (CHEN and MEITES, 1969).

Chapter 8

Oxytocin and Vasopressin

There can no longer be any doubt that oxytocin controls milk ejection, and it is generally accepted that milk ejection is mediated mainly by a humoral mechanism—the release of oxytocin in response to suckling (BISSET, 1968). It is also clear that suckling not only initiates the milk ejection reflex, but it is also necessary for maintenance of milk secretion in the lactating rat. The salient point would then be the problem whether synthesis of milk by the cells of the alveolar epithelium and the passage of milk from the cytoplasm of the alveolar epithelial cells into the alveolar lumen are dependent on oxytocin as assumed by SELYE (1934 a, b) and SELYE, COLLIP and THOMPSON (1934). It was concluded years ago that involution of the mammary glands in the rat after weaning is due not to accumulation of secreted milk, but to withdrawal of the suckling stimulus, and it was suggested that suckling stimulated also the secretion of prolactin from the anterior pituitary. Thus, PETERSEN (1944) as well as BENSON and FOLLEY (1956), BENSON (1960) and BENSON, COWIE and TINDAL (1958) advanced the idea that the oxytocin released by suckling constitutes a stimulus for prolactin release. This hypothesis was based on their success in retarding the involution of the lactating mammary gland by injections of oxytocin. Their theory was given weighty support by their finding that oxytocin stimulated mammary development only in intact—but not in hypophysectomized—mothers. Thus, the theory was developed that oxytocin, released from the nerve endings in the neurohypophysis by the milk-ejection reflex, reaches the adenohypophysis via the hypophysial portal vessels and stimulates it to release prolactin. This hypothesis is supported by the studies of McCANN, MACK and GALE (1959) who found that rats, bearing hypothalamic lesions in the supraoptic tract, suffered from a block of milk secretion and ejection which could be overcome by oxytocin.

MEITES and his group (MEITES, 1959 a; MEITES and NICOLL, 1959; MEITES and HOPKINS, 1961) demonstrated that a combination of prolactin with oxytocin is more effective than prolactin alone in retarding mammary involution, but they felt that two independent mechanisms controlled the release of oxytocin and prolactin by suckling, and that oxytocin retarded mammary involution not by releasing prolactin, but by a direct local action on the epithel of the mammary gland. The emptying of the alveoli by the action of oxytocin on the myoepithelium reduces the pressure of accumulated secretion against the epithelial cells and the blood capillaries, thus permitting synthesis of milk to proceed under the influence of prolactin (MEITES, NICOLL and TALWALKER, 1960; MEITES, NICOLL and TALWALKER, 1960).

Experiments by GALA and REECE (1965 a) and by our group (MISHKINSKY, DANON and SULMAN, unpublished results) both *in vivo* and *in vitro*, favor the concept that oxytocin can release prolactin from the rat anterior pituitary, and that any discrepancies are due to the fact that the effect is dose-dependent: Small doses of oxytocin stimulate prolactin release, but larger doses have in inhibiting effect. This aspect is still under intensive research, requiring the use of long-acting oxytocin preparations or repeated daily injections. Nevertheless, it is clear that whether or not oxytocin stimulates indirectly milk secretion by releasing prolactin, its direct action on the mammary gland is the cause of ejection of pre-formed milk from the alveoli.

For many years, the pharmacological activities of oxytocin and vasopressin were found to be overlapping. This was due to incomplete separation of the two hormones, and it was not until 1941 that ELY and PETERSON concluded that oxytocin was the natural hormone responsible for milk ejection, while the effect of vasopressin on the mammary gland was due to contamination by oxytocin. Moreover, a third factor was thought to be involved in the posterior pituitary; it was defined by TURNER, PIERCE and DU VIGNEAUD (1951) as arginine-vasopressin and found to have about one-fifth the potency of oxytocin when tested for milk-ejecting activity. Thus, it was concluded that milk-ejecting activity, like oxytocin activity, is an intrinsic property of vasopressin. It seems that although vasopressin may be released during suckling, it is the release of oxytocin which brings about milk ejection. The intravenous injection of this hormone will produce milk-ejection ("let-down") within a minute. This has as yet no practical application in dairy husbandry. Therapeutically it may be useful in sows which fail to eject milk after farrowing.

Some doubt has been expressed on whether oxytocin is the milk ejection factor in primiparous women. It is certainly active in pharmacological doses (2—4 IU intranasally) and increases milk yield. The milk-ejection reflex may be excited by various kinds of conditioned stimuli, but it is also subject to inhibition by emotionally stressful situations, in women as well as in cows (Fig. 13).

The reaction of oxytocin release in response to conditioned stimuli associated with machine milking in the cow was recently studied by CLEVERLEY (1967). Each cow was accustomed to a strict milking routine, which included the presence of the experimenter, for at least 10 days before blood sampling. Approach and entry of the milker stimulated the release of oxytocin. These results provide, for the first time, direct experimental evidence that the release of oxytocin can be conditioned to auditory and visual stimuli associated with the milking process. Similarly, washing the udder and foremilking also stimulated the release of oxytocin. Injections of 100 MU synthetic oxytocin into the external jugular vein near milking time elicited a rise in intramammary pressure equivalent to the natural ejection recorded during milking, with a latency of 40—50 sec. Recording was continued for 1 hr. after milking or oxytocin injection, and during this time the elevated intramammary pressure was maintained. These experiments show that oxytocin is released transiently and at low levels during milking in the cow. It can be released in response to conditioned stimuli, and/or direct physical stimuli applied to the udder.

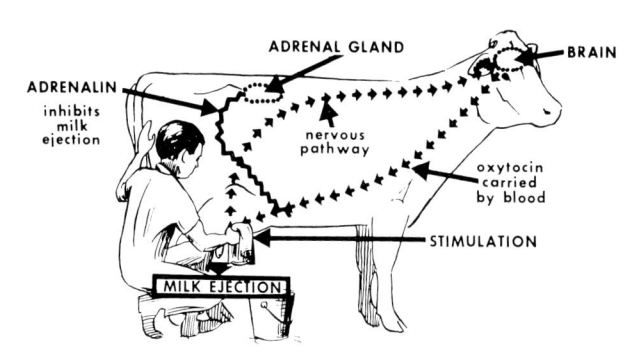

Fig. 13. Effect of oxytocin on milk yield: Milk ejection ("let-down") by oxytocin and milk retention by adrenaline following stress, pain or fright (by courtesy of F. P. NATZKE, A. M. MEEK and D. K. BANDLER, 1967, N. Y. State College of Agriculture, Cornell University, Ithaca, N. Y.). — 1. Udder washing stimulates nerves which carry the message to the brain to release oxytocin to the blood. Oxytocin reaching the udder causes milk ejection. 2. Pain or fright releases adrenalin which blocks the action of oxytocin (milk ejection)

In conclusion, it emerges that oxytocin (and vasopressin) can increase milk output, the main mechanism being the emptying of the myoepithelial elements of the mammary gland. Its central effect on the PIF or PRF is doubtful because it seems to be dose-dependent. This requires further study, in view of the fact that the posterior pituitary releases oxytocin during nursing, suckling and milking by direct stimuli as well as by conditioned stimuli.

Chapter 9

Parathormone and Calcitonin

Not many data exist on the effect of the parathyroid on lactation. Experiments by SMITH, STOTT and WALKER (1957) were carried out on parathyroid and thyroidectomized goats. They found an inhibition of lactation.

Parathormone appears to be necessary for full lactation, and its absence can result in interference with the calcium-concentrating mechanism of the mammary gland (COWIE and FOLLEY, 1961). No systematic study of the parathyroid hormone and/or calcitonin and their influence on mammogenesis or lactation has been carried out. There exists, however, one important clinical observation during lactation: Many species, especially cows, dogs and pigs, can develop lactation tetany, as their demand of calcium for milk production is excessively high. The condition is rare in women but, if observed, it points to a parathyroid deficiency.

The effect of calcitonin on the calcium and phosphorus levels of lactating cows has been studied by BARLET (1967). Injecting 20 RMC Units, he obtained a 20% reduction in blood calcium level and a 35% decrease in blood phosphates. These results indicate the possible importance of thyroid calcitonin in the development of lactation tetany.

Chapter 10

Insulin and Glucagon

Insulin does not seem to affect lactation *in vivo* unless given in relatively large doses. In this case the milk yield is reduced, apparently due to the severe hypoglycemia which may affect the supply of galactose. This detrimental effect was abolished by simultaneous administration of insulin and glucose (KRONFELD, MAYER, ROBERTSON and RAGGI, 1963). Smaller doses, which do not lower blood sugar levels substantially, were tried in guinea pigs (0.2 IU insulin/day) for 14 days after parturition. This treatment improved the milk yield and induced increased growth of the offspring (CORDELLI, CASTELLI and GORELLI, 1960).

Small doses of insulin, if given together with ovarian hormones, can apparently induce some development of the mammary ducts in hypophysectomized rats; presumably insulin corrects some of the metabolic deficiencies which result from pituitary ablation (AHRÉN and JACOBSOHN, 1956; JACOBSOHN, 1961). However, it has not been demonstrated that injections of insulin alone can elicit mammary growth in either intact or hypophysectomized animals. The stimulating effect of insulin on mammary tissue *in vitro* has recently been shown by Topper (unpublished results of Laurention Hormone Conference 1969).

Glucagon has not been studied with regard to its effect on lactation. It may, however, be safely assumed that small quantities stimulate lactation due to glucose mobilization. On the other hand, larger doses have been shown by Paloyan et al. (1965, 1967) to induce hypocalcemia, which would be detrimental to milk yield.

Chapter 11

Pineal Gland

Until recently very little was known about the effect of the pineal gland on mammary development and activity. Kitay and Altschule (1954), in their review of the literature on the physiology of the pineal gland, quote three references to the "galactogenic" action of pineal extracts, which in fact are reports of milk ejection occurring after the administration of these extracts. It is now known that they may contain arginine-vasotocin (Pavel and Petrescu, 1966). The question of a possible pineal endocrine effect on milk secretion was considered by Thieblot (1954), who concluded that, while no direct evidence of any effect on milk secretion was available, there was, however, proof that the pineal can inhibit the pituitary and its luteotropic activity, and that further studies on the pineal in relation to milk secretion were called for.

In a study carried out by Mishkinsky, Nir, Lajtos and Sulman (1966), female rats were pinealectomized 21 days after birth. When 26, 35, 50 and 100 days old, their inguinal mammary glands were excised and the degree of mammary development evaluated (MTI). Inguinal mammary glands of normal and sham-operated female rats were also investigated for comparison. As in the controls, the maximal development in the pinealectomized animals reached MTI stage 2, indicating that no change had occurred in their mammary development. In addition, in order to promote development of the mammary gland, 21-day-old rats weighing 40 ± 4 g, and adult female rats weighing 200 g, were primed for ten days with 10 µg estradiol/kg/day and

8 μg estradiol/rat/day respectively and then pinealectomized. Their mammary glands were examined six days later and again the MTI (stages 1—2) did not differ from that of the normal and sham-operated controls subjected to the same treatment.

As pinealectomy did not enhance mammary development nor the release of mammogenic or lactogenic hormone in female rats between the ages of 21 and 100 days, it became obvious that the pineal has no inhibitory effect on the development of the mammary gland.

In a further study attention was concentrated on pineal control of lactogenesis and maintenance of milk secretion in mother rats after parturition. This was done by comparing the milk yields of pinealectomized rats with those of untreated and sham-operate controls (NIR, MISHKINSKY, ESHCHAR and SULMAN, 1968). Female rats were pinealectomized when 21 days old and mated when four to six months old. Fourteen, 16, 18 and 20 days post-partum their milk yields were determined by the increase in the litter weights during suckling (after the litters had been adjusted to six pups each). The milk yields of the pinealectomized mothers were about 13 per cent lower (except on the 20th day) than those of the sham-operated controls, but not lower than those of the intact controls.

Our results suggest that removal of the pineal causes some depression of lactation in the rat, yet the differences in yields between the groups did not reach the accepted level of statistical significance ($P > 0.05$). Moreover, interpretation of the effects of pinealectomy was complicated by the unexpected tendency of the sham-operated controls to have higher milk yields than the intact controls. Further research is clearly necessary in order to determine the true significance of these effects before the role of the pineal in lactation can be established.

Chapter 12

Placental and Amniotic Fluid

In recent years it has been claimed that the human placenta, besides its gonadotropins, secretes also a mammotropic hormone, which has been called "Purified Placental Protein" (PPP) by FLORINI, TONELLI, BREVER, COPPOLA, RINGLER and BELL (1966), "Human Placental Lactogen" (HPL) by JOSIMOVICH and MACLAREN (1962),

"Chorionic Growth Hormone Prolactin" (CGP) by KAPLAN and GRUMBACH (1965 a, b), and "Placental Protein" (PP) by FRIESEN (1965). FLORINI's 70% pure PPP was reported (1966) to contain 2—4 IU/mg of prolactin, as assayed by the pigeon crop sac method.

Although EHRHARDT demonstrated prolactin activity in extracts of human placenta as early as 1936, considerable doubt was raised whether there exists a human placental lactogenic hormone (JOSIMO-VICH and MACLAREN, 1962). The existence of a prolactin-like substance (HPL) with immunological and lactogenic properties *in vivo* is now well established (FLORINI et al., 1966; FORSYTH, 1967), and its effect *in vitro* was shown by TURKINGTON and TOPPER (1966). It has neither hyperglycemic nor diabetogenic effect (KURCZ, 1967). HPL seems, therefore, to fulfill an essential function both in mammogenesis and colostrogenesis.

In a recent study by our group (MISHKINSKY, ZANBELMAN, KHAZEN and SULMAN, 1969) the mammotropic and prolactin-like effects of rat placenta homogenates, amniotic fluid and placental lactogen, as well as human placental lactogen and amniotic fluid, were measured by different assay-methods. Mammary development (MTI) was determined on days 2, 4, 7, 10, 14, 17 and 20 of pregnancy or pseudo-pregnancy (uterine-horn scratch).

Suspensions of rat placentas were prepared in normal saline by homogenizing placentas taken from pregnant rats at a later stage of pregnancy (16th day) with a cone-glass electric homogenizer soaked in ice. When removing the placenta, the amniotic fluid was collected and stored at $-20°$ C. The amniotic fluid or placental suspensions were injected subcutaneously once daily, for five days, to groups of 20—24 virgin rats, weighing 200 ± 10 g, and primed with estradiol (8 µg/rat/day) for 10 days. The doses chosen were the equivalent of 1 or 4 placentas/rat/day, and 0.5 or 2.0 ml amniotic fluid/rat/day; 24 hours after the last injection the rats were decapitated and their MTI evaluated. Rat placental lactogen (RPL) was prepared according to the method of JOSIMOVICH and MACLAREN (1962) for preparation of HPL: 100 rat placentas were homogenized in 2 vol. of chilled tris-maleate buffer at pH 8.6 and brought gradually to pH 4.5, centrifuged, lyophilized and weighed (yield was 20—22 mg/placenta).

The crop sac test was performed according to NICOLL (1967), and statistical evaluations were made by the "rank sum" method for ordinal data described by WHITE (1952).

The MTI values of different stages of pregnancy and pseudopregnancy in the rat are presented in Fig. 14. The MTI of pregnant rats increases rapidly throughout pregnancy. However, no mammotropic development was noticed in the adult pseudopregnant rats, nor was any clear-cut correlation established between the MTI of virgin pseudopregnant rats and the weights of their uteri (Fig. 15), though both phenomena are prolactin-dependent, and the uterine weight curve partly resembles the ascending curve of prolactin-activity dur-

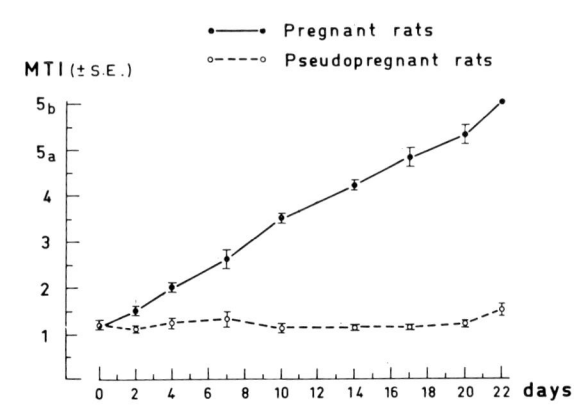

Fig. 14. Development of mammary gland in pregnant rats compared with lack of mammary differentiation in pseudopregnant rats

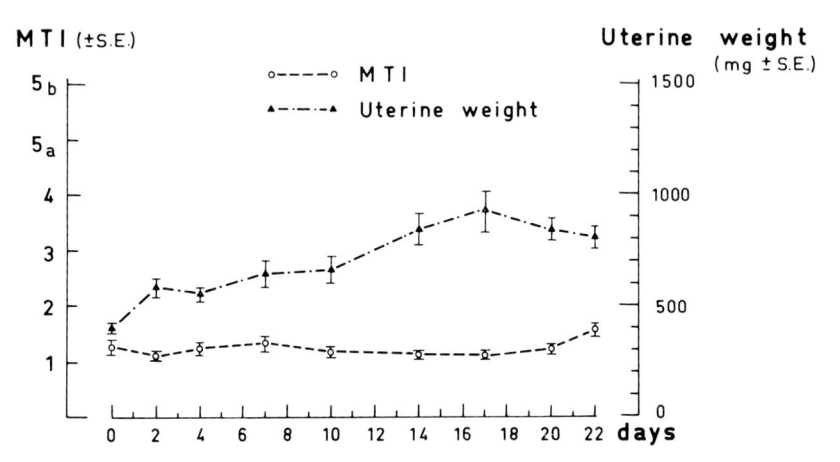

Fig. 15. Development of mammary gland in pseudopregnant virgin rats (lower curve) compared with decidual transformation and weight increase in uterus (upper curve)

ing pregnancy. Similar results were reported by DESJARDINS, PAAPE and TUCKER (1968).

The mammotropic activity of rat placental homogenates injected s.c. was evidenced only when the equivalent of 4 placentas was injected daily into each primed rat. It required this concentration for the difference between the MTI obtained after placenta homogenate injections and that of the controls to become significant ($P < 0.05$). Rat amniotic fluid had no evident mammotropic activity when injected s.c. into primed rats, nor when injected intradermally (0.6 ml native or $10 \times$ concentrated) over the pigeon crop. Rat placental lactogen, however, was found to have a clear prolactin-like effect when measured by the pigeon crop sac proliferation test.

Human placental lactogen (HPL) was found to be mammotropic in all doses tested (1—100 mg/rat/day). In the crop gland test it yielded also a prolactin-like effect in doses of 0.015—1 mg. On the other hand, human amniotic fluid was found to have a moderate mammotropic effect when 0.5 or 2.0 ml were injected s.c. into primed rats, but was devoid of prolactin-like activity when injected intradermally (0.06 ml native or $10 \times$ concentrated) over the pigeon crop. From these findings it emerged that rat and human amniotic fluids contain less than 0.01 IU/ml of prolactin or prolactin-like substance.

The study of HPL and RPL was undertaken by us because the experiments of WRENN, BITMAN, DE LAUDER and MENCH (1966) failed to prevent the decline in mammary development of pseudopregnant rats by various placental suspensions. Such experiments succeeded only when placental suspensions were injected into hypophysectomized-ovariectomized rats (RAY, AVERILL, LYONS and JOHNSON, 1955). Our findings agree with those of WRENN et al. (1966) and MATTHIES (1967), who found that the minimum effective dosage of rat fetus and placental extracts from the twelfth day of pregnancy, capable of stimulating the decidual response in hypophysectomized rats, were the equivalent of one placenta per day, injected s.c.

Summarizing, it is concluded that human and rat placenta contain a prolactin-like substance. The amniotic fluid of women contains only minute quantities of placental lactogen, and that of the rat does not contain any measurable amount at all (< 0.01 IU/ml).

Part II

Pharmacological Regulation of Lactation

Chapter 13

Hypothalamic Lactation Produced by Psychopharmaca

a) History

Tranquilizing drugs have been reported to interfere with a variety of endocrine functions, including lactation, in man and experimental animals. In 1954 WILKINS reported the development of gynecomastia following treatment with Rauwolfia alkaloids. WINNIK and TENNENBAUM (1955) showed that treatment with chlorpromazine (Largactil, Thorazine, Megaphen) might result in milk secretion in women. SULMAN and WINNIK (1956), and WINNIK and SULMAN (1956), studying this problem, described abnormal lactation in women and animals after treatment with chlorpromazine. Abnormal lactation was also reported to occur in women who received either reserpine or chlorpromazine (MARSHALL and LEIBERMAN, 1956; PLATT and SEARS, 1956). This abnormal lactation was confirmed in experimental animals as well. Treatment with reserpine resulted in milk secretion in rabbits (MEITES, 1957a, 1963; SAWYER, 1957; MEITES, NICOLL and TALWALKER, 1959, 1963) and in rats (DESCLIN, 1958; GRÖNROOS, KALLIOMÄKI, KEYRILÄINEN and MARJANEN, 1959). Administration of various phenothiazine derivatives also resulted in mammary differentiation and milk secretion in rats (KHAZAN, PRIMO, DANON, ASSAEL, SULMAN and WINNIK, 1962; TALWALKER, RATNER and MEITES, 1963; BEN-DAVID, DIKSTEIN and SULMAN, 1965). This endocrine disturbance is presumed to result from the action of tranquilizers on the reticular formation and hypothalamic regions which regulate the normal function of the pituitary (COURVOISIER, FOURNEL, DUCROT, KOLSKY and KOETSCHET, 1953; DELAY and DENIKER, 1953; PLUMER, EARL, SCHNEIDER, TRAPOLD and BARRETT, 1954). Thus the name "hypothalamic tranquilizers" has been given to these agents from which the conception of endocrine disturbances inducing lactation originated. They do not seriously affect "balanced tropins," i. e. tropins which are held in abeyance by the secondary hormone they produce, as shown in Fig. 16 (SULMAN, 1956 b). They do, however, seriously interfere with the function of the "unbalanced tropins" (LTH, MSH, ADH and oxytocin) as the latter do not produce secondary hormones which could counteract an overshoot produced by hypothalamic depression. Consequently, hypothalamic lactation results (Fig. 17) (SULMAN, 1959).

Many drugs and nonspecific agents have been shown to elicit mammary growth and secretion in estrogen-primed rats and rabbits if

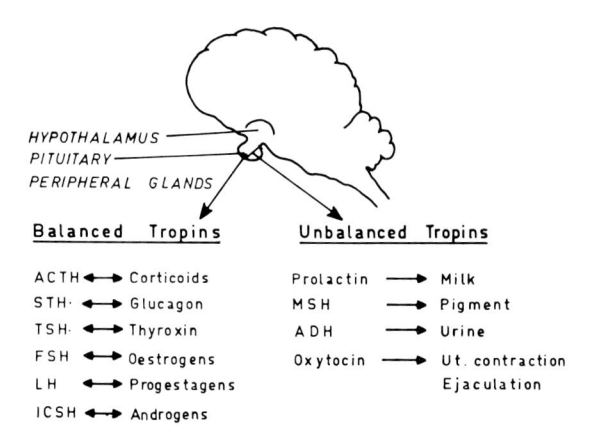

HYPOTHALAMUS
PITUITARY
PERIPHERAL GLANDS

Balanced Tropins	Unbalanced Tropins
ACTH ⟷ Corticoids	Prolactin ⟶ Milk
STH ⟷ Glucagon	MSH ⟶ Pigment
TSH ⟷ Thyroxin	ADH ⟶ Urine
FSH ⟷ Oestrogens	Oxytocin ⟶ Ut. contraction
LH ⟷ Progestagens	Ejaculation
ICSH ⟷ Androgens	

Fig. 16. Hypothalamic lactation: Balanced tropins have a positive and negative feedback mechanism. Unbalanced tropins react by overshoot when the hypothalamus is suppressed by phenothiazines

Hormonal reaction	Balanced tropins	Unbalanced hormones ↑
Week	1 2 3 4	Prolactin: Milk ↑
Hypothalamus	↓ ↓ ↓ ↓	MSH: Pigment ↑
Pituitary	— ↓ ↓ —	ADH: Urine ↓
Peripheral hormones	— — ↓ —	Oxytocin: Ut. Contract. ↓ Ejaculation ↓

Fig. 17. Reaction to hormonal depression by phenothiazine derivatives: Balanced tropins overcome depression of hypothalamus by feedback. Unbalanced hormones have no peripheral feedback. This results in stimulation of prolactin and MSH and in suppression of ADH and oxytocin

the priming dose is set high, and to maintain secretory activity in postpartum rats after litter removal (MEITES, 1963; MEITES, NICOLL and TALWALKER, 1963). The effective drugs in rats include—in addition to chlorpromazine and reserpine—meprobamate, adrenaline, noradrenaline, serotonin, atropine, dibenamine, eserine, pilocarpine, amphetamine, morphine, 3-methylcholanthrene, 3,4-benzpyrene, 9,10-dimethyl-1,2-benzanthracene and 10% formaldehyde. Stress, such as cold, heat, restraint, and electrical stimulation also induce mammary growth. The above "extra-hypothalamic" psychopharmaca will only provoke lactation if estrogen priming has been too high, e. g. 10 μg/rat/day for 10 days. We have shown that meprobamate, hydroxyine, ethinamate, phenobarbital and benactyzine are devoid of any mammotropic effect in rats primed for 10 days with sub-threshold doses (8 μg/day) of estrogen (BEN-DAVID, DIKSTEIN and SULMAN, 1965). On the other hand, when high threshold doses of estrogen (10 μg/day) were used for priming, many non-specific stressful stimuli, as reported above, can initiate secretion by the mammary gland.

b) Dissociation of Mammotropic and Sedative Effects

In a trial to dissociate the sedative from the mammotropic property of hypothalamic tranquilizers (KHAZAN, BEN-DAVID, MISHKINSKY, KHAZEN and SULMAN, 1966), two ways of approach were chosen in assaying the mammotropic effect in rats: Biological dissociation by combined treatment with tranquilizers and energizers, controlled by behavioral and electroencephalographic studies; chemical dissociation by changes in the molecule of phenothiazine tranquilizers which would eliminate the sedative effect. This effect was gauged by prolongation of the barbital sleep induction time in mice, and lowering of the body temperature (BTD) in rats (GREIG and MAYBERRY, 1951; CHILD, SUTHERLAND and TOMICH, 1961, 1962). The BTD is a standard test for hypothalamic depression, since the thermoregulatory centers of the thalamus are connected with the emotional centers of the hypothalamus.

c) Biological Dissociation through Combined Treatment with Tranquilizers and Psychostimulants

KHAZAN, BEN-DAVID, MISHKINSKY, KHAZEN and SULMAN (1966) investigated whether the mammotropic property of a tranquilizer is independent of its sedative effect and resistant to changes of mood produced by psychostimulants which block the sedative-depressant effects.

Groups of 10 adult female rats, weighing 200 ($\pm$ 10) g, were used in each series of experiments. They were primed by daily s.c. injections of 8 µg estradioal in oil for a period of 10 days. From the 11th to the 15th day the following daily treatments were administered s.c.: (1) saline as control, (2) perphenazine HCl 2.5 mg/kg, (3) tetrabenazine methane sulphonate 5 mg/kg, (4) combined treatment of etryptamine 5 mg/kg and perphenazine·HCl 2.5 mg/kg, (5) combined treatment of imipramine·HCl 10 mg/kg with tetrabenazine methane sulphonate 5 mg/kg, (6) combined treatment of ephedrine·HCl 5 mg/kg with perphenazine 5 mg/kg, (7) etryptamine 5 mg/kg alone, (8) imipramine ·HCl 10 mg/kg alone, and (9) ephedrine 5 mg/kg alone. From the 11th to the 16th day, the EEG and overt behavior of the rats were evaluated three times daily (Fig. 18). On the 16th day the rats were sacrificed and their MTI determined.

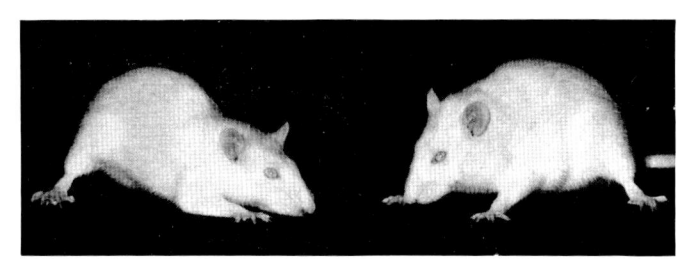

Fig. 18. Effect of perphenazine on overt behavior of rats after 5 days' treatment with: r: perphenazine 2.5 mg/kg s. c. alone—note tranquilizing effect; l: perphenazine 5 mg/kg s. c. combined with etryptamine 5 mg/kg s. c.—note alert behavior

The control rats showed normal proliferation of the mammary gland with an average MTI of 1—2. Perphenazine- and tetrabenazine-treated rats showed highly developed mammary glands (MTI 4—5). Combined treatment with etryptamine, imipramine or ephedrine, which blocked the sedative effects of perphenazine and tetrabenazine on the EEG and on overt behavior, nevertheless yielded a full mammotropic effect (MTI 4—5). The EEG and overt behavior were normal during the 5 days of observation (Fig. 19).

From the results obtained it can be concluded that the mammotropic and sedative effects of tranquilizers are not correlated, but are separate properties inherent in their molecules. The mammotropic properties are clearly independent and completely resistant to energizers which antagonize the tranquilizing effect.

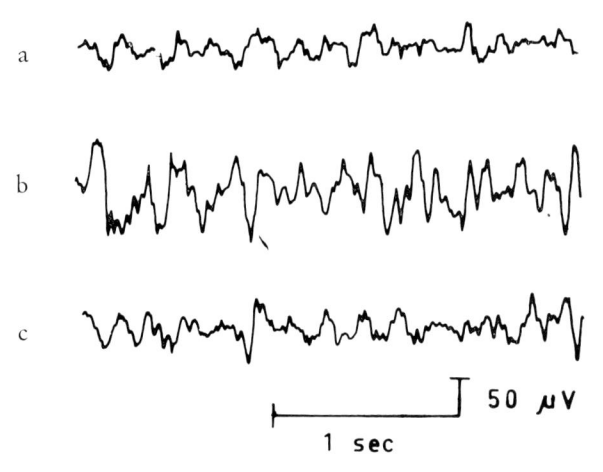

Fig. 19. Electrocortical tracings from unrestrained chronically implanted rats (a). Note the high voltage low frequency tracings induced by perphenazine (2.5 mg/kg) in (b), compared with the normal awake recording in (a). In (c) the desynchronizing action of etryptamine on the perphenazine effect is obvious

d) Chemical Dissociation by Molecular Changes of Tranquilizers

KHAZAN, BEN-DAVID, MISHKINSKY, KHAZEN and SULMAN (1966) used chemical changes for dissociation to evaluate the tranquilizing and mammotropic effects of three typical substances: a phenothiazine sulfoxide (RP-8580), 2-chlorpromazine and 2-bromopromazine (Fig. 20). RP-8580 showed no tranquilizing, but strong mammotropic effects. Chlorpromazine showed similar tranquilizing effects as 2-bromopromazine, but was much less mammotropic. From these results it can be concluded that the mammotropic and sedative effects of tranquilizers are separate properties inherent in their molecules. They may have a partially overlapping affinity to hypothalamic receptors, but the tranquilizing and mammotropic effects are by no means coupled with each other. Lack of parallelism between tranquilizing and mammotropic effects of 15 reserpine derivatives was described by BEN-DAVID, KHAZEN, KHAZAN and SULMAN, 1968) (Chapter 16). BEN-DAVID (1968 a) recently found that persistent influence of a drug must be maintained in order to induce artificial hypothalamic lactation. This became evident when a single high dose

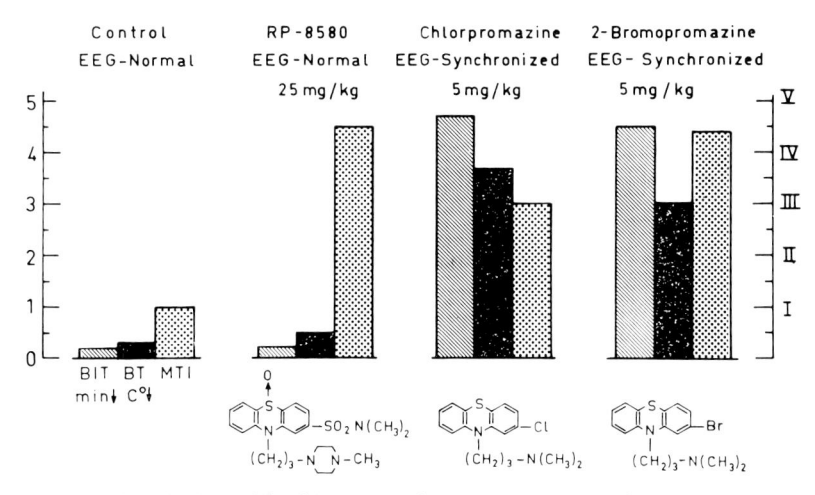

Fig. 20. Dissociation of barbiturate induction time (BIT), body temperature drop (BT) and mammotropic index (MTI), obtained with three different phenothiazine derivatives: RP-8580, yielding only high MTI; chlorpromazine, yielding high sedative reaction but low MTI; 2-bromopromazine, yielding high BIT and MTI but low BT

of fluphenazine-HCl (the short-acting form of the drug) failed to induce mammary differentiation comparable to that obtained by one injection of a high dose of fluphenazine enanthate (Chapter 14). The latter long-acting form of the drug converted the non-developed gland (control) into a well-developed gland with copious milk secretion. Thus it seems that the depressant activity of a phenothiazine drug on the hypothalamic prolactin-inhibiting factor (PIF)—which results in oversecretion of pituitary prolactin—can be accomplished only by continuous exposure to the drug for a minimum of 4 days.

e) Studies in Hypophysectomized Rats

Direct evidence that perphenazine stimulates pituitary prolactin secretion by suppressing the hypothalamic PIF is to be found in the *in vitro* work of DANON, DIKSTEIN and SULMAN (1963), who showed that perphenazine, added to pituitary-hypothalamus co-culture, abolished the inhibitory influence of hypothalamic fragments upon prolactin secretion by the pituitary. Nevertheless, *in vivo* no direct evidence is available concerning the mode of action of perphenazine. BEN-DAVID (1968 c), therefore compared the mammotropic effect of perphenazine in intact and hypophysectomized adult rats, some of which also received replacement therapy, in order to clarify the role of pituitary and peripheral hormones in perphenazine-induced lactation.

Experiments were carried out without any estradiol priming and started 7 days after hypophysectomy. Priming with estrogen was avoided in this study, because estrogen may induce mammary duct growth and some end-bud growth (LYONS, 1958) and stimulate prolactin secretion from the pituitary (RAMIREZ and McCANN, 1964). Animals received s.c. daily injections of perphenazine for 5 days, and they were sacrificed ony day after the last injection, and their mammotropic index determined.

Fig. 21 shows that increased doses of perphenazine treatment resulted in higher development of the mammary glands in the intact rats. A daily dose of 5 mg/kg of perphenazine for 5 days was able to change undeveloped glands into the well developed and secretory type. Moreover, the more the lobulo-alveolar system was developed, the less fat tissue was found in the perphenazine-treated intact rats. However, the same doses of perphenazine, when given to hypophysectomized animals, failed to induce any mammotropic effect even at the high dose of 10 mg/kg. Results of the mammotropic effect of 5 days' treatment with perphenazine (5 mg/kg/d), estradiol (0.01 mg/kg/d), hydrocortisone (10 mg/kg/d) and ovine prolactin (1 mg = 24 IU/kg/d), individually and combined, showed that perphenazine treatment alone in intact rats resulted in a MTI of 4.7, while such development was never achieved in hypophysectomized animals with any combination of treatment, comprising perphenazine and estradiol or hydrocortisone or ovine prolactin. In intact rats, the additional hormonal treatment improved the lactogenic effect achieved by perphenazine, but no initiation of milk secretion was found in any of the hypophysectomized rats receiving the same combination of treatment.

The failure of perphenazine to induce mammary gland development in the hypophysectomized animals at any dose level clearly

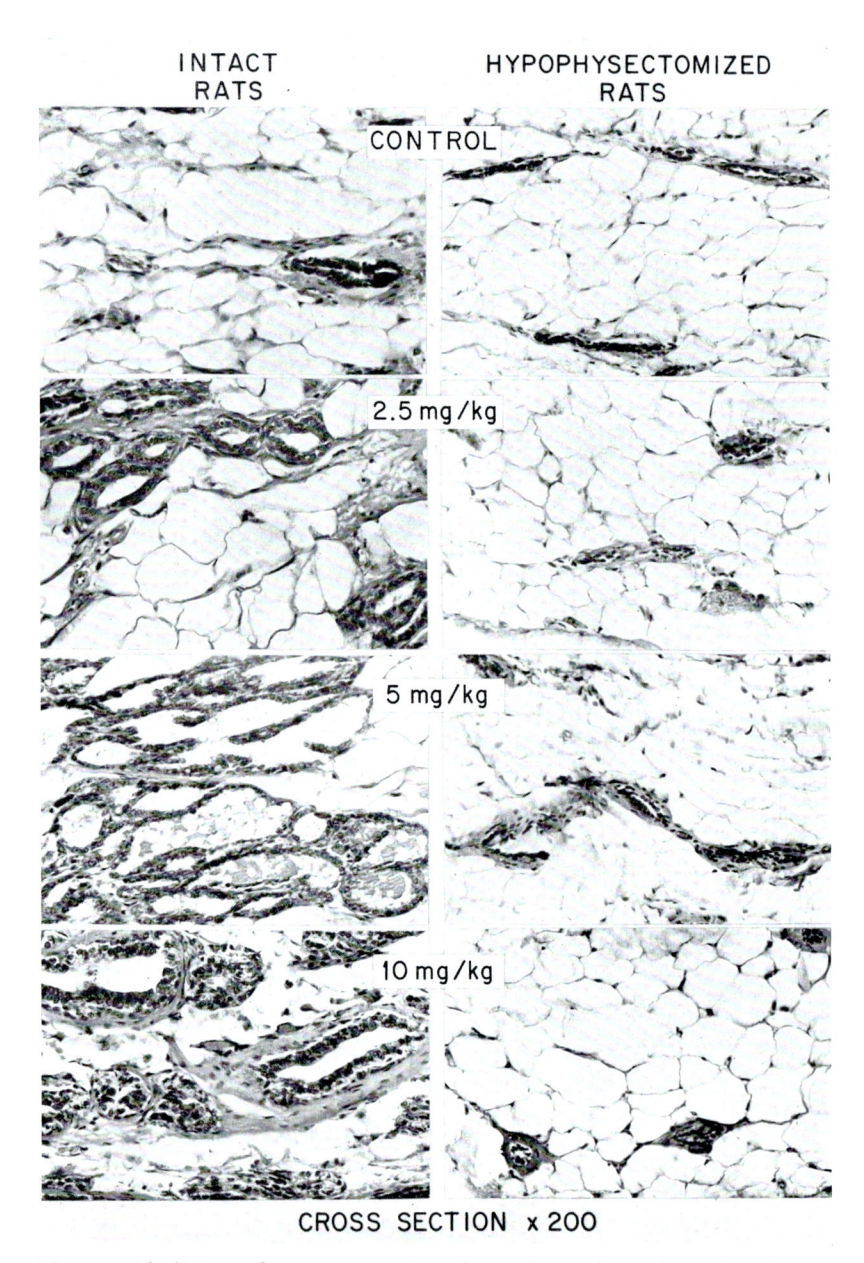

Fig. 21. Abolition of mammotropic effect of perphenazine after hypophysectomy: l: mammotropic reaction of intact rats after 5 days' treatment with 2.5, 5 and 10 mg/kg perphenazine respectively: MTI 5; r: complete absence of mammotropic reaction in hypophysectomized rats: MTI 1.5 after same treatment

indicates that the presence of the hypophysis is essential for this process, and that a drug endowed with mammotropic effect acts by stimulating secretion of mammotropic hormones from the pituitary.

Further evidence has been accumulated to show that perphenazine acts mainly via the hypothalamus, through stimulation of pituitary prolactin rather than other pituitary tropins. Moreover, the aforementioned results show that, in hypophysectomized rats, additional "replacement" treatment with steroid hormones was unable to produce a mammotropic effect similar to that achieved in intact animals by perphenazine alone. It is interesting that addition of ovine prolactin to the treatment failed substantially to improve the mammotropic effect in the hypophysectomized animals, a fact which could be explained by species differences in the source of prolactin (LEVY and SAMPLINER, 1962) or by the use of a subthreshold dose (24 IU/kg/d for 5 days), or by both together.

f) Conclusions

Drugs which depress hypothalamic activity, such as derivatives of phenothiazine, reserpine, butyrophenone, etc., can induce lactation. This effect cannot be prevented by hypothalamic energizers. The effect is specific and cannot be obtained by tranquilizing or sedative drugs which are not active at the hypothalamic level. Such drugs may evoke lesser degrees of lactation in as far as they affect the hypothalamus along with the cortical or internuncial synapses. In this case, over-priming with estrogen can unmask and potentiate their lactogenic effect. The sedative effect *per se* is not decisive for any lactogenic activity, indeed a separation of both properties seems possible. The mechanism of the lactogenic effect of all hypothalamic tranquilizers is based on the depression of the hypothalamic PIF; it depends on the presence of the hypothalamus and the pituitary.

Chapter 14

Mammotropic Effect of Phenothiazine Derivatives

a) History

Phenothiazine derivatives have a broad spectrum of pharmacological effects: antiemetic, analgetic, antihistaminic, adrenergic blocking, anticholinergic, antipsychotic, antipruritic, antitussive, antinauseant, antispasmotic, tranquilizing and mammogenic (HOLLISTER, 1964).

Galactorrhea in female patients, and breast development in males, after chlorpromazine treatment, were reported by various authors (WINNIK and TENNENBAUM, 1955; POLISHUK and KULCSAR, 1956; SULMAN and WINNIK, 1956; WINNIK and SULMAN, 1956; ROBINSON, 1957; HOOPER, WELCH, POINT and SHACKELFORD, 1961). Most authors related these effects to an interference with the normally-functioning endocrine release system of the hypothalamus. This view is strengthened today by a more exact knowledge of the mechanism of lactation, according to which depression of the hypothalamus causes suppression of the hypothalamic prolactin-inhibiting factor (PIF), thus enhancing prolactin secretion from the anterior pituitary.

Preliminary reports on the structure-activity relationship (SAR) of some phenothiazines with regard to their mammotropic effect were published by our Department (KHAZAN et al., 1962; BEN-DAVID, DIKSTEIN and SULMAN, 1965; KHAZAN, BEN-DAVID, MISHKINSKY, KHAZEN and SULMAN, 1966; KHAZEN, MISHKINSKY, BEN-DAVID and SULMAN, 1968 a, b) and by others (MEITES, 1957 a; BHARGAVA and CHANDRA, 1964; WELSH, 1964; ROSSI, 1964; LAUNCHBURY, 1964), mostly as secondary parameters of chemical and pharmacological research.

b) Mammotropic Index (MTI)

We undertook to screen all available phenothiazine and related structures in clinical and research use (tranquilizers and non-tranquilizers), in order to establish the rules of the SAR of their mammotropic effect. The two parameters used were the index of mammary development, "mammotropic index" (MTI), and body temperature drop (BTD) after subcutaneous injection of the substances. Measurement of these two parameters was performed as described in Chapter 26 — b.

Female albino rats of the Hebrew University "Sabra" strain, weighing 200 ± 10 g, were primed for 10 days s.c. with 8 µg/rat/day estradiol. On the 11th—15th day the drug to be tested was injected s.c. in doses of 5—40 mg/kg/day. On the 16th day the animals (5—10 per drug) were sacrificed and the MTI determined. Maximum development of the mammary glands of the controls (estradiol-primed for 10 days and saline-injected for the following 5 days) was found to reach MTI 2, with a mean MTI of 1.2. It is essential to keep strictly to a low priming dose (in our rats 8 µg/rat/day of estradiol) and to use female rats of 200 ± 10 g, because animals of larger weight or primed with bigger doses may tend to give a higher MTI and thus yield false positive results.

c) Body Temperature Drop (BTD)

Groups of 10 female rats (200 ± 10 g) were injected s.c. with a single dose of the phenothiazines. Temperatures were recorded at 0, 1, 2, 3, 4 and 5 hours with an electrocouple thermometer (Electric Universal Thermometer TE_3, "Ellab Instruments", Copenhagen). Control animals were injected with 0.2 ml saline s.c.

d) Screening of Phenothiazine Derivatives

Table 1 compares the MTI with the BTD effected by 87 phenothiazine derivatives at a dosage level of 5 and 40 mg/kg. It emerges that the most mammotropic phenothiazine derivatives were piperacetazine, perphenazine, propericiazine, thiopropazate, pipamazine, TPO-33, SKF 7100-A, 2-bromopromazine, fluphenazine, acetophenazine and SKF 7221-I.

It seems that a straight 3-C side chain originating at N_{10} is of primary importance because it may structurally form half a hexane ring, and that a heterocyclic ring at its end considerably increases mammotropic activity (cf. Chapter 19).

e) Screening of Phenothiazine Sulfoxides and Sulfodioxides

Our attention was drawn to the sulfoxide derivatives for three main reasons:

1. Only very mild oxidation is needed to convert phenothiazine into its lactogenic sulfoxide derivative, while more drastic oxidation yields the inactive sulfone derivative (FORREST and PIETTE, 1964).

2. The reversibility of the phenothiazine sulfoxide oxidation is very high: acetylpromazine and RP-8909 are oxidized by air and produce their sulfoxides very easily, while the respective sulfoxides are reduced to the parent substances only by exposure to light (YAMAMOTO and FUJISAWA, 1963).

3. Mammalian drug metabolism will normally stop at the sulfoxide level (SO). Sulfones, which contain SO_2, are, therefore, not found in biological material like blood, urine and tissue.

Table 1. *Mammotropic index of 87 phenothiazine derivatives in oestrogen-primed female rats (200 ± 10 gm). Oestradiol priming: 8 μg/rat/day, for 10 days; phenothiazine treatment: 5—40 mg/kg/day for 5 days. Preparations are listed in descending order of their lactogenic effect at 5 mg/kg*

Generic and proprietary name	R₁	R₂	Daily dose mg/kg	Mammotropic index	Body temperature drop °C
Piperacetazine (Quide, Pitman-Moore)	O=C—CH₃	—(CH₂)₃—N⟨piperidine⟩—CH₂—CH₂OH	5 40	4.8 5.2	1.8 3.9
Perphenazine (Trilafon, Schering)	—Cl	—(CH₂)₃—N⟨piperazine⟩—CH₂—CH₂OH	5 40	4.7 5.9	3.3 3.3
Propericiazine (Neuleptil, Specia) (SKF 20716)	—CN	—(CH₂)₃—N⟨piperidine⟩—OH	5 40	4.6 5.2	2.3 2.4
Thiopropazate (Dartal, Searle)	—Cl	—(CH₂)₃—N⟨piperazine⟩—CH₂CH₂—O—C(=O)—CH₃	5 40	4.6 5.0	1.1 2.0

Screening of Phenothiazine Sulfoxides and Sulfodioxides

Name	R₂	R₁₀	dose	a	b
Pipamazine (Mornidine, Searle)	—Cl	—(CH₂)₃—N⟨piperidine-4-C(=O)NH₂⟩	5 / 40	4.5 / 4.4	1.1 / 2.4
TPO-33 (Sandoz)	O=S—CH₃	—CH₂—CH₂—⟨piperidine NH⟩	5 / 40	4.4 / 3.0	0 / 0.3
SKF 7100-A (Smith, Kline & French)	—CF₃	—(CH₂)₃—N⟨piperidine-4-C(=O)N(CH₃)₂⟩	5 / 40	4.4 / 4.5	0 / 1.8
2-Bromopromazine (NIMH)	—Br	—(CH₂)₃—N(CH₃)₂	5 / 40	4.4 / —	1.7 / 7.4
Fluphenazine (Prolixin, Squibb; Permitil, White)	—CF₃	—(CH₂)₃—N⟨piperazine-N—CH₂—CH₂OH⟩	5 / 40	4.2 / 5.6	0 / 3.4
Acetophenazine (Tindal, Schering)	O=C—CH₃	—(CH₂)₃—N⟨piperazine-N—CH₂—CH₂OH⟩	5 / 40	4.2 / 3.5	0.9 / 2.1
SKF 7221-I (Smith, Kline & French)	—CF₃	—(CH₂)₃—N⟨piperidine-4-C(=O)—O—C₂H₅⟩	5 / 40	3.8 / 3.8	1.1 / 2.0

6*

Table 1 (continued)

Generic and proprietary name	R$_1$	R$_2$	Daily dose mg/kg	Mammotropic index	Body temperature drop °C
SKF 7136-A (Smith, Kline & French)	—CF$_3$	—(CH$_2$)$_3$—N⟨N⟩—SO$_2$—N(CH$_3$)$_2$	5 / 40	3.8 / 4.2	1.3 / 0.8
Carphenazine (Proketazine Wyeth)	O=C—C$_2$H$_5$	—(CH$_2$)$_3$—N⟨N⟩—CH$_2$—CH$_2$OH	5 / 40	3.8 / 3.4	0.6 / 1.6
Pasaden (Chemiewerk Homburg)	—CH$_3$	—(CH$_2$)$_3$—N⟨N⟩—CH$_2$—CH$_2$OH	5 / 40	3.8 / 1.9	0 / 1.1
Ethylbutrazine (Sergetyl, Vaillant) (Ethymemazine, Specia) (RP 6484)	—C$_2$H$_5$	—CH$_2$—CH—CH$_2$—N(CH$_3$)$_2$ | CH$_3$	5 / 40	3.6 / 4.2	0.3 / 2.2
SKF 6063-A (Smith, Kline & French)	—Cl	—CH$_2$—CH$_2$—C(=O)—N(H)—(CH$_2$)$_3$—N(CH$_3$)$_2$	5 / 40	3.6 / 3.1	0.4 / 0.4
Butyrylperazine (Randolectil, Bayer)	O=C—C$_3$H$_7$	—(CH$_2$)$_3$—N⟨N⟩—CH$_3$	5 / 40	3.5 / 5.2	1.1 / 4.8

Name	R	Side chain	Dose		
Thiethylperazine (Torecan, Sandoz)	—SC₂H₅	—(CH₂)₃—N⟨piperazine⟩N—CH₃	5 40	3.4 3.9	0 0
Acepromazine (Plegicil, Byla)	—C(=O)—CH₃	—(CH₂)₃—N(CH₃)₂	5 40	3.2 4.3	1.6 4.2
Trifluorperazine (Stelazine, SKF)	—CF₃	—(CH₂)₃—N⟨piperazine⟩N—CH₃	5 40	3.2 5.4	0.1 0.4
Thioridazine (Mellaril, Sandoz)	—S—CH₃	—CH₂—CH₂—⟨piperidine⟩N—CH₃	5 40	3.2 3.0	1.1 0.6
Methophenazine (Frenelon, Un. Works, Pharm. & Diet. Prod., Budapest)	—Cl	—(CH₂)₂—N⟨piperazine⟩N—(CH₂)₂—O—C(=O)—(3,5-dimethoxyphenyl)	5 40	3.2 2.7	0.7 0
SKF 7887-A₂ (Smith, Kline & French)	—Cl	—(CH₂)₃—⟨piperidine with N(CH₃)₂⟩	5 40	3.2 1.8	0 0.9
SKF 8448-I (Smith, Kline & French)	—Cl	—(CH₂)₃—NH—CH₂—CF₃	5 40	3.2 2.7	0 0.6

Table 1 (continued)

Generic and proprietary name	R_1	R_2	Daily dose mg/kg	Mammo-tropic index	Body temper-ature drop °C
SKF 5910-I₂ (Smith, Kline & French)	—Cl	—CH₂—CH(CH₂—N(CH₃)₂)(CH₂—N(CH₃)₂)	5 40	3.2 3.4	0 0.4
UCB 2493 (Chlorosucos UCB, Brussels)	—Cl	—CH₂—CH(CH₃)—CH₂—N—N—(CH₂)₂—O—CH₂—CH₂OH (piperazine)	5 40	3.2 4.4	0.3 1.8
Cyamepromazine (R. P. 7204, Specia)	—CN	—CH₂—CH(CH₃)—CH₂—N(CH₃)₂	5 40	3.2 5.2	3.0 6.0
TPN 12 (Sandoz)	—SO₂—CH₃	—CH₂—CH₂—(N-methylpiperidinyl)	5 40	3.0 4.0	1.3 2.5
Chlorpromazine (Largactil, Specia Thorazine, SKF)	—Cl	—(CH₂)₃—N(CH₃)₂	5 40	3.0 3.6	2.8 9.4

Screening of Phenothiazine Sulfoxides and Sulfodioxides

Compound	R	Side chain	Dose		
SKF 6058-I₂ (Smith, Kline & French)	—Cl	—(CH₂)₃—N(CH₃)(CH₂)₃—N(CH₃)₂	5 / 40	3.0 / 3.1	0 / 1.6
2-Chlorophenothiazine (SKF 3227)	—Cl	—H	5 / 40	3.0 / 2.5	0 / 0.8
Prochlorperazine (Compazine, SKF) (R.P. 6140 Specia)	—Cl	—(CH₂)₃—N⟨⟩N—CH₃	5 / 40	3.0 / 4.4	1.4 / 0.6
Chlorproethazine (R.P. 4909, Specia)	—Cl	—(CH₂)₃—N(C₂H₅)₂	5 / 40	2.8 / 3.4	0 / 2.5
Perazine (Taxilan, Promonta)	—H	—(CH₂)₃—N⟨⟩N—CH₃	5 / 40	2.8 / 3.5	0 / 0.1
SKF 6064-J (Smith, Kline & French)	—Cl	—C₆H₄—CH₂—N(CH₃)₂	5 / 40	2.8 / 2.4	1.2 / 1.2
SKF 6108-I₂ (Smith, Kline & French)	—CF₃	—CH₂—CH(OH)—CH₂—N⟨⟩N—CH₃	5 / 40	2.8 / 3.0	0 / 0.2

Table 1 (continued)

Generic and proprietary name	R_1	R_2	Daily dose mg/kg	Mammo- tropic index	Body temper- ature drop °C
SKF 8745 (Smith, Kline & French)	—Cl	—(CH$_2$)$_3$—N⟨piperazine⟩N—C(=O)—NH$_2$	5 40	2.8 3.5	2.3 1.9
Thioperazine (Majeptil, Specia) (Vontil, SKF) (RP 7843)	—SO$_2$N(CH$_3$)$_2$	—(CH$_2$)$_3$—N⟨piperazine⟩N—CH$_3$	5 40	2.6 2.4	0 0
Alimemazine R.P. 6549, Specia (Temaril, SKF)	—H	—CH$_2$—CH(CH$_3$)—CH$_2$—N(CH$_3$)$_2$	5 40	2.6 2.9	0.9 1.6
SKF 5757-G (Smith, Kline & French)	—Cl	—(CH$_2$)$_3$—N⟨piperidine⟩—CH$_3$	5 40	2.6 4.4	0 0
SKF 7009-I (Smith, Kline & French)	—CF$_3$	—(CH$_2$)$_3$—N⟨thiomorpholine⟩	5 40	2.6 3.7	0.2 0.5
SKF 6051-A$_2$ (Smith, Kline & French)	—Cl	—(CH$_2$)$_3$—NH(CH$_2$)$_3$—N(CH$_3$)$_2$	5 40	2.6 4.3	0.6 2.2

Screening of Phenothiazine Sulfoxides and Sulfodioxides

Compound	R₁	R₂	Dose		

R.P. 6710 (Specia)	—Cl	—CH₂—CH—CH₂—N(piperazine)N—CH₃
		                     |
		                     CH₃
				5
				40
					2.5
					3.8
						0.8
						1.4

7-OH-Chlorpromazine (NIMH)	—Cl	—(CH₂)₃—N(CH₃)₂ and 7-OH	5 40	2.5 3.6	— —
3-Hydroxypromazine (NIMH)	3-OH	—(CH₂)₃—N(CH₃)₂	5 40	2.5 4.3	0.7 2.6
8-Hydroxychlorpromazine (NIMH)	—Cl 8-OH	—(CH₂)₃—N(CH₃)₂	5 40	2.5 2.6	0 0.3
2-Methoxyphenothiazine (NIMH)	—OCH₃	—H	5 40	2.5 3.6	1.0 1.1
Diethazine (Diparcol, Specia)	—H	—(CH₂)₂—N(C₂H₅)₂	5 40	2.4 2.9	0 0
Dipyripromazine (Promoton, Galma)	—Cl and 3,7—N(piperidine with NH₂—CO)	—(CH₂)₃—N(CH₃)₂	5 40	2.4 2.8	2.0 5.0

Table 1 (continued)

Generic and proprietary name	R_1	R_2	Daily dose mg/kg	Mammo-tropic index	Body temper-ature drop °C
Mepazine (Pacatal, Promonta)	—H	—CH$_2$—(piperidine N—CH$_3$)	5 40	2.4 3.2	0 3.2
SKF 6693-A$_2$ (Smith, Kline & French)	—Cl	—(CH$_2$)$_3$—NH—(piperidine N—CH$_3$)	5 40	2.4 3.2	0 0.5
SKF 5611-A (Smith, Kline & French)	—Cl	—CH$_2$—(cyclohexyl N(CH$_3$)$_2$) Cis isomer	5 40	2.4 2.5	0.7 0.3
SKF 3072-A (Smith, Kline & French)	4-Cl	—(CH$_2$)$_3$—N(CH$_3$)$_2$	5 40	2.4 3.5	0.7 1.1
SKF 7838 (Smith, Kline & French)	—Cl	—(CH$_2$)$_3$—NH—S(=O)$_2$—C$_6$H$_4$—CH$_3$	5 40	2.4 3.4	0.3 1.4

Screening of Phenothiazine Sulfoxides and Sulfodioxides 79

Name	R	R'	Dose		
Mesoridazine (NC-123, Sandoz)	$\overset{O}{\underset{\|}{S}}-CH_3$	–CH₂–CH₂– [2-methylpiperidine]	5 / 40	2.2 / 3.9	0 / 0.8
SKF 6262-A₂ (Smith, Kline & French)	–Cl	$\overset{O}{\underset{\|}{C}}$–CH₂–CH₂–N[piperazine N–CH₃]	5 / 40	2.2 / 2.8	0 / 0
Methoridazine (KS-33, Sandoz)	–OCH₃	–CH₂–CH₂– [2-methylpiperidine N–CH₃]	5 / 40	2.2 / 3.6	0.4 / 1.0
Promazine (Sparine, Wyeth)	–H	–(CH₂)₃–N⁺(CH₃)₂	5 / 40	2.6 / 2.2	0.8 / 3.4
Compound 22 (Winthrop)	–H	–CH₂–CH₂–CH₂–N[piperidine-OH]	5 / 40	2.2 / 3.6	1.3 / 3.1
SKF 2680-J (Smith, Kline & French)	–Cl	–(CH₂)₃–N⁺(CH₃)₃	5 / 40	2.2 / 3.3	1.0 / 1.1
SKF 12116-A (Smith, Kline & French)	–Cl	$\overset{O}{\underset{\|}{C}}$–CH₂–CH₂–N(C₂H₅)₂	5 / 40	2.0 / 3.0	0 / 0.9

Table 1 (continued)

Generic and proprietary name	R_1	R_2	Daily dose mg/kg	Mammo-tropic index	Body temper-ature drop °C
SKF 5612-A (Smith, Kline & French)	—Cl	—CH$_2$—⟨cyclohexyl-N(CH$_3$)$_2$⟩ Trans isomer	5 40	2.0 3.3	1.5 1.1
KS-75 (Sandoz)	H —OC(CH$_3$)$_2$	—(CH$_2$)$_2$—⟨piperidine-N—CH$_3$⟩	5 40	2.0 2.6	0.9 1.6
Methylene blue	—H 3,7-N(CH$_3$)$_2$	—H	5 40	1.9 2.9	1.3 3.5
Profenamine Parsidol, R.P. 3326 (Specia)	—H	CH$_3$ —CH$_2$—CH—N(C$_2$H$_5$)$_2$	5 40	1.9 2.0	0.1 3.0
Methdilazine (Tacaryl, Mead-Johnson)	—H	H CH$_2$ —CH$_2$—C—CH$_2$ CH$_2$—N—CH$_3$	5 40	1.8 2.6	2.5 2.7

Screening of Phenothiazine Sulfoxides and Sulfodioxides

Compound	R	Side chain	Dose				
Methopromazine (Mopazine, Specia)	—OCH$_3$	—(CH$_2$)$_3$—N(CH$_3$)$_2$	5	1.8	1.0
			40	1.4	3.0
Propionylpromazine (Combelen, Bayer)	$\underset{}{\overset{O}{\|\|}}$C—C$_2H_5$	—(CH$_2$)$_3$—N(CH$_3$)$_2$	5	1.8	1.0
			40	3.6	2.0
SKF 13813-A (Smith, Kline & French)	$\underset{}{\overset{O}{\|\|}}$C—CH$_3$	—C(=O)—CH$_2$—CH$_2$—N(CH$_3$)$_2$	5	1.8	0.9
			40	2.6	1.1
SKF 4523 (Smith, Kline & French)	—H	—C(=S)—N(C$_2$H$_5$)$_2$	5	1.8	0.2
			40	3.3	1.4
Dixyrazine (Esucos UCB, Brussels)	—H	—CH$_2$—CH(CH$_3$)—CH$_2$—N(piperazine)N—(CH$_2$)$_2$—O(CH$_2$)$_2$OH	5	1.7	0.6
			40	4.2	2.2
3-Chlorophenothiazine (SKF 18722)	3-Cl	—H	5	1.6	0.1
			40	2.2	1.0
Levomepromazine (Nozinan, Specia)	—OCH$_3$	—CH$_2$—CH(CH$_3$)—CH$_2$—N(CH$_3$)$_2$	5	1.6	1.9
			40	1.6	4.6

Table 1 (continued)

Generic and proprietary name	R_1	R_2	Daily dose mg/kg	Mammotropic index	Body temperature drop °C
Dimethoxanate (Clothera, Ayerst)	—H	—C(=O)—O—(CH$_2$)$_2$—O—(CH$_2$)$_2$—N(CH$_3$)$_2$	5 40	1.6 1.8	0 1.6
Phenothiazinyl-propionitril	—H	—CH$_2$—CH$_2$—CN	5 40	1.6 1.8	0.3 1.4
Dimetiotazine (R.P. 8599 Specia)	—SO$_2$—N(CH$_3$)$_2$	—CH$_2$—CH(CH$_3$)—N(CH$_3$)$_2$	5 40	1.6 2.5	0.6 1.0
SKF 6045-A$_2$ (Smith, Kline & French)	—Cl	—(CH$_2$)$_3$—N(CH$_3$)—(CH$_2$)$_2$—N(CH$_3$)$_2$	5 40	1.6 3.2	0.5 0.9
10-phenothiazinyl-propionic acid (self synthesis)	—H	—(CH$_2$)$_2$COOH	5 40	1.6 2.2	0 0
Aminopromazine (Lispamol, Specia)	—H	—CH$_2$—CH(N(CH$_3$)$_2$)—CH$_2$—N(CH$_3$)$_2$	5 40	1.5 2.5	0 0.3

Screening of Phenothiazine Sulfoxides and Sulfodioxides 83

Compound	R	R'	Dose			
Promethazine (Phenergan, Specia)	–CH₂–CH–N(CH₃)₂ | CH₃	–H	5	1.4	0.7	
			40	2.3	2.1	
Propriomazine (Largon, Wyeth) (Dorevane, Clin)	–CH₂–CH–N(CH₃)₂ | CH₃	O=C–CH₂CH₃	5	1.4	0.2	
			40	3.2	1.0	
Pyrathiazine (Pyrrolazote, Upjohn)	–CH₂–CH₂–N⟨ ⟩	–H	5	1.4	0	
			40	2.4	1.0	
6-OH-Chlorpromazine (NIMH)	–(CH₂)₃–N(CH₃)₂ and 6-OH	–Cl	5	1.4	—	
			40	2.1	—	
1-Chloro-phenothiazine (SKF 24200)	–H	1-Cl	5	1.4	0	
			40	2.2	0.3	
Lauth Violet (Thionin, Merck)	–H	3-amino, 7-imino-	5	1.4	0.4	
			40	2.6	1.0	
Toluidine Blue Tolonium (Blutene, Abbott)	–H	3-N(CH₃)₂ 5-Cl 7-NH₂ 8-CH₃	5	1.4	1.0	
			40	2.3	2.0	
Phenothiazine (Specia)	–H	–H	5	1.2	0.2	
			40	1.8	0.6	
Saline (control)	—	—	0.5 ml	1.2	0	

Table 2. *Mammotropic index of 15 phenothiazine sulfoxides and sulfodioxides in oestrogen-primed female rats (200 ± 10 gm). Oestradiol priming: 8 μg/rat/day for 10 days; treatment: 5—200 mg/kg/day, for 5 days. Preparations are listed in descending order of their lactogenic effect at 5 mg/kg*

Generic and proprietary name	R₁	R₂	Daily dose mg/kg	Mammotropic index	Body temperature drop °C
Perphenazine sulfoxide (Schering, USA)	—Cl	—(CH₂)₃—N⟨piperazine⟩N—CH₂—CH₂OH	5 40 200	3.8 4.3 5.0	0 0 0.7
Piperacetazine sulfoxide (self synthesis)	O=C—CH₃	—(CH₂)₃—N⟨piperidine⟩—CH₂—CH₂OH	15 40 200	3.6 4.4 5.0	— —
R.P. 8.580 (Specia)	—SO₂—N(CH₃)₂	—(CH₂)₃—N⟨piperazine⟩N—CH₃	5 40 200	3.3 4.9 5.0	0 0.3 0.7

Compound	R	Side chain	Dose		
Methdilazine sulfoxide (Mead Johnson)	—H	—CH$_2$—CH—CH$_2$—N(CH$_3$)(CH$_2$) (with CH$_3$, CH$_2$ branches)	5 40 200	2.6 3.0 4.4	0 1.0 1.3
SKF 4611-A sulfodioxide (Smith, Kline & French)	—Cl	—(CH$_2$)$_3$—N(CH$_3$)$_2$	5 40 200	2.6 3.3 3.9	0 0.1 1.2

Promethazine sulfoxide (Phenergan sulfoxide, Specia)	—H	—CH$_2$—CH—N(CH$_3$)$_2$
		      |
		     CH$_3$
			5
			40
			200
				2.5
				2.8
				2.8
					0.8
					2.1
					1.9

SKF 5419-A Triflupromazine sulfoxide, (Smith, Kline & French)	—CF$_3$	—(CH$_2$)$_3$—N(CH$_3$)$_2$	5 40 200	2.4 3.2 4.7	0 1.4 2.7
SKF 17910-A (Smith, Kline & French)	—Cl	O=C—(CH$_2$)$_2$—N(CH$_3$)$_2$	5 40 200	2.2 2.5 4.0	— — —
SCH 13361 (Schering)	—Br	—(CH$_2$)$_3$—N⟨piperazine⟩N—CH$_2$CH$_2$OH	5 40 200	2.2 2.6 3.8	— — —

Table 2 (continued)

Generic and proprietary name	R₁	R₂	Daily dose mg/kg	Mammotropic index	Body temperature drop °C

SKF 7145-A sulfoxide (Smith, Kline & French)	—Cl	—(CH₂)₃—N—CH₃
		                   |
		                   H
			5
			40
			200
				1.8
				2.5
				3.0
					0
					0.2
					0.5

Chlorpromazine sulfoxide (R.P. 5293, Specia) (SKF 4260-A)	—Cl	—(CH₂)₃—N(CH₃)₂	5 40 200	1.9 2.8 4.0	0.5 0 0.5
Promazine sulfoxide (Wy 1193, Wyeth)	—H	—(CH₂)₃—N(CH₃)₂	5 40 200	1.8 2.8 3.4	0 0 2.8
Phenothiazine sulfoxide	—H	—H	5 40 200	1.6 2.2 3.0	0.9 0 1.2

Oxomemazine sulfodioxide (Doxergan, Specia)	—H	—CH₂—CH—CH₂—N(CH₃)₂
		            |
		            CH₃
			5
			40
			200
				1.4
				2.6
				4.0
					0.8
					1.0
					1.9

SKF 5332-I sulfoxide (Smith, Kline & French)	—OCH₃	—(CH₂)₃—N(CH₃)₂	5 40 200	1.4 — —	— — —

Testing of 15 sulfoxide derivatives of phenothiazine revealed that oxidation decreased considerably the sedative and tranquilizing effect as measured by body temperature drop. The mammotropic activity, however, was much less affected. This opened a way to synthesis of lactogenic drugs practically free of tranquilizing side effects. The results are presented in Table 2. The most mammotropic phenothiazine sulfoxides were perphenazine sulfoxide, RP-8580, and methdilazine sulfoxide.

In Tables 1 and 2 the phenothiazine derivatives studied are listed in descending order of their lactogenic effect at 5 mg/kg. This arrangement conspicuously shows that special substitutions at C_2 and N_{10} of the phenothiazine molecule are essential for eliciting strong mammotropic activity. Substitution of C_2 by Cl, Br, CF_3, SCH_3, SC_2H_5, CN or $CO(CH_3)$ increased the mammotropic effect. Likewise substitution at N_{10} by piperidine, pyrrolidine and piperazine rings provided highly mammotropic compounds (cf. Chapter 19). The preparations, which yielded a mammotropic index of 2.0 or less, may be regarded as completely ineffective compounds. Moreover, we showed that no parallelism exists between the mammotropic and tranquilizing effects of several phenothiazines, e. g. perphenazine sulfoxide and chlorpromazine sulfoxide. Both compounds showed only very slight tranquilizing activity or none at all, but they still proved to be strongly mammotropic. The tables present also the body temperature drop as a measure of tranquilizing activity.

Table 2 shows the lactogenic effect of 15 phenothiazine sulfoxides. Comparison of their lactogenic activity with that of their phenothiazine parent compounds (containing S instead of SO) shows a clearly diminished but still good lactogenic activity which can be enhanced to maximum effect by increased dosage. For practical purposes this would be no drawback, since sulfoxides have no tranquilizing effect unless their dosage is increased beyond a 20-fold concentration (HOTOVY and KAPFF-WALTER, 1960).

ROSSI (1964) pointed out that phenothiazine derivatives of the piperazine series generally possess a greater antipsychotic activity per milligram than those of the dimethylamine or piperidyl series. Within these three series, differences in psychotherapeutic activity are also attributable to the nature of the substitution at position 2 on the phenothiazine nucleus. As a rule, the chlorine-substituted derivatives are less active than the trifluoromethylsubstituted compounds. Mercaptomethyl substitution appears to imbue the compound with an activity approximately equal to that of the chlorine substitution. Phenothiazines, lacking a substitution in position 2, or having an acetyl group, appear to possess the lowest anti-psychotic potency within their respective series. Sulfoxide formation nearly abolishes the anti-psychotic effect, without greatly diminishing the lactogenic effect. It is interesting to note that most of the rules pertaining to the psychotherapeutic activity hold also—though not necessarily—for the lactogenic effect. We know now that the common denominator for both actions is a 3—4 ring system which, in its structural arrangement, is similar to that of the steroids and therefore hypothalamo-tropic and hypophysio-tropic (cf. Chapter 19).

f) Stimulation of Synthesis and Release of Prolactin by Perphenazine

As previously described, perphenazine (Trilafon)—an amino derivative of chlorphenothiazine—was found to be a potent lactogenic agent in intact virgin rats (BEN-DAVID, DIKSTEIN and SULMAN, 1965) but entirely inactive in hypophysectomized animals (BEN-DAVID, 1968 c). Perphenazine was also reported to stimulate luteotropic activity (BEN-DAVID, 1968 b) and to stimulate prolactin secretion *in vitro* when added to pituitary-hypothalamus co-culture (DANON, DIKSTEIN and SULMAN, 1963). Hence it was postulated that perphenazine exerts its lactogenic effect through stimulation of pituitary prolactin secretion. This stimulation is thought to be effected by interference with the action of the prolactin-inhibiting factor (PIF) released from the hypothalamus (PASTEELS, 1961 a; GROSVENOR, MCCANN and NALLER, 1965; SCHALLY, KUROSHIMA, ISHIDA, REDDING and BOWERS, 1965; BEN-DAVID, 1968 c). However, the processes taking place in the hypothalamus and pituitary subsequent to the administration of perphenazine, leading to the ultimate release and biosynthesis of prolactin, have not been elucidated.

BEN-DAVID and CHRAMBACH (1969), therefore, studied the simultaneous effect of daily injections of perphenazine (5 mg/kg/d) for 1—5 days on mammary differentiation, pituitary weight and on pituitary and prolactin levels in female rats, using a modification of the pigeon-crop DNA method (BEN-DAVID, 1967). Blood prolactin was extracted as described in Chapter 26—d.

The development and histological patterns of the mammary glands of control animals (prior to treatment) and animals after varying numbers of daily injections of 5 mg perphenazine/kg/day are shown in Figs. 22 and 23. The histological patterns reveal remarkable mammary differentiation in direct proportion to the duration of perphenazine treatment. Initiation of lobulo-alveolar development occurred after 2 days of treatment, and mammary secretion appeared in the glands after 3 days of treatment. After 4 days of daily perphenazine administration, and, similarly, after 5 days of treatment, mammary secretion was abundant.

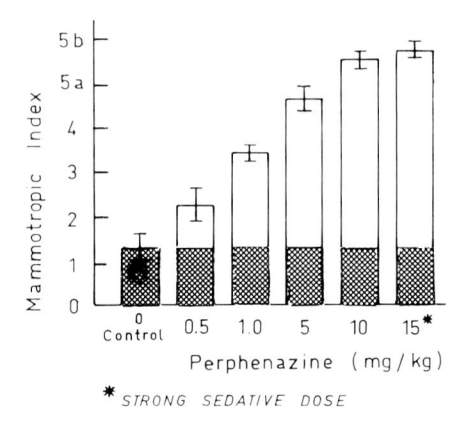

Fig. 22. Mammotropic effect (MTI) of increasing doses of perphenazine (0.5—15 mg/kg/day), given during 5 days to adult female rats. The MTI exceeds 3 after a daily dose of 1 mg perphenazine, 5 or 10 mg are optimum dosages, 15 mg have a strong sedative effect

The relative increase in the wet weight of pituitaries per 100 g body weight, following various periods of perphenazine treatment, is shown in Fig. 24. A significant increase of 32.4% ($P<0.05$) was found 24 hours after a single injection of 5 mg perphenazine/kg, and a significant average increase of 40% ($P<0.01$) was found after 2, 3, 4, and 5 days of treatment. Since the sedative action of the drug produced diminished food intake and loss of body weight in the test animals, a control series of non-treated rats, subjected to fasting for varying periods ranging between 1 and 5 days, was studied. In these animals there was a parallel decline in both pituitary and body weights, so that the quotient of pituitary weight divided by body weight was practically unchanged. Thus, the reported increases in pituitary weight per 100 g bodyweight represent not only increases in protein mass but,

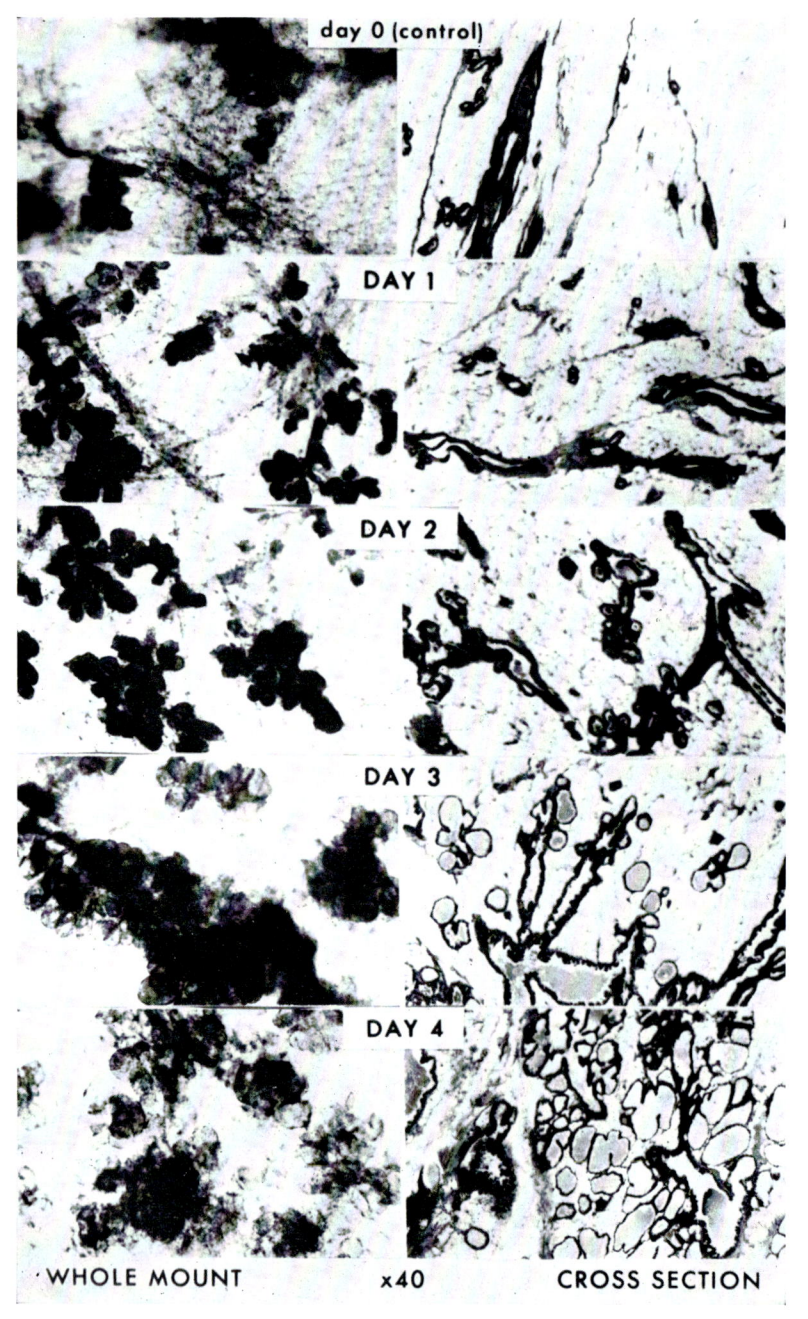

Fig. 23. Whole mounts and cross sections of mammary glands of virgin female rat which has received daily injections of 5 mg/kg perphenazine. Note the appearance of secretion starting after three days of treatment and the highly developed lobulo-alveolar system after four days of treatment

more specifically, the absence of any decline in the pituitary weight of perphenazine-treated animals together with the loss in total body weight.

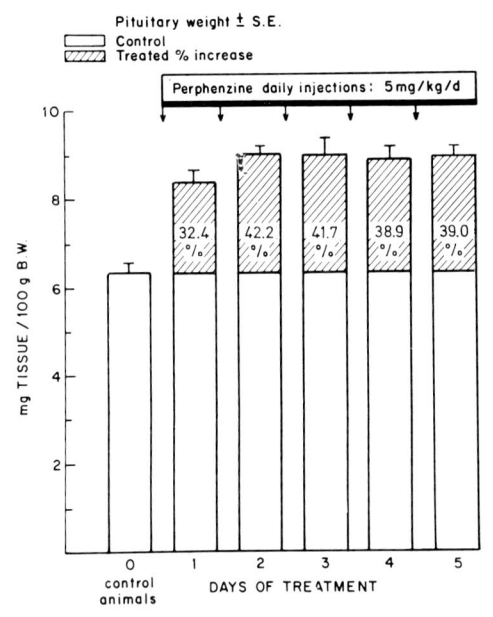

Fig. 24. Effect of perphenazine administration on the relative wet weight of rat pituitaries per 100 g body weight. The abscissa denotes the number of consecutive days of s. c. perphenazine administration. The weights are expressed as mg/100 g body weight ± S. E.

Fig. 25 shows prolactin activity in pituitaries and in serum of rats treated daily with perphenazine for periods up to 5 days. During this 5-day treatment, the initial release of prolactin from the pituitary into the blood and the subsequent production of prolactin in the pituitary can be distinguished kinetically. Twenty-four hours after a single injection of perphenazine, prolactin activity of the pituitary decreased to 48% ($P < 0.05$) of that of the untreated control animal. Concurrently, prolactin activity of the blood increased from 0.5 to 3.3 mU/ml ($P < 0.001$). By the 3rd day of treatment, pituitary prolactin per mg of tissue leveled off at 55% above the rate of the controls; the weight of the pituitary increased simultaneously with the relative amount of prolactin (activity per total weight of tissue).

This shows that the mammotropic action of perphenazine is mediated by pituitary prolactin, since the drug does not cause mammary growth in hypophysectomized animals (BEN-DAVID, 1968 c). The action of the drug on the pituitary gland is indirect; it is directed primarily at the hypothalamus, since implantation of perphenazine

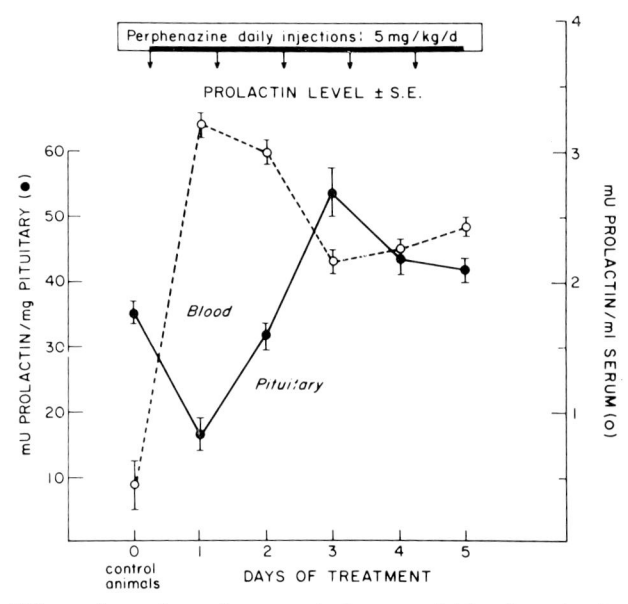

Fig. 25. Effect of perphenazine on pituitary and blood prolactin activity. The abscissa denotes the number of consecutive days of perphenazine administration. The vertical lines indicate the standard error. Note the immediate increase in blood prolactin heralded by its depletion from the pituitary

into the median eminence causes lactation in rabbits (MISHKINSKY, LAJTOS and SULMAN, 1966). Action of the drug on the pituitary gland is generally considered to be mediated by the hypothalamic prolactin-inhibiting factor (PIF) (PASTEELS, 1962; SCHALLY et al., 1965; GROSVENOR et al., 1965). Direct evidence that perphenazine stimulated prolactin secretion from the pituitary by suppressing the PIF is found in the work of DANON et al. (1963). It was shown that the addition of perphenazine to a pituitary-hypothalamus combined organ culture abolished the inhibitory effect of the hypothalamic fragments upon prolactin secretion by the pituitary. The blocking action of perphenazine on the PIF appears to be reversible and temporary, as diestrus in rats—a concomitant of its administration—terminates 7 days after treatment with the drug, with resumption of a normal estrus cycle (BEN-DAVID, 1968 b).

Perphenazine treatment seems to affect mainly pituitary prolactin rather than other hormones. This hypothesis is supported by the fact that perphenazine maintains full mammotropic activity in propylthiouracil-treated rats (BEN-DAVID, DIKSTEIN and SULMAN, 1966) and also in ovariectomized or adrenalectomized rats when administered immediately after the operation (BEN-DAVID, 1968 b, c; BEN-DAVID and SULMAN, 1969). Moreover, other phenothiazine derivatives, such as chlorpromazine, are reported to cause inhibition rather than stimulation of FSH and LH secretion (SULMAN and WINNIK, 1956; WINNIK and SULMAN, 1956; ALLOITEAU, 1957).

Whether perphenazine, mediated by PIF, regulates prolactin release or production in the pituitary remains to be established. The biphasic kinetics of simultaneous prolactin assays in the pituitary and in the blood depicted in Fig. 25 support the hypothesis that the action of the drug involves both release and production of prolactin. It remains to be shown to what extent the initial release-phase also comprises hormone production; and whether the production of the hormone involves biosynthesis or activation of a precursor protein.

The following facts suggest that perphenazine treatment results in "turning on" the pituitary: Following one day of treatment with the drug, pituitary weight significantly increases (Fig. 24), in spite of the pronounced depletion of prolactin (Fig. 25). Subsequent to the release of the hormone, its activity increases with time and reaches a steady state of secretion significantly above the level characteristic for the control gland (Figs. 24 and 25). It seems plausible, in view of the apparent bi-phasic state of the pituitary, that the second phase of perphenazine action predominantly involves the biosynthesis of the hormone.

Activation of the pituitary by perphenazine appears to be similar to that obtained by pituitary transplantation (BOOT, ROEPCKE and MUEHLBOCK, 1964), and found in pituitary isografts in rats (NIKITOVITCH-WINER and EVERETT, 1958) and mice (BOOT et al., 1964). These isografts induced mammary differentiation (BARDIN, A. G. LIEBELT and R. A. LIEBELT, 1962; DAO and GAWLAK, 1963), and lactation in the host animals (COWIE, TINDAL and BENSON, 1960; HARAN-GERA, personal communication). Activation in both cases has been interpreted as being due to the elimination of the inhibitory effect of PIF by spatial separation of the hypothalamus from the pituitary. As pointed out above, perphenazine can similarly be considered a blocking agent of the PIF.

In summary, treatment with perphenazine results in a prompt release of prolactin from the pituitary into the blood, followed by a rate of production and release of the hormone which first increases and later levels off at a stage significantly above that of the controls. The elevated steady-state production and release of the hormone is accompanied by increased weight of the pituitary and by mammary gland growth and secretion.

g) Intracerebral Implantation of Phenothiazine Derivatives

In order to locate the mammotropic effect of phenothiazine derivatives, intracerebral stereotaxic implantations were carried out by MISHKINSKY, LAJTOS and SULMAN (1966).

Thirty-eight adult female rabbits weighing 3.5—4.5 kg were ovariectomized and primed with estradiol (0.1 mg s.c. daily) for 10 days. On the 11th day the left abdominal mammary gland was removed under pentobarbital anesthesia, and a stainless steel tube, bearing a perphenazine crystal, was lowered stereotaxically into different regions of the brain: the frontal lobe (6 animals), the thalamus (5 animals), the occipital lobe (5 animals), the median eminence (7 animals), and the pituitary anterior lobe (5 animals). In the control experiments, empty tubes were similarly implanted into the median eminence (5 rabbits); in addition, tubes bearing the phenothiazine derivative aminopromazine (Lispamol) which does not initiate lactation were inserted into the same site (5 rabbits). The implants were prepared by heating small perphenazine crystals (about 10 mg) on the tip of a 0.4 mm-diameter stainless steel tube. The tube was then thoroughly cleaned, leaving only the small perphenazine crystal visible on it.

Implantation into the median eminence was achieved by inserting the tubes at an angle of 6° to the midline in order not to damage the superior sagital sinus. All other implantations were carried out vertically on either side of the midline, at a distance of 1.5 mm. After implantation, the tube was fixed to the skull of the rabbit with dental cement and 2 anchoring screws.

On the 16th day the rabbits were killed and their abdominal mammary glands taken for autopsy, stained with hematoxylin and eosin, examined and graded on the mammotropic index scale (MTI) of 1—5, used in our Department (Chapter 26). At autopsy, the brains were carefully dissected and fixed in 4% formaldehyde solution and the steel tubes removed. They were then sectioned by the paraffin method into 7 µ slides and stained with hematoxylin and eosin. The localization of the implantation sites (Fig. 26) was verified from the serial sections.

Fig. 26. Transverse frontal section of rabbit hypothalamus at P_1, showing an implanted site of the posterior basal median eminence at the bottom. The implant (black) promoted prolactin secretion, empty tubes (control) did not

The results are summed up in Fig. 27. No mammary development was noticed in rabbits with perphenazine implanted into their frontal lobe, thalamus, occipital lobe or pituitary. In a few cases we observed mammary involution, but full development could be observed only where perphenazine had been implanted into the median eminence of the hypothalamus (MTI 5). These implantations increased mammary gland development and brought it to the maximum stage Vb of the rabbit MTI scale. Neither aminopromazine (Lispamol) (Fig. 28) nor empty tubes evoked any mammary development.

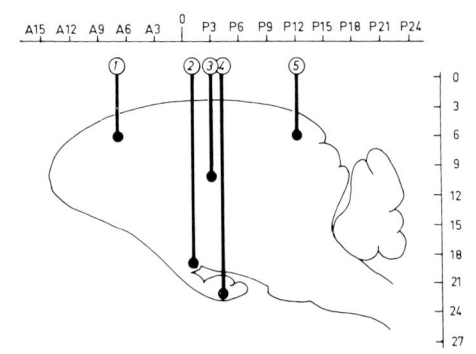

Fig. 27. Sites of perphenazine implantations in rabbit brain: *1* frontal lobe; *2* median eminence; *3* thalamus; *4* pituitary anterior lobe; *5* occipital lobe. (A, P, H are the stereotaxic coordinates.) Only median eminence implantation (2) promoted mammary development

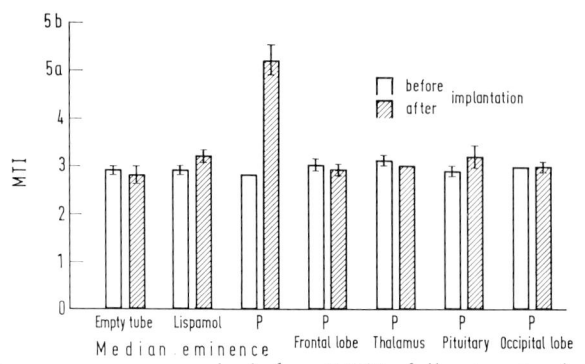

Fig. 28. Average mammotropic index (MTI) following implantation of perphenazine (P) crystals into different sites of the brain of ovariectomized, estrogen-primed rabbits (0.1 mg per day for 10 days). Only perphenazine implantation in the median eminence produces lactation. Lispamol (aminopromazine)—a non-lactogenic phenothiazine derivative—gives a negative reaction, as does the empty tube control (cf. also Chapters 14 and 19)

These results indicate that in rabbits the neurons of the posterior basal tuberal region of the hypothalamus are involved in the inhibitory mechanism which governs prolactin secretion by production or release of a PIF. Perphenazine implantations, like electrolytic lesions, apparently decrease the PIF level of the hypothalamus, causing release of prolactin from the anterior pituitary and activation of the mammary glands. It is assumed that perphenazine interferes with a balance established within the hypothalamus between the PIF and PRF mechanisms, by eliminating the inhibitory one. This interference follows the same rules of specificity as those established in section f of this chapter, since no mammary development was obtained when aminopromazine implants of the same diameter as that of the perphenazine-bearing electrodes were used for control. These results, thus, well confirm the hypothesis that a PIF is located in the posterior basal tuberal region of the hypothalamus.

h) Mammotropic Effect of Fluphenazine Enanthate in the Rat

Fluphenazine enanthate is, to date, the main long-acting phenothiazine tranquilizer widely used by psychiatrists. Laboratory experiments (EBERT and HESS, 1965) and clinical experience (KINROSS-WRIGHT and CHARALAMPOUS, 1965) suggest that its tranquilizing effect lasts for 10—28 days with few side effects. Lately it has been shown that both fluphenazine HCl (KHAZEN, MISHKINSKY, BEN-DAVID and SULMAN, 1968 a) and fluphenazine enanthate (BEN-DAVID, 1968 a) evoke mammotropic activity, which appears to be one of their main side effects. MISHKINSKY, KHAZEN and SULMAN (1969), therefore, studied mammary development in rats for thirty days after subcutaneous injection or hypothalamic implantation of fluphenazine HCl and fluphenazine enanthate, in order to elucidate the mechanism of action of these two substances.

We used female albino rats weighing 200 ± 10 g each. They were primed for 10 days s.c. with 8 μg/rat/day estradiol. On the 11th day the animals were divided into five groups and injected s.c. as follows: In groups 1 (35 rats) and 2 (28 rats), each animal received a single injection of fluphenazine IICl, 5 or 25 mg/kg, respectively. In groups 3 (35 rats) and 4 (29 rats), each animal received a single injection of fluphenazine enanthate, 5 or 25 mg/kg, respectively; the animals in group 5 (35 rats) received a single injection of 0.2 ml sesame oil, and served as controls. A further 98 rats were primed for 10 days as above, and were implanted on the 11th day stereotaxically in the median eminence of their hypothalamus with glass tubings (Rudmond's disposable pipettes), bearing the following substances: group 6 (25 rats)—fluphenazine base; group 7 (27 rats)—fluphenazine enanthate; and group 8 (36 rats)—the control group—pure cocoa butter. The fluphenazine base and fluphenazine enanthate were melted prior to the experiment in cocoa butter at a ratio of 1:4 by weight, and

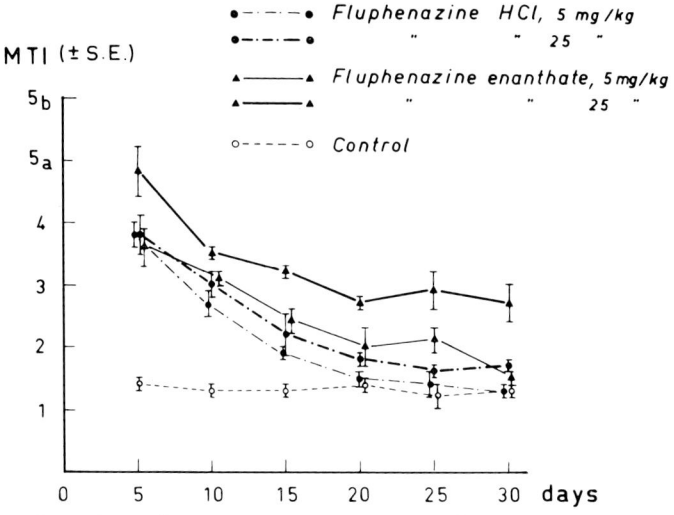

Fig. 29. Duration of mammotropic reaction after s. c. injection of fluphenazine salt compared with that produced by its ester. Average mammotropic index (MTI) of estradiol-primed rats, followed up for 30 days after single s. c. injections of fluphenazine HCl and fluphenazine enanthate

were tamped into the glass tubing. Subsequently, the external surface was carefully cleaned and the excess material removed. About onesixth of the animals in each of the 8 groups were sacrificed at 5-day intervals from the 5th day until the 30th day after treatment and their MTI determined.

From Fig. 29 it can be seen that single injections of fluphenazine HCl and fluphenazine enanthate provoked moderate mammotropic activity, the MTI being about 4 on the 5th day after the single injection of either of the two substances and declining, after the 15th day, to MTI 1.7 in the fluphenazine HCl-injected rats and to MTI 2.7 in the fluphenazine enanthate-injected rats. The slope of mammotropic involution was not quite as steep in the rats which received fluphenazine enanthate as in those which received fluphenazine HCl. Implants of fluphenazine base caused a significantly higher mammotropic effect than those of fluphenazine enanthate (Fig. 30). Moreover, there was a tendency towards involution of the developed mammary pads 15 days after fluphenazine base implantation, at a time when fluphenazine enanthate developed its mammotropic effect lasting up to the 30th day. Beyond this date its effect declined.

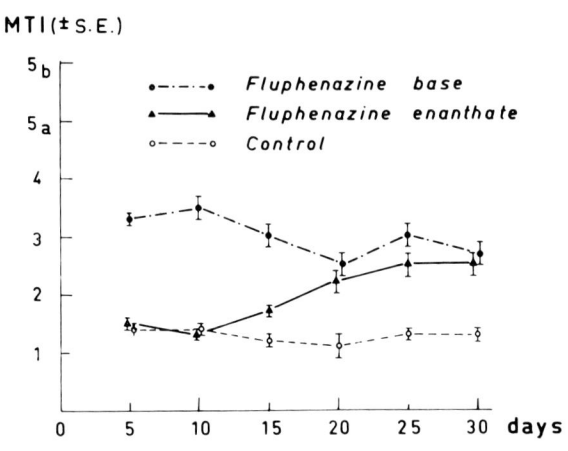

Fig. 30. Development of mammotropic effect after stereotaxic implantation in the hypothalamic median eminence. Average mammotropic index (MTI) of estradiol-primed rats, followed up for 30 days after stereotaxic implantation of 0.6 mg fluphenazine base and fluphenazine enanthate. Note the delayed and inferior mammotropic reaction elicited by the enanthate

Fluphenazine enanthate—the heptanoic acid ester of fluphenazine—was developed as a long-acting phenothiazine derivative to control severe cases of schizophrenia by monthly or bi-monthly injections. Although all phenothiazines have side effects, and fluphenazine enanthate, like other phenothiazine derivatives, may produce extrapyramidal reactions, PARKES, BROWN and MONCK (1962) showed that only a few of the patients receiving fluphenazine enanthate injections suffered any side effects—mainly drowsiness, lethargy, blurred vision or hypotension. Metabolic or endocrine side effects, such as weight-gain or lactation, have only rarely been reported.

Our interest in the mechanism of action of fluphenazine enanthate was aroused by previous studies which had established the structure-activity relationship of phenothiazines (KHAZEN, et al., 1968 a) and phenothiazine-like substances (KHAZEN, MISHKINSKY, BEN-DAVID and SULMAN, 1968 b). We concluded that there was a possible hypothalamic site sensitive to the phenothiazine moiety, and the question arose whether esterification would impair its lactogenic activity. It emerged during the 30-day follow-up that fluphenazine enanthate has a definite mammotropic activity when injected, but less so when implanted in the brain, probably due to its inability to undergo de-esterification in that organ. These results are supported by the findings of EBERT and HESS (1965), who demonstrated that fluphenazine enanthate metabolized in the rat into fluphenazine and fluphenazine sulphoxide, but that only free fluphenazine was to be found in the brain.

The implanted quantity of both fluphenazines (0.6 mg base per rat) was rather small, yet sufficient to yield a clear mammotropic effect without sedation. EBERT and HESS (1965), injecting higher doses (30 mg/kg), showed in their drug-metabolism study that only 0.2% of the radioactive fluphenazine enanthate was found in the brain immediately after injection, and this decreased to 0.04% after 21 days. This dose was sufficient to produce tranquilization and inhibit conditional avoidance responses for the whole three-week period, the animals reverting to normal thereafter. This led EBERT and HESS (1965) to conclude that the active agent ultimately deposited in the brain was fluphenazine base, a hypothesis which is confirmed in the present investigation by stereotaxic implantation of the two substances.

It seems that the slowing-down of the sedative effect parallels that of the mammotropic effect, supporting the view of BEN-DAVID (1968 a) that a persistent influence of the drug must be maintained for a minimum of 4 days in order to induce artificial hypothalamic lactation.

i) Comparative Study of Phenothiazine Effect on Hypothalamic Prolactin-Inhibiting Factor and MSH-Releasing Factor

It has recently been reported that acid extracts of rat hypothalamus, injected intravenously into recipient rats, induce an acute decrease in the melanocyte-stimulating hormone (MSH) concentration in the pituitary. The active agent in the hypothalamus has been referred to as a melanocyte-stimulating hormone-releasing factor (MSH-RF). It has been shown to be effective at low doses, the degree of the effect being related to the dose injected (TALEISNIK, ORIAS and OLMOS, 1966).

Nine phenothiazine derivatives—fluphenazine, trifluoperazine, perphenazine, thiopropazate, chlorpromazine, triflupromazine, methoxypromazine, prochlorperazine, promazine—were found to cause melanophore dispersion in the intact frog; mepazine was shown to be ineffective. When tested in hypophysectomized animals, none of the drugs showed any effect. The release of MSH is effected by direct action on the hypophysis or indirectly via the hypothalamus (SCOTT and NADING, 1961).

It is striking that the same phenothiazine derivatives which are most active in releasing prolactin are also potent MSH releasers. *In vitro* experiments on the mechanism of prolactin and MSH release by phenothiazine derivatives have recently been started in our Department.

Pituitaries and hypothalami were taken from normal ewes at the slaughter house within 5 min of slaughtering and put on ice. The organs were thawed immediately before the experiment, and frontal slice sections were prepared from the pituitary (1.5 mm thick) and from the hypothalamus (3 mm thick). One-half of a pituitary slice (about 100 mg) was cut into ten small fragments and distributed on 5 small Erlenmeyer flasks, containing 2 ml of medium M-199 each. The hypothalamic slices (about 300 mg each) were cut into 30 pieces and distributed among five other Erlenmeyer flasks containing two fragments each from the second half of the pituitary slice. After 4 hrs. shaking in a Dubnoff shaker at 37° C in a 5% CO_2/O_2 atmosphere, both sets of flasks were tested for prolactin and MSH effect. Prolactin concentration was determined with the local pigeon crop sac test at various dilutions, and MSH was tested on the tree frog Hyla arborea (SULMAN, 1952, 1956 a).

Our preliminary results show that the hypothalamus of the ewe inhibits prolactin release from its pituitary but enhances the production of MSH. These experiments will now be extended in order to find a method of separating and concentrating the two hypothalamic factors involved. Concurrently, mediators, releasers and inhibitors of this reaction will be studied *in vivo* and *in vitro*.

j) Conclusions

Screening of phenothiazine derivatives and their sulfoxides shows that both are lactogenic if the chemical structure includes a straight three-membered carbon side-chain at N_{10}. Sulfoxide formation nearly abolishes the sedative effect (down to 5%), without affecting lactogenic activity, operating via depression of the hypothalamic PIF. Prolactin is thus released from the pituitary.

Proof of this could be provided by direct assay of prolactin in blood and its depletion from the pituitary. Another approach is the stereotaxic implantation of phenothiazine derivatives into the hypothalamus. Derivatives, which are active when injected s.c., are also active by implantation; inactive derivatives are as negative as "empty" implants. The center which reacts to such implantation is located in the tuberal region of the posterior hypothalamus (median eminence). Fluphenazine enanthate is of special interest since its hypothalamic implantation is less lactogenic than implantation of fluphenazine base. This is due to the fact that it cannot enter the "receptor" of the PIF region unless broken down into unesterified fluphenazine. After subcutaneous injection, however, fluphenazine enanthate is de-esterified locally and has higher lactogenic activity than fluphenazine hydrochloride.

The structure-activity relationship (SAR) rules pertaining to the lactogenic effect of phenothiazine derivatives hold also for their melanophore activity in frogs. This points to the existence of a similar hypothalamic receptor for PIF and MIF.

Chapter 15

Mammotropic Effect of Phenothiazine-like Compounds

In Chapter 14 we have shown that a 3-C side chain—which may structurally form half a hexane ring—on a heterocyclic nucleus is far more important for the mammotropic effect than the ring system as such. This would explain why the phenothiazine nucleus by itself is not mammotropic (KHAZEN, MISHKINSKY, BEN-DAVID and SULMAN, 1968 a). It has also been established that sulfoxidation of a phenothiazine nucleus minimizes its tranquilizing effect without substantially affecting its lactogenic potency (BEN-DAVID, DIKSTEIN and SULMAN, 1965; KHAZAN, BEN-DAVID, MISHINSKY, KHAZEN and SULMAN, 1966). Modern psychiatry tends towards use of phenothiazine-like substances in preference to the classic phenothiazines, because of the latter's chronic side-effects, i. e. photo-sensitivity, liver damage, eye melanosis, etc. Many related heterocyclic systems have, therefore, been synthetized, and these were tested by us.

Since the first synthesis of phenothiazine by BERNTHSEN in 1883, many modifications have been made in its nucleus, mainly by replacing the sulphur and nitrogen bridges between the two benzene rings by methylene or ethylene groups. STEINER and HIMWICH (1963), in their review on modified phenothiazine systems and imipramine derivatives, concluded that substitutions in one or both bridges between the benzene rings of phenothiazine tranquilizers reduce the antidepressant effect. STEINER's subjects were humans, and his parameters for tranquility were based on subjective observations, since the EEG pattern after administration of phenothiazines did not differ from that induced by phenothiazine-like substances. This general finding was confirmed both for phenthiazine sulfoxides (KHAZEN, MISHKINSKY, BEN-DAVID and SULMAN, 1968 a) and for the modified phenothiazines (KHAZEN, MISHKINSKY, BEN-DAVID and SULMAN, 1968 b).

The striking result, i. e. that the unsubstituted phenothiazine nucleus has no mammotropic effect, led us to investigate other similar nuclei, the main groups tested being the thioxanthenes and the iminodibenzyls.

KHAZEN et al. (1968 b) screened 18 ring-systems related to phenothiazine (Fig. 31), all of which contain a system of three consecutive rings: Rings A and C are mainly benzenic (with three exceptions), whereas ring B is composed of six or seven components. We found that, while the phenothiazine system is not required for achieving a mammotropic effect, and that a straight side chain—consisting of 3 carbons—may compensate for adverse changes in the phenothiazine nucleus, profound changes in it drastically reduce the mammotropic effect. Such substances may therefore find wider use only in psychiatry (LAUNCHBURY, 1964).

The mammotropic index and body temperature drop were evaluated as in Chapter 14. The results which are compiled in Table 3 show that the phenothiazine-like chemicals tested are less mammotropic but also less sedative than the corresponding phenothiazines. Two groups—thioxanthenes and iminodibenzyls—furnished some slightly mammotropic derivatives, but their low MTI proved that sulphur and nitrogen bridges in ring B are vital for a higher mammotropic effect. The most mammotropic phenothiazine-like substances were

Table 3. *Mammotropic effect of 52 phenothiazine-like compounds in oestrogen-primed female rats (200 ± 10 g). Oestrogen priming: 8 μg/rat/day for 10 days; treatment: 5 and 40 mg/kg/day for 5 days. Preparations are listed in descending order of their lactogenic effect at 5 mg/kg*

Generic and proprietary name	Nucleus type (Fig. 31)	Structure R₁	Structure R₂	Daily dose mg/kg	Mammotropic index (MTI)	Body temperature drop °C
P-4657 B (Pfizer)	1	—SO₂N(CH₃)₂	=CH—(CH₂)₂—N⟨piperazine⟩—CH₃	5 / 40	4.2 / 4.4	0.1 / 0.4
Clopenthisol (Ciatyl, Tropon) (Sordinol, Lundbeck)	1	—Cl	=CH—(CH₂)₂—N⟨piperazine⟩—(CH₂)₂—OH	5 / 40	3.3 / 4.4	1.8 / 3.5
Chlorprothixene (Taractan, Roche)	1	—Cl	=CH—(CH₂)₂—N(CH₃)₂	5 / 40	3.0 / 3.0	1.7 / 4.8
Sandoz 11—18	1	—H	=⟨cyclohexylidene⟩—N—CH₃ and 2-Cl	5 / 40	2.6 / 3.9	1.0 / 2.3
Sandoz BP 400	1	—H	=⟨cyclohexylidene⟩—N—CH₃	5 / 40	2.6 / 2.9	0.1 / 1.2
Methixene (Tremaril, Wander)	1	—H	—CH₂—⟨piperidine⟩—CH₃	5 / 40	2.6 / 2.0	0.7 / 0

Compound						
Sandoz 11—665 (BP 400 Sulfoxide)	1	—H	N—CH₃ and Sulfoxide	5 40	2.4 2.6	0.1 1.2
Sandoz 10—798	1	—H	=CH—CH₂—N(CH₃)₂	5 40	2.2 2.3	0 0.6
Sandoz 11—445	1	—H	=CH—CH₂—N⟨ ⟩	5 40	1.4 2.2	0.2 0.2
Oxyperpendyl (Pervetral, Homburg)	2	—H	—(CH₂)₃—N⟨ ⟩N—(CH₂)₂—OH	5 40	3.3 3.7	1.2 7.5
SQ-15283 (Squibb) (Selvigon, Homburg)	2	—H	O=C—O(CH₂)₂—O—(CH₂)₂—N⟨ ⟩	5 40	2.9 2.2	0 0

Isothipendyl (Andantol, Homburg)	2	—H	—CH₂—CH—N(CH₃)₂
			         |
			         CH₃
				5
				40
					2.8
					2.2
						1.3
						1.7

Prothipendyl (Dominal, Homburg)	2	—H	—(CH₂)₃—N(CH₃)₂	5 40	1.8 1.7	2.1 9.5
Sandoz 12—039	3	—H	N—CH₃ and Sulfoxide	5 40	2.4 2.2	0.3 2.0

Table 3 (continued)

Generic and proprietary name	Nucleus type (Fig. 31)	Structure R_1	Structure R_2	Daily dose mg/kg	Mammotropic index (MTI)	Body temperature drop °C
SKF 9062-A$_2$ (Smith, Kline & French)	4	—CF$_3$	—(CH$_2$)$_3$—N⟨ ⟩N—(CH$_2$)$_2$—OH	5 / 40	2.0 / 3.7	0 / 0.1
Brilliant Cresyl Blue (National Aniline)	4	—CH$_3$	—H and 8=NH, 2—N(CH$_3$)$_2$	5 / 40	1.6 / 2.4	0 / 1.2
Gallocyanin (I.G. Farben)	4	—H	—H and 1—OH, 2=O, 4—COOH, 8—N(CH$_3$)$_2$	5 / 40	1.6 / 2.5	0.6 / 0
Eosin Blue (Edward Gurr)	5	—NO$_2$	and 1,9—Br and 2,8=O and 7—NO$_2$ (phenyl-COONa)	5 / 40	1.9 / 2.2	0.8 / 0.7
Methantheline (Banthine, Searle)	5	—H	O=C—O—(CH$_2$)$_2$—N(C$_2$H$_5$)$_2$ (phenyl-COONa)	5 / 40	1.7 / 2.4	0.9 / 1.9
Merbromine (Mercurochrome, Dobler)	5	—Br	and 1—HgOH and 2—ONa and 8=O and 7—Br	5 / 40	1.2 / 3.3	0 / 0

Quinacrine (Atabrine, Winthrop)	6	—Cl	—N—CH—(CH$_2$)$_3$—N(C$_2$H$_5$)$_2$ 　│ 　H　CH$_3$ and 7—OCH$_3$	5 40	2.0 3.2	0 0
Acridon (Fluka)	6	—H	—H and 9=O	5 40	2.0 1.8	0.3 0.9
Proflavine (BDH)	6	—NH$_2$	—H and 6—NH$_2$	5 40	1.7 2.4	0.1 1.5
Acriflavine (Pfizer)	6	—NH$_2$	—CH$_3$ and 6—NH$_2$	5 40	1.5 3.3	0.8 6.4
Litracen (N 7049, Lundbeck)	7	—H	=CH(CH$_2$)$_2$—NHCH$_3$ and 9,9-dimethyl	5 40	2.4 3.1	0.8 0.3
Melitracen (N 7001, Lundbeck)	7	—H	=CH—(CH$_2$)$_2$—N(CH$_3$)$_2$ and 9,9-dimethyl	5 40	2.2 2.4	0 0
Alizarin (National Aniline)	7	—OH	=O and 2—OH, 9=O	5 40	1.4 2.2	0.6 1.0
Trimepramine (Surmontil, Specia)	8	—H	—CH$_2$—CH—CH$_2$—N(CH$_3$)$_2$ 　　　│ 　　　CH$_3$	5 40	2.4 3.5	0 0.6
Imipramine (Tofranil, Geigy)	8	—H	—(CH$_2$)$_3$—N(CH$_3$)$_2$	5 40	2.4 2.8	0 2.7
G-34586 (Anafranil, Geigy)	8	—Cl	—(CH$_2$)$_3$—N(CH$_3$)$_2$	5 40	1.8 1.9	0 0.4

Table 3 (continued)

Generic and proprietary name	Nucleus type (Fig. 31)	Structure R₁	Structure R₂	Daily dose mg/kg	Mammotropic index (MTI)	Body temperature drop °C
R.P. 8307 (Specia)	8	—H	—(CH₂)₃—N⟨piperidine⟩—(CH₂)₂—OH	5 / 40	2.2 / 2.8	0 / 0.3
G-22150 (Geigy)	8	—H	—CH₂—CH(CH₃)—N(CH₃)₂	5 / 40	2.0 / 2.8	0.4 / 0.6
G-32883 (Geigy)	8	—H	—CONH₂	5 / 40	1.9 / 2.6	0 / 0.7
GP-33006 (Geigy)	8	—H	—(CH₂)₃—N⟨piperidine⟩—(CH₂)₂OH	5 / 40	2.7 / 3.5	0.1 / 0.4
Nor-Imipramine (Pertofran, Geigy)	8	—H	—(CH₂)₃—NHCH₃	5 / 40	1.4 / 2.9	0.4 / 2.7
Carbamazepine (Tegretol, Geigy)	9	—H	—CONH₂	5 / 40	2.2 / 1.8	0 / 2.9
G-31406 (Geigy)	9	—H	—(CH₂)₃—H(CH₃)₂	5 / 40	2.2 / 3.2	0 / 1.1

Opipramol (Insidon, Geigy)	9	–H	–(CH₂)₃–N⌒N–(CH₂)₂OH	5 / 40	1.5 / 2.0	0 / 0.7
Deptropine (Brontine, Brocades)	10	–H	(structure with N–CH₃ and O)	5 / 40	1.6 / 2.0	0.4 / 2.4
Nortriptyline (Aventyl, Lilly)	10	–H	=(CH₂)₃–NHCH₃	5 / 40	1.6 / 1.6	1.3 / 2.2
Amitriptyline (Elavil, Merck) (Elatrol, Assia-Zori)	10	–H	=CH–(CH₂)₂–N(CH₃)₂	5 / 40	1.6 / 2.4	0.2 / 2.1
Cyproheptadine (Periactin, Merck)	11	–H	(piperidine N–CH₃)	5 / 40	2.6 / 2.8	0 / 2.3
Protriptyline (Vivactil, Merck)	11	–H	–(CH₂)₃–NHCH₃	5 / 40	1.7 / 2.5	0.4 / 1.5
MK-130 (Merck)	11	–H	=CH–(CH₂)₂–N(CH₃)₂	5 / 40	2.4 / 4.4	0.1 / 1.7
Curatin (Doxepin, Pfizer)	12	–H	=CH–(CH₂)₂–N(CH₃)₂	5 / 40	2.0 / 1.2	0 / 1.1
AW-14,2333 (Wander)	13	–H	(piperazine N–CH₃)	5 / 40	3.0 / 3.3	1.1 / 5.6

Table 3 (continued)

Generic and proprietary name	Nucleus type (Fig. 31)	Structure R₁	Structure R₂	Daily dose mg/kg	Mammotropic index (MTI)	Body temperature drop °C
SUM-3170 (Wander)	14	—Cl	—N⌬N—CH₃	5 40	5.0 5.5	6.3 6.0
Dibenzodiazepine (Noveril, Wander)	15	—H	=O and 11—CH₃ and 5—(CH₂)₂—N(CH₃)₂	5 40	2.0 2.4	0.3 1.0
Clothiapine (HF-2159, Dorsey)	16	—Cl	—N⌬N—CH₃	5 40	4.0 5.2	3.4 6.4
Clothiapine sulfoxide (self synthesis)	16	—Cl	—N⌬N—CH₃	15	3.7	—
Sandoz BC 105	17	—H	=⌬N—CH₃	5 40	3.1 2.8	0.1 1.4
Sandoz 11—233	18	—H	=⌬N—CH₃	5 40	2.0 2.5	0.8 1.9
Saline (Control)	—	—	—	0.5 ml	1.3	0

P-4657 B and clopenthisol (thioxanthene nucleus), oxyperpendyl (azaphenothiazine nucleus), Sum-3170 (dibenzooxazepine nucleus) and clothiapine (dibenzothiazepine group).

It seems that the spheric orientation of the molecule is a most important factor for its effect on the central nervous system: the aza-

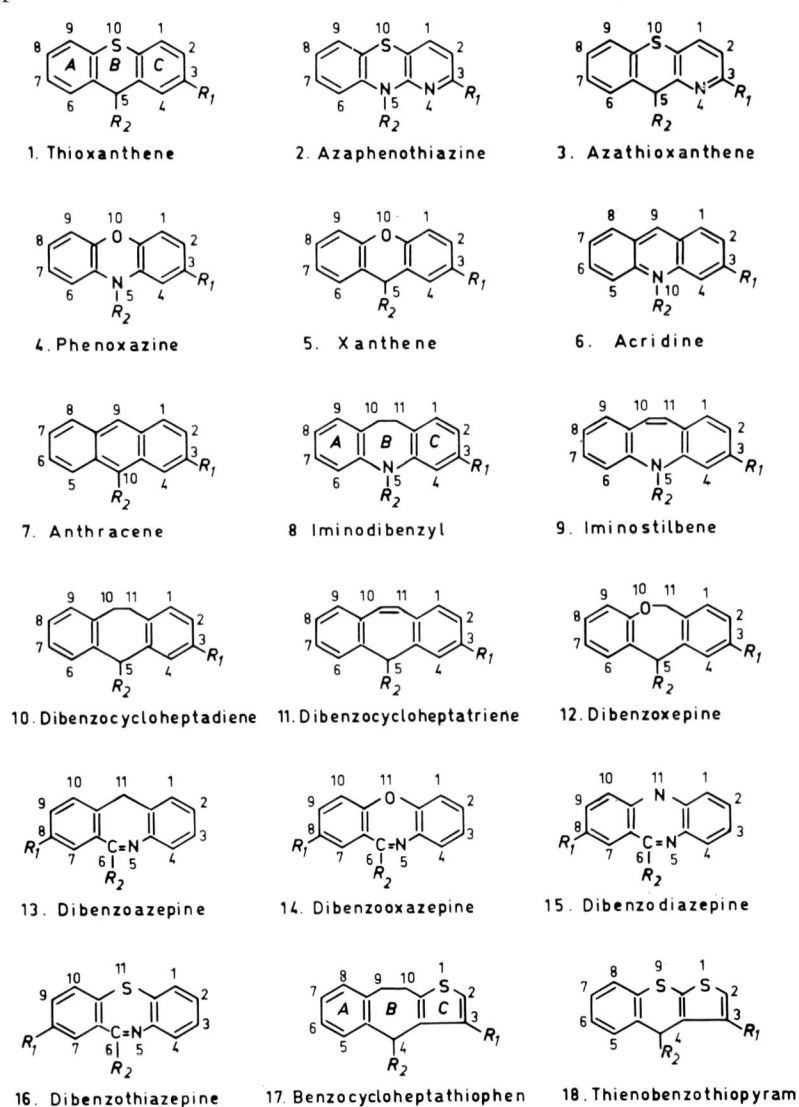

Fig. 31. Eighteen phenothiazine-like nuclei tested for their mammotropic effect. Only Nos. 1, 2, 14 and 16 yielded active compounds

phenothiazines containing nitrogen in the C ring had a markedly lower mammotropic effect. These findings suggest that chemicals bearing A and C benzene rings are the most active. It also seems that the mammotropic effect mediated by psychopharmaca requires a nitrogen atom in the nucleus: none of the substances lacking such a nitrogen atom had any mammotropic effect whatever.

Another important fact emerges from Table 3, *viz.* that there is no direct relationship between the mammotropic and the tranquilizing effects of the tricyclic nuclei tested, although in many cases both effects run somewhat parallel. It may be concluded that the diphenyl moiety is the minimum prerequisite for a mammotropic effect, and that phenothiazine is superior in this respect to all its variations. None of the derivatives tested had significantly high MTI values. The very popular derivatives of azaphenothiazine were non-mammotropic: oxyperpendyl, a substance closely related to perphenazine, was found to have a very slight mammotropic effect, due to lack of a polar group at C_3 and insertion of a second nitrogen in place of C_4. Much of the loss of the mammotropic effect in this case can be attributed to the spheric orientation of the nucleus bearing a second nitrogen.

The sulphur and nitrogen bridges may be replaced by carbons as long as the polar group at C_3 and the two benzene rings remain intact: dibenzothiazepine and dibenzooxazepine—when bearing a polar chlorine atom at C_3 and intact A and C benzene rings—are mammotropic, while substitution of the chlorine by a hydrogen atom markedly reduces the mammotropic effect. Neither the anthracene system nor the dibenzocycloheptanes were lactogenic: when the hetero-atom was shifted to the C ring (Table 3: BC-105), the substance had some mammotropic effect, but a very similar drug—cyproheptadine (Periactin)—without any hetero-atom had no mammotropic effect at all. This may be explained by the quick accumulation of the sulphur-containing rings in the brain (STEINER and HIMWICH, 1964). Thus it emerges that small changes in the three-carbon side chain, like demethylation of the terminal N, which proved to be very important for abolishing the mammotropic effect in phenothiazines, are of lesser importance in phenothiazine-modified substances.

In regard to sedation, KHAZEN et al. (1968 b) could in some instances confirm by body temperature drop the observation of PETERSON and NIELSON (1964), that central depression largely depends on the presence of a double bond between the nucleus and the R_2 side chain, and on polar substitution at C_3.

Another theory (ABOOD, 1955) relates the sedative effect of the phenothiazine-like substances to their reactivity with the phosphorylative complex of mitochondria in the hypothalamus. This complex greatly resembles the diphenyl moiety, due to its three-dimensional molecular structure. This may suggest that the phenothiazine-like substances, like many other diphenyls, achieve their depressive effect by inhibiting oxidative phosphorylation. This, however, does not hold true for the mammotropic effect.

To sum up this series of conclusions, one must admit that tricyclic variants of phenothiazine are no more lactogenic than the mother compound. Still, these compounds, although they help us considerably in understanding the mechanism of lactogenic drug action, do not supply us with a complete explanation. Only a precise receptor theory, which will be discussed in Chapter 19, can provide full understanding.

Chapter 16

Mammotropic Effect of Reserpine Derivatives

The endocrine effects of reserpine are well known. They are mediated by the hypothalamus (KHAZAN and SULMAN, 1961; KHAZAN, SULMAN and WINNIK, 1961; KHAZAN, BEN-DAVID and SULMAN, 1963).

Gynecomastia was reported by WILKINS (1954) to have developed in male patients following administration of Rauwolfia alkaloids. Rauwolfia alkaloids and reserpine were also reported by GAUNT et al. (1954, 1963) to interfere with estrus cycles in mice, and with ovulation and menstruation in monkeys (DE FEO and REYNOLDS, 1956). PLATT and SEARS (1956), WOOLMAN (1956), DURLACH (1957), MEITES (1957 a), ROBINSON (1957), TINDAL (1959), KHAZAN et al. (1962), KANEMATSU, HILLIARD and SAWYER (1963 a, b) and GATCHEW (1966) found reserpine to have a mammotropic and lactogenic effect in rats, rabbits and women. All these authors are agreed that prolactin hypersecretion and discharge from the anterior hypophysis are responsible for the mammotropic and lactogenic effects observed after administration of reserpine and phenothiazine derivatives. This has been confirmed by KHAN and BERNSTORF (1964) and was later also proved

Table 4. *Mammotropic effect of 16 reserpine analogues compared with their sedative effect*

Oestradiol priming: 8 μg/rat/day for 10 days s.c.
Reserpine treatment 0.5—20 mg/kg/day for 5 days s.c.
Female rats 200 (± 10) g
Scoring from 0 to 5:
Overt behaviour: 0 = alert → 5 = reserpine syndrome
Body temperature drop: 0 = 1/2 °C → 5 = 2.5 °C
Decrease in barbital sleep induction time: 0 = 0 min → 5 = 5 min
Mammotropic index: 1 = normal mammary gland
5 = lactating mammary gland

No.	Generic and proprietary name	R$_{10}$	R$_{11}$	R$_{17}$	R$_{18}$	Dose mg/kg	Overt behaviour	Body temperature drop	Decrease in barbital Sleep induction time	Mammotropic index MTI	Recommendable for lactogenic use
							Scoring of sedative effect				
0	Control: Saline solution	—	—	—	—	0.5–1 ml	0	0	0	1.5	—
1	Deserpidine Harmonyl-Abbott	—	—	OCH$_3$	O–CO– (3,4,5-tri-OCH$_3$ phenyl)	0.5 / 1	2.5 / 3.5	2.1 / 3.2	4.0 / 5.0	4.5 / 5.0	±
2	Reserpine Serpasil-Ciba	—	OCH$_3$	OCH$_3$	O–CO– (3,4,5-tri-OCH$_3$ phenyl)	0.5 / 1	3.5 / 5.0	3.2 / 3.2	4.0 / 4.2	4.2 / 5.0	±

#	Name				Structure/Notes	Dose						
3	Rescinnamine Moderil-Pfizer	—	OCH$_3$	OCH$_3$	O—CO—CH=CH—⟨phenyl with OCH$_3$, OCH$_3$, OCH$_3$⟩	1 5	3.5 4.0	1.2 2.2	2.2 2.5	4.0 5.0	±
4	Su-10,704 Ciba (maleate)	—	OCH$_3$	OCH$_3$	O—CH$_2$—CH$_3$ in beta position	0.5 1	0 5.0	0 2.2	0 4.9	2.0 4.6	+
5	Su-9064 Ciba	—	OCH$_3$	OCH$_3$	O—CH$_3$ in alpha position	0.5 1 5 10	0 1.1 2.3 3.6	0 2.0 2.2 5.0	0 4.0 4.5 5.0	3.3 3.8 4.0 4.6	+
6	Tetrabenazine Nitoman-Roche	OCH$_3$	OCH$_3$	—	Ring B and E missing O and CH$_2$—CH—(CH$_3$)$_2$ on ring D	0.5 1 5 10	0 1 2.3 3.5	0 1 1.6 2.2	0 2 4.0 4.8	2.8 3.2 3.6 4.2	+
7	Su-7064 Ciba	—	OCH$_3$	OCH$_3$	⟨tetrahydropyran⟩ in beta position	0.5 10	0 3.1	0 2.2	0 2.3	2.8 3.6	+
8	Su-11,279 Ciba	—	OCH$_3$	OCH$_3$	O—CH$_2$—CH$_2$—O—CH$_3$, in alpha position at C$_{16}$: CO—O—CH$_2$—CH$_2$—O—CH$_3$	1 10 20	0 0 0	0 0 1.8	0 1.0 1.6	3.3 3.4 4.0	++
9	Su-9673 Ciba (Sulfate)	—	OCH$_3$	OCH$_3$	O—CH$_2$—CH$_2$—CH$_3$	0.5 1 5	0 0 2.5	0 0 2.1	0 1.3 2.6	1.8 2.3 3.4	±

Table 4 (continued)

No.	Generic and proprietary name	R_{10}	R_{11}	R_{17}	R_{18}	Dose mg/kg	Overt behaviour	Body temperature drop	Decrease in barbital Sleep induction time	Mammotropic index MTI	Recommendable for lactogenic use
10	Su-10,092 Ciba (maleate)	—	OCH$_3$	OCH$_3$	O—CH$_2$—CH$_2$—CH$_3$	0.5 / 1 / 5	0 / 1 / 2.5	0 / 0 / 2.0	0 / 1.5 / 2.5	2.0 / 2.2 / 3.2	±
11	Benzquinamide Quantril-Pfizer	—	—	—	Ring B and E missing O—CO—CH$_3$ and CO—N(C$_2$H$_5$)$_2$ on ring D	5 / 10 / 20	1.5 / 2.3 / 2.5	0 / 1.1 / 1.6	1.6 / 3.2 / 4.0	2.6 / 3 / 3.2	—
12	Yohimbine Quebrachine	—	—	OH	—	1 / 5	1.1 / 1.5	1.4 / 4.2	2.2 / 3.7	2.2 / 2.6	—
13	Ajmaline Cardiorythmine-Servier	—	—	—	Ring E connected to ring C OH and C$_2$H$_5$ on ring D, CH$_3$ on ring B	1 / 5	0 / 2.3	0 / 1.1	1.2 / 2.5	1.6 / 2.6	—
14	Su-9300 Ciba (hydrochloride)	—	OCH$_3$	OCH$_3$	O—CH$_2$—CH$_3$ in alpha position	0.5 / 1	0 / 1.5	0 / 0	0 / 3	1.8 / 1.7	—
15	Methoserpidine Decaserpyl-Roussel	OCH$_3$	—	OCH$_3$	O—CO— with 2,4,6-tri(OCH$_3$) benzene	1 / 5	1.1 / 1.5	0 / 2.0	2.5 / 3.6	1.2 / 1.6	—
16	Ajmalicine Raubasin-C.H. Boehringer	—	—	—	C$_{17}$ replaced by O C$_{19}$ methylated C$_{16}$—C$_{17}$: Δ	1 / 5	1.1 / 1.5	0 / 2.0	0 / 3.6	1.8 / 2.2	—

by radioimmunoassay (BRYANT and GREENWOOD, 1968) and by bioassay (BEN-DAVID and CHRAMBACH, 1969).

KHAZAN, BEN-DAVID, MISHKINSKY, KHAZEN and SULMAN (1966) showed that dissociation between the mammotropic and sedative effects of tranquilizers is possible, both being caused by independent mechanisms. BEN-DAVID, KHAZEN, KHAZAN and SULMAN (1968) studied 16 reserpine derivatives (Table 4) for correlation of the endocrine and the psychopharmacologic properties. The analogues differed from each other through varying positions of the methoxy groups and different radicals attached to the etherial linkage at position 18 at both configurations alpha (epi) and beta (normal). One of the analogues studied—Su-11,279—was the only one belonging to the 3-iso-reserpate series. Solutions for injection were made with a non-toxic reserpine solvent (propylene glycol 25%, acetic acid 1.14%, aqua bidest. ad 100%) which could be injected with 0.5—1 ml of normal saline.

Four parameters were compared:

a) *The mammotropic effect* (MTI) was studied, for each dose level, in groups of 5 to 10 adult female rats, each weighing 200 ± 10 g. The animals were primed for 10 days by daily s.c. injections of 8 µg estradiol in oil. From the 11th to the 15th day, daily s.c. injections of the reserpine analogues were given, and on the 16th day the animals were sacrificed. The right inguinal mammary pads were removed and their MTI determined (scoring 1—5).

b) *The body temperature drop* (BTD) was studied in groups of 10 female rats (200 ± 10 g), which were injected s.c. with a single dose of the reserpine analogues. Rectal temperatures were recorded at 0, 1, 2, 3, 4 and 5 hours with an electrocouple thermometer (Electric Universal Thermometer TE$_3$ "Ellab Instruments", Copenhagen). Control animals were injected with 0.2 ml saline s.c. Scoring: $1 = -0.5\,°C$, $5 = -2.5\,°C$.

c) *The effect on general overt behavior* (GOB) was studied in groups of 10 female rats, weighing 125 ± 10 g each. Immediately after injection, the rats were put into individual jars to permit accurate evaluation of the effect. Ptosis was judged by the degree of dropping of the eyelid, the ability of the rats to reopen their eyes after shaking of the jars, and the time which elapsed till the rats closed their eyes again. Exploratory behavior was evaluated by putting the rats on an upside-down jar and measuring the time lapse till the rats calmed down. Irritability was judged by the rats' reaction towards pricking of their backs with a blunt hypodermic needle just above the attachment of the tail (Scoring 1—5, Fig. 32).

d) *The barbital sleep induction time* (BIT) was studied in groups of 10 female mice, weighing 30 ± 2 g each. Two hours after i.p. injection of the different analogues, the animals were injected with 500 mg/kg of barbital sodium. Control mice received barbital sodium alone. The time elapsing between barbital injection and hypnosis, measured in minutes, was considered the barbital induction time (BIT). An animal was judged to be hypnotic if it lost the righting reflex for one minute or longer. Scoring: 1 = 1.5 min., 5 = 8 min.

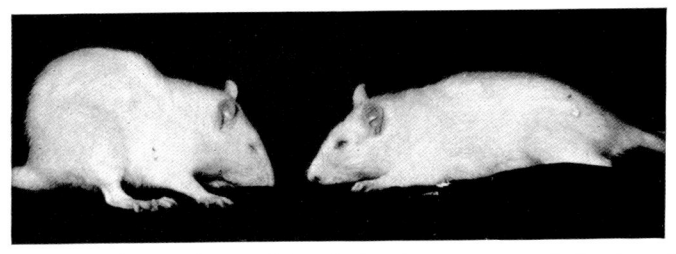

Fig. 32. Effect of reserpine and its derivatives on overt behavior of rats: l: reserpine syndrome (eye ptosis, arched position combined with irritability); r: normal control rat

Table 4 shows the comparative effectiveness of the reserpine analogues at different dose ranges used, arranged in order of their lactogenic activity (MTI: 1—5). At higher doses, all the reserpine analogues studied exerted strong sedative hypothalamic effect for less than 8 hours, except for Su-8,064 (10 mg/kg) and Su-10,704 (1 mg/kg) whose sedative effect lasted for more than 8 hours but was still shorter than the classical reserpine-syndrome which continues for 24 hours. Deserpidine, reserpine, rescinnamine and Su-10,704 were found to be the most effective analogues with regard to both their sedative and their mammotropic effects, when given at a 1 mg/kg dose level. Su-9,064, tetrabenazine, Su-7,064, Su-9,673, and Su-10,092 had weaker effects. Surprisingly, Su-11,279 was found to lack all sedative effect even at a dose as high as 20 mg/kg; on the other hand, it showed a high degree of mammotropic activity. Benzquinamide, yohimbine, ajmaline, Su-9,300 and methoserpidine had a very low lactogenic effect.

These results furnish an interesting example for the SAR between closely related chemical compouonds, comparable to earlier results with phenothiazine derivatives (cf. Chapter 14). It can be concluded that, when an ethyl radical in beta configuration is attached to position 18 on a reserpine molecule (Su-10,704), a most potent analogue will result, with regard to both sedative and mammotropic effects. If this ethyl radical is attached in alpha configuration (Su-9,300), the potency of the analogue drops sharply.

Attachment of a 2-methoxy-ethyl at position 18, in alpha configuration, to a 3-isoreserpate (Su-11,279) abolished the sedative effect, with the mammotropic effect remaining fairly high. This most interesting finding induced us to give Su-11,279 the highest scoring for practical use as a lactogenic agent. In general, scoring for recommendability of a substance for lactogenic use must be based on maximum lactogenic effect combined with minimal sedative effect. For this reason, Su-10,709, Su-9,064, tetrabenazine and Su-7,064 range higher than deserpidine, reserpine or rescinnamine in spite of the excellent lactogenic activity of the three latter substances.

Reserpine analogues with a methyl group in alpha or beta configuration were studied by SCHLITTER and PLUMMER (1964) for their sedative effects. The authors found that the beta-analogues had a higher sedative potency than the alpha-analogues. Highest potency was attained when an ethyl radical replaced the methyl in both positions. GALA and REECE (1963), in their study of Su-9,064, reported that this analogue was able to release lactogenic hormone from the anterior pituitary, similar to reserpine, without, however, being tranquilizing or hypotensive at the reserpine dosage. High doses were reported to induce the full reserpine-syndrome. The authors rightly concluded that the lactogenic effect of reserpine is not due to its hypotensive or tranquilizing action, but is inherent in the basic structure of the molecule.

PURSHOTTAM (1962) reported that syrosingopine, another reserpine derivative, had a hypotensive effect equaling that of reserpine, but was less sedative and less active in suppressing pituitary gonadotropin release. PALLADINO, BUTCHER and FUGO (1968) demonstrated that yohimbine suppressed the PIF and produced luteotropic hormone in pregnant rats. Whe showed that the mammotropic and the tranquilizing properties are unrelated, independent of each other, and mediated through two selective actions of the chemical agents on different CNS receptors, mediators or mechanisms (Assael, Gabai, Winnik, Khazan and Sulman, 1960; Khazan, Ben-David, Mishkinsky, Khazen and Sulman, 1966). Su-11,279 furnishes a classic example for such "dissociation," as it shows no sedative property but high mammotropic activity.

Summarizing it seems that reserpine and most of its derivatives have a strong lactogenic effect. The absence of all lactogenic effect in methoserpidine is particularly interesting because it differs from the highly lactogenic reserpine only in that the methoxy group is shifted from C_{11} to C_{10}. This fact shows the importance of a polar substitution at C_{11}. It would seem that the hypothalamus has a specific array of receptors which are sensitive to the action of cyclic compounds when they have a polar group at a corresponding spot, which is for the three-cyclic phenothiazines at C_2, for the four-cyclic steroids at C_3, and for the five-cyclic reserpine derivatives at C_{11}. Any compound fitting into this set-up is lactogenic but not necessarily tranquilizing, as has been shown. This problem will be further discussed in Chapter 19 devoted to receptor theories. Studies on the SAR of reserpine analogues supply further evidence for a possibility of complete dissociation between the mammotropic and sedative activities of hypothalamic tranquilizers.

Chapter 17

Mammotropic Effect of Butyrophenones

The clinical use of butyrophenones as neuroleptics originated at the research laboratories of Janssen Pharmaceutica in Beerse (Belgium) about ten years ago. It arose from a systematic investigation of the relationship between the chemical structure of 4-phenyl-piperidines and propiophenones and their pharmacological properties. Members of the higher homologue, the butyrophenones, proved to have a clear antipsychotic neuroleptic potency, and quite a few of them are used nowadays in psychiatry, obstetrics and anesthesiology (Janssen et al., 1959; Haase and Janssen, 1965). They posses chlorpromazine-like properties (Janssen, 1967 a, b), and were brought to our attention through their mammotropic activity, which may be a possible side-effect at unusually high clinical dose levels.

MISHKINSKY, KHAZEN, GIVANT and SULMAN (1969) have so far tested 23 representatives of this group for their mammotropic activity —as measured by mammary gland development, their sedative effect— as measured by body-temperature drop, and their cataleptic effect— as measured by catatonic behavior and ptosis.

All parameters were scored by numbering them from 1 to 5, to allow easy comparison of effects, and the aggregate of the scores of each group was divided by the number of animals. The assay of the mammotropic index (MTI) is described in Chapter 26, and the assay of body-temperature drop (BTD) in Chapter 14. The other two parameters were assayed as follows:

Cataleptic Effect (CPE)

Cataleptic effect was determined in female rats weighing 150 ± 10 g (5 rats per drug per dose) 60 minutes after s.c. injection. The test procedure consisted of putting the rat gently on its hind legs, in a head-upward position, with one of its forelegs placed on cork plates 3 or 9 cm thick. In this test we followed the method of WIRTH, GÖSSWALD, HÖRLEIN, RISSE and KREISKOTT (1958) who described different stages of cataleptic behavior in rats induced with phenothiazine compounds. The following five stages for the cataleptic effect were scored:

1. Transient cataleptic reaction lasting less than 30 seconds;
2. Cataleptic position, forelegs 3 cm raised for 30 seconds;
3. Cataleptic position, forelegs 9 cm raised for 30 seconds;
4. Prostrate posture with reaction to pricking;
5. Prostrate posture without reaction to pricking.

Failure to correct the imposed posture for 30 seconds was considered a positive cataleptic reaction. The test was carried out at each dose level on both right and left forelegs.

Ptosis (PTS)

Ptosis was evaluated by the degree of dropping of the upper eyelid. For scoring, the following five stages were used:

1. Intermittent lid closure;
2. One side semi-closed;
3. Both sides semi-closed;
4. One side completely closed;
5. Both sides completely closed.

Table 5. *Mammotropic index (MTI), body temperature drop (BTD), cataleptic effect (CPE) and ptosis (PTS) of 23 butyrophenones in oestradiol-primed female rats (200 ± 10 g). Oestradiol priming: 8 µg/rat/day for 10 days. Butyrophenone treatment: 1, 5 and 40 mg/kg/day for 5 days. Preparations are listed in descending order of their lactogenic effect (MTI) at 1 mg/kg*

Formula F—⟨phenyl⟩—C(=O)—CH$_2$—CH$_2$—CH$_2$—R$_1$

Code No. and generic anme	Substitution of R$_1$	Daily dose (mg/kg)	Mammo-tropic index (MTI)	Body temperature drop (BTD)	Cataleptic effect (CPE)	Eyelid ptosis (PTS)
R-2498 Trifluperidol (Janssen)	[4-hydroxy-4-(3-trifluoromethylphenyl)piperidine]	1 / 5 / 40	4.0 / 5.0 / 4.6	0.6 / 2.0 / 3.6	2.0 / 3.0 / 4.0	1.5 / 2.5 / 3.5
R-1658 Moperone (Janssen)	[4-hydroxy-4-(4-methylphenyl)piperidine]	1 / 5 / 40	3.9 / 4.0 / 4.3	0.9 / 2.1 / 4.9	2.5 / 3.0 / 3.5	1.0 / 2.0 / 3.0

R-5147 Spiroperidol (Janssen)		1 5 40	3.8 5.0 4.4	1.1 2.4 2.2	2.8 4.0 4.0	1.0 2.0 3.0
R-4749 Droperidol (Janssen)		1 5 40	3.8 3.5 3.4	0.9 1.5 2.2	1.5 2.5 4.0	3.0 3.0 3.5
R-9298 Clofluperol (Janssen)		1 5 40	3.7 4.4 4.0	1.1 1.7 1.7	2.5 3.0 3.5	0 0.5 1.0

Table 5 (continued)

Code No. and generic name	Substitution of R₁	Daily dose (mg/kg)	Mammo-tropic index (MTI)	Body temperature drop (BTD)	Cataleptic effect (CPE)	Eyelid ptosis (PTS)
R-4584 Benzperidol (Janssen)		1 5 40	3.4 3.8 3.8	1.3 1.3 1.7	2.5 3.0 4.0	1.5 2.5 3.5
R-1625 Haloperidol (Janssen)		1 5 40	3.3 4.7 4.8	1.1 2.3 3.0	2.5 2.5 4.0	1.0 1.0 3.0
R-6238 Pimozide (Janssen)	instead of O=	1 5 40	3.2 3.2 4.1	0.7 0.9 1.4	0 1.0 1.5	0 0 0.5

R-2028 Fluanisone (Janssen)	[structure]	1 5 40	3.0 3.5 4.0	1.7 2.6 4.6	1.0 2.0 5.0	1.0 2.0 3.0
R-3201 Haloperidide (Janssen)	[structure]	1 5 40	3.0 3.0 3.3	1.0 2.1 3.3	1.5 3.0 3.0	0.5 2.5 2.5
R-4457 (Janssen)	[structure]	1 5 40	3.0 3.0 3.6	0.8 2.4 3.8	1.0 2.0 3.0	0 1.5 2.0
R-3345 Floropipamide (Janssen)	[structure]	1 5 40	2.8 2.9 2.8	1.7 3.2 4.7	0 1.0 2.0	0 0 1.0

Table 5 (continued)

Code No. and generic name	Substitution of R_1	Daily dose (mg/kg)	Mammo-tropic index (MTI)	Body temperature drop (BTD)	Cataleptic effect (CPE)	Eyelid ptosis (PTS)
R-2963 Methylperidide (Janssen)		1 5 40	2.7 2.8 3.3	0.6 2.0 3.7	3.0 3.0 3.5	2.0 2.5 3.5
R-2962 Amiperone (Janssen)		1 5 40	2.6 3.0 3.4	0.8 1.4 3.8	2.0 3.0 3.0	1.5 2.0 3.0
R-1892 Butyropipazone (Janssen)		1 5 40	2.5 2.4 3.0	1.5 3.6 5.0	2.0 3.0 4.0	0.5 1.5 3.0

Mammotropic Effect of Butyrophenones

Compound	Dose				
FR-02 (Sandoz)	1	2.4	1.0	1.0	0
	5	2.6	2.9	1.0	0.5
	40	3.6	2.2	3.0	1.0
R-1647 Anisoperidone (Janssen)	1	2.0	0.6	0	0
	5	2.5	2.4	1.5	2.5
	40	2.8	3.5	2.0	3.0
R-4082 Floropipeton (Janssen)	1	2.0	0.7	0	0
	5	2.2	1.6	2.5	1.0
	40	3.8	3.6	5.0	3.0
R-4006 (Janssen)	1	2.0	0.1	0	0
	5	2.8	3.1	2.0	1.0
	40	3.6	4.8	2.5	1.0
R-1929 Azaperone (Janssen)	1	2.0	1.9	1.0	0
	5	2.8	4.0	1.1	0.5
	40	4.2	5.0	4.0	3.0

Structural features:
- FR-02 (Sandoz): spiro piperidine with N–CH$_3$ imide and C$_4$H$_9$ substituent
- R-1647 Anisoperidone (Janssen): 4-phenyltetrahydropyridine, F replaced by CH$_3$O—
- R-4082 Floropipeton (Janssen): O=C–CH$_2$–CH$_3$ attached to piperidinyl-piperidine
- R-4006 (Janssen): dispiro dipiperidine
- R-1929 Azaperone (Janssen): N-(2-pyridyl)piperazinyl-piperidine

Table 5 (continued)

Code No. and generic name	Substitution of R₁	Daily dose (mg/kg)	Mammo-tropic index (MTI)	Body temperature drop (BTD)	Cataleptic effect (CPE)	Eyelid ptosis (PTS)
R-3264 (Janssen)	–CH₂–NH–C(=O)–OC₂H₅ on 4-phenylpiperidine	1 5 40	1.9 2.9 3.4	2.1 3.8 5.0	1.5 1.0 2.5	0.5 0.5 1.5
FR-33 (Sandoz) R-7158 (Janssen)	N–CH₃ spiro piperidine-succinimide	1 5 40	1.6 1.8 3.5	2.5 3.3 5.0	1.5 2.0 4.0	0 0.5 2.5
R-3248 Aceperone (Janssen)	–CH₂–NH–C(=O)–CH₃ on 4-phenylpiperidine	1 5 40	1.5 2.3 3.7	0.6 2.0 3.0	3.0 3.0 3.0	1.5 2.5 3.0
Solvent (Control)	ethanol + 0.3 N HCl āā → 0.5 ml	0.5 ml	1.2	0	0	0

The degree of ptosis was evaluated simultaneously with the cataleptic behavior, 60 minutes after s.c. injection of the drug tested.

The 4 parameters measured and evaluated after s.c. injections into female rats of three different doses of the 23 butyrophenones are presented in Table 5. Eleven butyrophenones were found to be mammotropic at a dose of 1 mg/kg (MTI above 3.0). This result did not improve at a dosage of 5 mg/kg, but at 40 mg/kg ten more butyrophenones gave a mammotropic reaction.

All butyrophenones tested caused a significant lowering in body temperature (i. e. more than 0.6° C) even at the small dose of 1 mg/kg. MTI and BTD showed a very close correlation to each other, apparently due to a similar mechanism of action, though they differed quantitatively.

The cataleptic and ptotic reactions were also closely interrelated, though without any connection with the mammotropic effect. Ten of the butyrophenones tested proved to be cataleptic at the small dose of 1 mg/kg, seven others were cataleptic at 5 mg/kg and five became cataleptic at the highest dose of 40 mg/kg. One substance—pimozide—did not show any significant cataleptic effect within 60 minutes; its onset of action is probably very slow.

The ptosis reaction followed the pattern of the cataleptic butyrophenones except that—in addition to pimozide—clofluperol, floropipamide, FR-02 and R-4006 also showed a low degree of ptosis.

Many of these results were to be expected, as butyrophenones cause a marked hypothalamic depression, similar to the typical chlorpromazine and reserpine effects. Earlier experience has shown that the MTI and the BTD are most reliable parameters of specific hypothalamic depression, the MTI measuring the effect on the prolactin-inhibiting factor (PIF). The BTD is a standard test for hypothalamic sedation, since the thermoregulatory centers of the thalamus are connected with the emotional centers of the hypothalamus (LISSEN and PARKES, 1957; PETERSON and NIELSON, 1964). Serotonin increases temperature, whereas noradrenaline decreases it (FELDBERG, 1965). Butyrophenone derivatives, like phenothiazine derivatives, deplete both serotonin and noradrenaline and thus deprive the hypothalamus of its heat regulatory mechanism.

We concentrated mainly on the mammotropic effect of the butyrophenones, and were struck by the fact that they promote mammotropic development at lower doses than do the phenothiazines. This is probably due to their surface-spreading tendency, forming a film on biological structures, and thus becoming potent membrane permeability blockers (JANSSEN, 1967 a, b; GUTH and SPIRTES, 1964).

All neuroleptics are competitive antagonists of adrenaline, noradrenaline and dopamine, which act as neurotransmitters in the central nervous system (JANSSEN, 1967 a, b). It is postulated that haloperidol and other potent neuroleptic drugs block the dopaminergic neurones by forming a permeability-decreasing mono-layer on the membranes surrounding the cleft on the dopaminergic synapse. In this way, the dopamine molecules fail to reach the receptor site, the nerve impulses cease to be transmitted, and the animal becomes depressed. This may also be the mechanism of the BTD in the rat, since it seems that temperature is the outcome of a fine balance between the release of catecholamines and serotonin in the hypothalamus (FELDBERG, 1965), and it has been suggested that butyrophenones antagonize the influence of catecholamines, thus decreasing body temperature. The dopamine molecules are either stored in the dopamine pools (HORNYKIEWICZ, 1960) or metabolized by monoamine oxidase (MAO) in the cytoplasm. Depressing the dopaminergic neurones by butyrophenones requires much smaller doses than are needed when using chlorpromazine (JANSSEN, 1967 a, b).

The relation between the ability of the butyrophenones to antagonize catecholamines and their mammotropic effect was studied by COPPOLA, LEONARDI, LIPPMANN, PERRINE and RINGLER (1965), who suggested an indirect mechanism for prolactin release by depletion of noradrenaline from the PIF center in the median eminence. This theory was recently adopted by VAN MAANEN and SMELIK (1968), who localized the monoaminergic prolactin-inhibitory system in the tuberal part of the hypothalamus and postulated that it reacted to dopamine. The two hypotheses about the site of action of the butyrophenones may well explain the mechanisms of their mammotropic and body temperature lowering effects.

The mammotropic indices for the 23 butyrophenones tested are listed in Table 5 in descending order of their lactogenic effect at a dose of 1 mg/kg. Like other potent mammotropic neuroleptics—such as fluphenazine, trifluperazine and reserpine—the potent mammotropic butyrophenones are tertiary amines, derivatives of 4-substituted piperidine or piperazine rings. Replacement of the ketonic moiety by an aromatic radical does not seem to impair the mammotropic effect.

Small polar substituents, in para-position to the piperidine ring, increase the neuroleptic potency without, however, reducing the mammotropic effect, but lengthening of this group does impair lactogenic activity. This fact is also described in reference to the lactogenic phenothiazines (cf. Chapter 19).

The evaluation of the cataleptic effect of mammotropic drugs is obviously important, as ideal mammotropic activity should lack all concomitant cataleptic effect. The latter phenomenon, first observed by COURVOISIER, DU CROT and JULOU (1957) is caused by an activation of cholinergic processes and it is antagonized by anticholinergic and antidepressant drugs (ZETLER, 1968). It seems to be linked to the mechanism of the BTD, since hypothermia, too, can be caused by centrally acting cholinergics and may be prevented or reversed by anticholinergics (MAICKEL, STERN and BRODIE, 1963). Catalepsy is obviously related to ptosis, and is also known as "the catalepsy-ptosis observation test in rats" (JANSSEN, NIEMECEERS and SCHELLEKENS, 1965).

Summing up we feel that for the purpose of lactogenic activity a substance should have a high MTI and low BTD, CPE and PTS scores, as e. g. pimozide. The only three other substances approaching this goal among the butyrophenones tested were trifluperidol (R-2498), droperidol (R-14749) and clofluperol (R-9298).

Chapter 18

Mammotropic Effect of Miscellaneous Psychotropics Drugs

The mammotropic effect obtained with derivatives of phenothiazine (Chapter 14), tricyclic ring systems (Chapter 15), reserpine (Chapter 16), and butyrophenones (Chapter 17) prompted us to study some psychopharmaca which were assumed to act on the hypothalamus (unpublished results).

Thirty-six compounds were tested. Only four of them were found to induce mean MTI values of 3 to 3.5 at non-toxic doses: guanoclor (Vatensol, Pfizer), meclozine (Bonamine, Pfizer), methyldopa (Aldomet, Merck), tripelennamine (Pyribenzamine, Ciba). The other thirty-two compounds did not induce mammary development beyond MTI 2.8 at non-toxic doses. These inactive drugs were: amphenidone (Dornwal, withdrawn preparation), azacyclonol (Frenquel, Merrell), captodiamine (Covatin, Lundbeck), clomiphene (Mer-41, Clomid, Merrell), chlordiazepoxide (Librium, Roche), chlorothiazide (Diuril, Merck), diazepam (Valium, Roche), digitoxine (Nativelle), dyphylline (Hyphylline, Continental Labs.), fucidin (Leo Labs.), guanoxan (Envacar, Pfizer), histamine dichloride (Merck), HPt-430 (C. H. Boehringer, Ingelheim), HR-1978 (Wander, Berne), 5-hydroxytryptophane (Sandoz), hydroxyzine (Atarax, Pfizer), iproniazide (Marsilide, Roche), isocarboxazide (Marplan, Roche), Kö 592 (C. H. Boehringer, Ingelheim), lanatoside C (Cedilanid, Sandoz), melatonin (Upjohn), meprobamate (Miltown, Wallace), methamphetamine (Pervitin, Temmler), monzal (Thomae, Germany), morphine HCl and narceine HCl (Sandoz), phencarbamide (Escorpal, Bayer), D-phenylalanine (Merck), scillaren A (Transvaalin, South Africa), thalidomide (Contergan, withdrawn preparation), theophylline (Aminophylline, Searle), tybamate (Solacen, Wallace).

The cardiac glycosides which are known to induce gynecomastia in man (LEWINN, 1953) did not induce full lobulo-alveolar mammary development in the MTI test. The gynecomastia-inducing activity of digitalis preparations is attributed to their metabolic conversion into estrogen-like derivatives. Since hypothalamus-mediated mammotropism is assumed to result from protracted effects on the hypothalamus-pituitary axis, short-term administration of cardiac glycosides would not elicit mammotropic action. This was confirmed by us in animal experiments. We also had negative results with morphine and methamphetamine, which can produce lactation in highly estrogen-primed rats. Doubtless, our lower estrogen-priming allows more specific reactions, probably because of the subthreshold suppression of the PIF center. Investigations in our laboratory by BEN-DAVID (unpublished) reveal that estrogen-priming is not required for a mammotropic reaction if specific lactogenic phenothiazine compounds acting on the hypothalamus are injected into rats. In future studies of the mammotropic effects in *non*-estrogen-primed rats we hope to be able to elucidate reports claiming the non-specificity of many mammotropic effects of drugs (NICOLL, TALWALKER and MEITES, 1960; MEITES, NICOLL and TALWALKER, 1963).

The evidence presented in this and preceding chapters enables us to state that mammotropism is a quite specific property of drugs, more or less closely related to hypothalamotropic actions. This structure-activity relationship is discussed in Chapter 19.

Chapter 19

Structure-Activity Relationship of Mammotropic Drugs and Receptor Theory

The data reported in Chapters 13—18 can now be summarized to define the requirements for a mammotropic compound. Mammotropic activity is a typical side-effect of the derivatives of phenothiazine, reserpine, the butyrophenones and many hypothalamic psychopharmaca, which does not run parallel to the tranquilizing activity of these groups. In order to find the most mammotropic phenothiazine derivative, some hundred chemicals bearing a phenothiazine nucleus were screened in our laboratory, and the following six requirements for optimal lactogenic activity were established:

1. Changes in the phenothiazine nucleus impair the mammotropic effect, but phenothiazine sulfoxides do not lose much of their mammo-

MTI	Structure	Ring Type
4-5		PHENOTHIAZINE RING
4-5		PHENOTHIAZINE - SULFOXIDES *RP 8580*
3		PHENOTHIAZINE - SULFONES *DOXERGAN*
2-3		AZAPHENOTHIAZINES *PERVETRAL* *SELVIGON*
2-3		THIAXANTHENES *TARACTAN* *BP 400*
2-3		IMIPRAMINE DERIVATIVES *TOFRANIL* *PERTOFRAN*
2		AMITRYPTILIN DERIVATIVES *ELAVIL* *PERIACTIN*

Fig. 33. Rules of mammotropic effect of phenothiazine dericatives: 1. Changes in the phenothiazine ring reduce the mammotropic effect—sulfoxides are optimal

tropic activity (Fig. 33). This indicates that the action of the S moiety is not essential for the mammotropic effect.

2. Substitution at C_2 by Cl, Br, CF_3, SCH_3, SC_2H_5, CN or $CO(CH_3)$ increases the mammotropic effect; substitution at C_2 by H or OCH_3 decreases it (Fig. 34). This indicates that a polar group at C_2 is required for the mammotropic effect.

MTI	Structure	Generic & Proprietary name
4.7	[phenothiazine with Cl at C_2; $CH_2-CH_2-CH_2-N\underset{\smile}{}N-CH_2-CH_2OH$]	PERPHENAZINE *TRILAFON*
4.6	[phenothiazine with CF_3; $CH_2-CH_2-CH_2-N\underset{\smile}{}N-CH_2-CH_2OH$]	FLUPHENAZINE *PROLIXIN*
4.6	[phenothiazine with CN; $CH_2-CH_2-CH_2-N\underset{\smile}{}-OH$]	PROPERICIAZINE *NEULEPTIL*
4.2	[phenothiazine with $COCH_3$; $CH_2-CH_2-CH_2-N\underset{\smile}{}N-CH_2-CH_2OH$]	ACETOPHENAZINE *TINDAL*
3.0	[phenothiazine with Cl; $CH_2-CH_2-CH_2-N(CH_3)_2$]	CHLORPROMAZINE *LARGACTIL*
2.8	[phenothiazine with H; $CH_2-CH_2-CH_2-N\underset{\smile}{}N-CH_3$]	PERAZINE *TAXILAN*
2.4	[phenothiazine with H; $CH_2-CH_2-CH_2-N(CH_3)_2$]	PROMAZINE *SPARINE*
1.8	[phenothiazine with OCH_3; $CH_2-CH_2-CH_2-N(CH_3)_2$]	METHOPROMAZINE *MOFAZINE*

Fig. 34. Rules of mammotropic effect of phenothiazine derivatives: 2. Substitution at C_2 by Cl, Br, CF_3, SCH_3, SC_2H_5, CN or $COCH_3$ increases mammotropic effect; substitution at C_2 by H or OCH_3 decreases mammotropic effect

Structure-Activity Relationship of Mammotropic Drugs   135

3. The distance between N_{10} and N_{14} must be filled by three C (Fig. 35). This allows a configuration resembling a halved hexane ring.

4. The side chain C_{11}—C_{13} must be devoid of any substitution. This points to its possible entrance into a receptor ready to receive the halved hexane ring (Fig. 36).

MTI	Structure	Generic & Proprietary name
4.8		PIPERACETAZINE *QUIDE*
2.2		SKF 5757-G
1.5		PYRATHIAZINE *PYRROLAZOTE*

Fig. 35. Rules of mammotropic effect of phenothiazine derivatives: 3. The distance between N_{10} and N_{14} must be filled in by 3 C

5. Heterocyclic rings (e. g. piperidine, pyrrolidine or piperazine) following C_{11}—C_{13} increase the mammotropic effect (Fig. 36). This differentiates the receptor of the phenothiazine molecule from that of the steroid molecule.

6. Incorporation of C_{12} or C_{13} into the piperidine ring decreases the mammotropic effect (Fig. 37). This confirms the requirement of a halved hexane ring for the hypothalamic mammotropic receptor.

Mammotropic substances can thus be divided into two groups wich require similar but different hypothalamic receptors (Fig. 38):

A. steroids (estrogens, including stilbestrol and its derivatives);

B. cyclic compounds composed of 4—5 rings containing N (phenothiazines and their tricyclic derivatives if not prolonged by esterification, reserpine derivatives, certain butyrophenones, Fig. 39).

Both groups have the following features in common (Fig. 40):

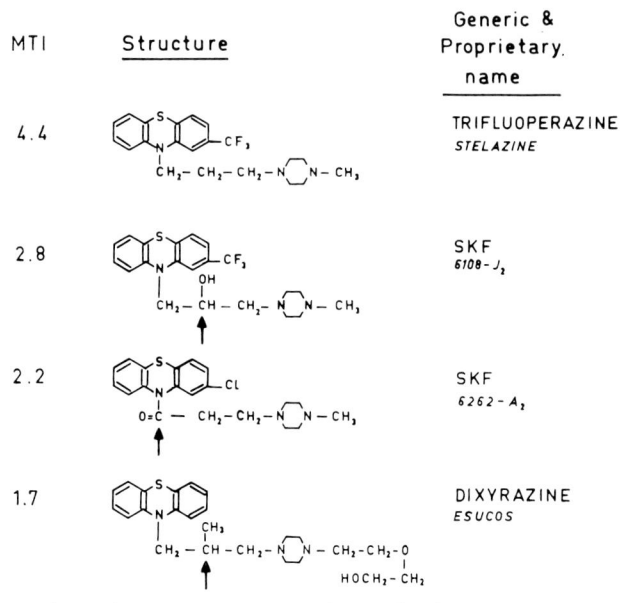

Fig. 36. Rules of mammotropic effect of phenothiazine derivatives: 4. An unbranched aliphatic side chain of C_{11}—C_{13}, connecting N_{10} with N_{14} is a prerequisite for mammotropic effect

Fig. 37. Rules of mammotropic effect of phenothiazine derivatives: 5. A heterocyclic ring (piperidine, pyrrolidine or piperazine) following C_{11}—C_{13} increases the mammotropic effect

a) an aromatic ring A with a polar group at the same site (C_3 for steroids, C_2 for phenothiazines, C_{11} for reserpine derivatives);

b) 3—4 hydrogenated B—E rings.

The receptor for steroids can be blocked by steroids; it is related to the hypothalamic prolactin-releasing factor since clinical experience

Fig. 38. Rules of mammotropic effect of phenothiazine derivatives: 6. Incorporation of C_{12} or C_{13} into the piperidine ring decreases the mammotropic effect

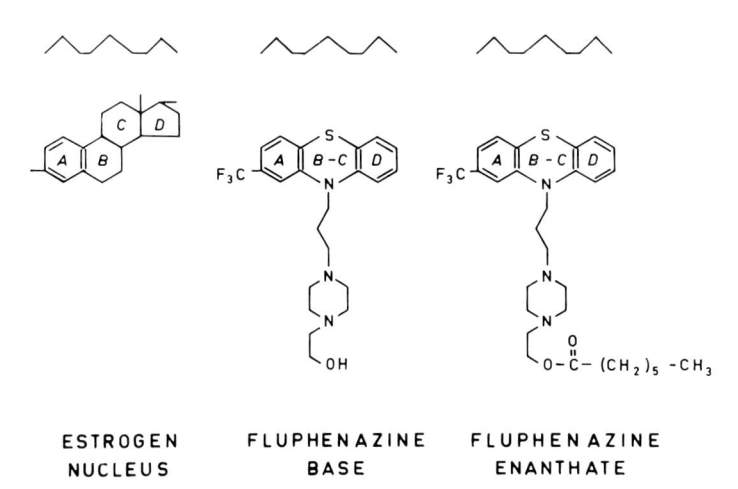

Fig. 39. Hypothetic estrogen receptor which can partly receive fluphenazine base but hardly cope with fluphenazine enanthate

has shown that it is stimulated by small doses of estrogen and suppressed by larger ones, e. g. in post-partum lactation. The receptor for the cyclic N-compounds cannot be blocked by steroids; it is linked to the hypothalamic prolactin-inhibiting factor, since clinical experience has shown that it is susceptible to psychopharmaca but not to estrogens. This experience is well known to any clinician who has tried the suppressive effect of estrogens in Chiari-Frommel Syndrome or Ahumada-del Castillo Syndrome.

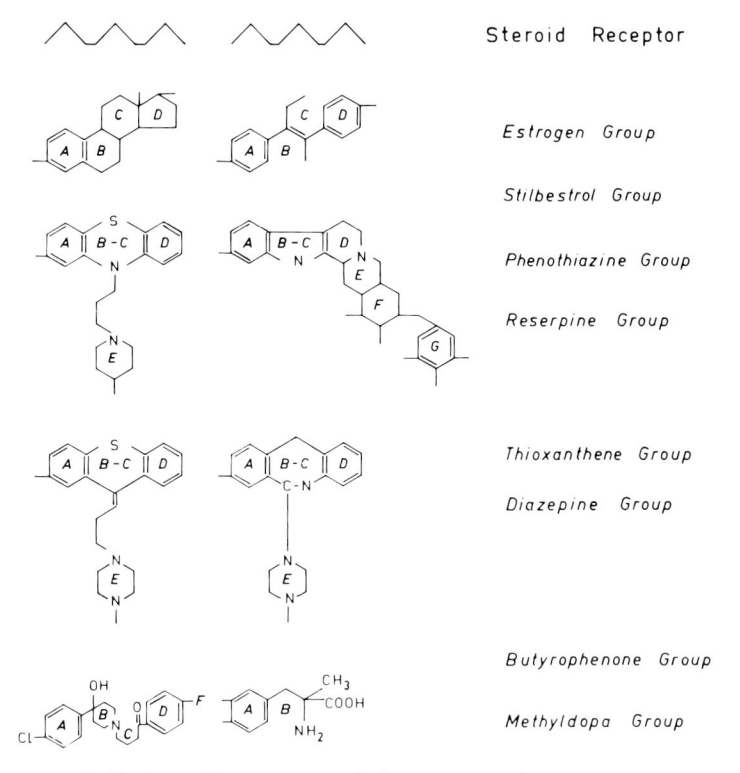

Fig. 40. Similarity of structure of lactogenic substances: 1 estrogens; 2 stilbestrol; 3 phenothiazines; 4 reserpines; 5 thioxanthenes; 6 diazepines; 7 butyrophenones; 8 methyldopa. All drugs found to be powerful prolactin releasers have the following steroid features in common: a) one aromatic ring with a substitution at C_3; b) at least two additional rings, or c) structural arrangements which simulate two additional rings. This structure makes them fit into the hypothalamic estrogen receptor which promotes lactation

The fact that side chains protruding from the tree-cyclic systems should consist of an unbranched 3-C-membered chain—which can fold in order to imitate half a hexane ring—seems of special importance since it proves the existence of a tripartite hypothalamic receptor sensitive to cyclic structures (Figs. 39, 40).

The validity of this conception has been accepted in pharmacology since the introduction of sympathomimetic (adrenergic) amines emanating from alkyl derivatives of ethylamine. Experience with these alkyl adrenergic amines has shown that addition of a methyl or ethyl group to the beta-C-ethylamine moiety does not yield active preparations. However, substitution by a triple-carbon (propyl) or a larger chain, which allows folding, results in preparations which pharmacologically mimic the adrenergic effects of adrenaline or noradrenaline. These new derivatives are known as alkyl or aliphatic pressor amines (1,3 dimethylamine, Forthane: $3C+2C$; N,1-dimethylhexylamine, Oenethyl: $4C+2C$; Heptamine or Tuamine: $5C+2C$; Octylamine: $6C+2C$; Nonylamine: $7C+2C$). It is obvious that an aliphatic adrenergic amine composed of less than $3C+2C$ cannot fold with its head piece to fill the receptor site which normally receives the phenyl moiety of adrenaline.

These rules were confirmed by the results obtained with fluphenazine enanthate (Chapter 14). Whereas the nonesterified mother substance—fluphenazine—had high lactogenic activity (MTI 3.3—4.6), the ester when given in the same dose was inactive (MTI 1.2—2). This surprising result induced us to review the properties required of an active lactogenic drug, and we arrived at the following conclusions:

a) The enanthate esterification deprives fluphenazine of its lactogenic activity because it adds bulk to the five-ring system which prevents it from entering a hypothetic hypothalamic receptor. It, therefore, needs to be split up at its site of injection in order to reach the hypothalamus (Fig. 39).

b) The hypothalamic receptor must have at least three main sites for natural ring systems which stimulate lactation in women and animals, i. e. the steroids (Fig. 39).

c) Most substances described as potent lactogenic stimulants in this report have a ring formation which imitates steroid rings (Fig. 40).

d) The six rules for active phenothiazine derivatives (Figs. 33—38) fit extremely well into this theory. If they are redrawn, Fig. 40 will present the conclusion that at a typical position of an aromatic ring, which is identical for all lactogenic substances (estrogen: C_3, stilbestrol: C_3, phenothiazine: C_2, reserpine: C_{11}), there must be a substitution which can act as a polar group within the body.

e) This rule is supported by the puzzling fact (Chapter 16) that reserpine with a methoxy group at C_{11} is lactogenic, whereas its isomer methoserpidine—with the methoxy group at C_{10}—is inactive as a lactogen, but as effective a tranquilizer as reserpine itself.

f) Fig. 40 also furnishes and explanation why thioxanthene derivatives, such as P-4657-B, and the Wander compounds, such as SUM 3170 and HF 2159 (Chapter 15), are lactogenic. Even haloperidol and its derivatives (Chapter 17) could be conceived in such a way that they fit into the hypothalamic steroid acceptor, as might methyldopa (Chapter 18).

Thus we arrive at the conclusion that study of the structure-activity relationship in this research furnishes us with a simple rule for future synthesis of lactogenic substances. These should have at least 3 rings—the first one aromatic—to fit the hypothalamic lactation receptor. The aromatic ring should have a polar group or a substitution at C_3 to resemble the steroids. It should have one or two piperidine, piperazine, pyrrolidine or even azepine rings attached. Side chains should be avoided, and up to 6 rings in all could be expected to be active lactogens. Synthesis of ring compounds to fit these rules and to obtain highly lactogenic substances which would specifically affect the LTH centers of the hypothalamus, but exert no tranquilizing or sedative effect, is now being planned in our laboratory.

Another aspect of the above receptor hypothesis is the proof recently furnished by us that estradiol and phenothiazine derivatives compete for the same brain receptor (MISHKINSKY, GIVANT, EYLATH and SULMAN, unpublished data). Labeled and unlabeled phenothiazine derivatives (0.05—50 mg/kg) were given by intravenous injection to ovariectomized rats and their cerebral uptake studied. When tritiated estradiol (50 c/mM) was injected i.v. (1 µg/kg) 30 seconds after the phenothiazines, its uptake in the anterior pituitary and median eminence depended upon the quantity of phenothiazines given before. The dose-response curve revealed only 10—50% of estradiol uptake after administration of perphenazine, aminopromazine or phenothiazine. This proves that phenothiazine derivatives compete quantitatively with estradiol for a common steroidal receptor in the hypothalamus and anterior pituitary. Control measurements with other brain sectors did not show any significant uptake of steroids or of phenothiazine derivatives (Fig. 41). The reverse experiment, giving estradiol before the phenothiazine, did not reveal any significant competition.

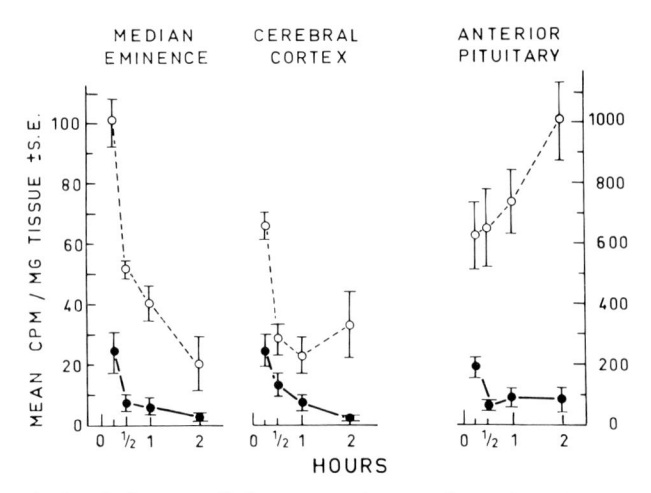

Fig. 41. Block of the estradiol receptor by perphenazine given s. c. 30 sec before i. v. injection of estradiol. The rats were killed $1/4$, $1/2$, 1 and 2 hours after estradiol injection. Median eminence block: 70—90%; Anterior pituitary block: 70—90%; Cerebral cortex block: 30—45% only. Note the specific uptake and block of both substances in median eminence and pituitary which has its optimum at $1/4$—$1/2$ hr. but lasts beyond 2 hrs. Pituitary uptake of estradiol is, as a rule, 10 times higher than that of median eminence. ●———● Estradiol after Perphenazine; ○- - - -○ Estradiol alone

Chapter 20

Pituitary-Ovary Axis in Hypothalamic Lactation

Perphenazine was found to be an effective agent for elucidation of the mechanism of hypothalamic lactation (BEN-DAVID, DIKSTEIN and SULMAN, 1965). It stimulates pituitary prolactin secretion *in vitro* in hypothalamus-pituitary co-culture (DANON, DIKSTEIN and SULMAN, 1963) and, *in vivo* via the synthesis and release of pituitary prolactin in intact animals (BEN-DAVID and CHRAMBACH, 1970). Similarly, perphenazine (VELARDO, 1958), reserpine and chlorpromazine (BARRACLOUGH, 1957) have also been reported to induce pseudopregnancy, indicating stimulation of the secretion of luteotropic hormone which produces progesterone (VELARDO, DAWSON, OLSEN and HISAW, 1953). Estrogen, too, stimulates prolactin secretion (RAMIREZ and McCANN, 1964). Thus, all ovarian hormones may play a role in mammary gland growth. LYONS showed already in 1958 that, in the hypophysectomized rat, ovarian and hypophysial hormones are required for normal mammogenesis. Only later did TALWALKER and MEITES (1961) demonstrate that, in the hypophysectomized-ovariectomized rat anterior pituitary hormones are mammogenic by themselves if given in sufficient quantity. It can, thus, not be determined whether perphenazine treatment induces lactation via the hypothalamus and the anterior pituitary—by stimulating the release of pituitary mammotropic hormones (LTH and STH)—or also via the hypothalamus, the anterior pituitary and the ovaries, i. e. by stimulating ovarian hormone secretion as well. In order to elucidate the role of the ovaries in perphenazine-induced lactation, experiments were carried out (BEN-DAVID, 1968 b) in intact and ovariectomized, estradiol-primed and non-primed adult rats and in juvenile rats, in which simultaneous determination of the DNA content of the mammary glands and histological studies of the mammary glands, ovaries, uterus, and vaginal smears were performed after 5 days' treatment with perphenazine. In bilaterally ovariectomized rats treatment was started 7 days after operation. For priming, 8 µg estradiol was injected for 10 days prior to perphenazine treatment.

In normal (non-primed) adult rats, full mammary gland development (MTI:5) was obtained by 5 daily injections of 10 mg perphenazine/kg. In ovariectomized adult rats the mammotropic effect of perphenazine was always lower than in the intact rats, the MTI not exceeding 3.8. This lesser response was, however, overcome by first priming the ovariectomized animals with small doses (8 µg/d) of estradiol for 10 days. In estradiol-primed rats a daily dose of 10 mg perphenazine/kg produced a maximal lactogenic response (MTI:5) in both intact and ovariectomized animals.

The DNA content of the mammary glands increased with the perphenazine dosage, but not RNA. A sharp increase in both DNA/mg tissue and total DNA content was obtained when the perphenazine dose was increased to 5 mg/kg/day. However, the DNA/mg tissue rose less sharply than total DNA content when higher doses of perphenazine were employed. In juvenile rats, perphenazine treatment alone (10 mg/kg/day, for 5 days) failed to induce significant mammary growth (MTI: 2) even though the absolute weight of the pituitary and that per 100 g body weight were substantially ($P<0.05$) increased by 20% and by 29% respectively. However, in estradiol-primed juvenile rats, similar perphenazine treatment induced considerable mammary growth (MTI: 4.8). Perphenazine administered to juvenile rats caused no significant change in the weight of ovaries or uterus, nor did it induce precocious opening of the vagina.

The changes in the mammary glands, ovaries, and uterus of intact adult non-primed rats receiving perphenazine for 5 days, starting at early diestrus, are illustrated in Fig. 42. The mammary glands were found to be highly developed and the lactogenic response increased with the dose. Milk was present in the mammary glands of all animals receiving the excessive dose of 15 mg/kg. Examination of the ovaries showed a greater number of corpora lutea (3—6) in the ovaries of treated animals than in the ovaries of the control rats (0—2), an observation confirmed by sectioning the ovaries.

No animal among the perphenazine-treated rats had a fluid distended uterus (which usually indicates estrogenic activity), as against 40% of the control animals. Histological investigation of the uteri of treated rats (Fig. 42) showed that basal vacuoles ("lucid zones") appeared in the endometrium which contained many more nuclei than that of the control. Moreover, diestrus persisted in the treated rats until about 7 days after the end of treatment.

Histological studies and determination of the DNA content of the mammary glands after administration of various doses of perphenazine showed that the degree of mammary growth was directly proportional to the doses employed. While total DNA content of the left inguinal mammary pad showed a continuous sharp increase parallel with the increase in perphenazine dosage up to 15 mg/kg, only a limited rise in DNA per 100 mg mammary tissue was obtained with doses higher than 5 mg/kg, indicating that, on reaching their maximum degree of development, the mammary cells started to produce milk rather than to divide. At the same time, however, larger doses of the drug caused further growth of mammary tissue, as indicated by the increase in total DNA.

In estradiol-primed rats a comparison between the lactogenic effect of perphenazine in intact and ovariectomized rats showed no difference in the mammary response at any of the doses given, while in non-

144   Pharmacological Regulation of Lactation

primed rats there appeared to be greater mammary response in the intact animals. Moreover, in the non-primed ovariectomized rats the maximal mammary response did not exceed MTI 3.8, even when a higher dose (10 mg/kg/day) was employed. It may be assumed, therefore, that, in order to achieve full lactogenic response after perphenazine treatment, the mammary tissue must be primed with small

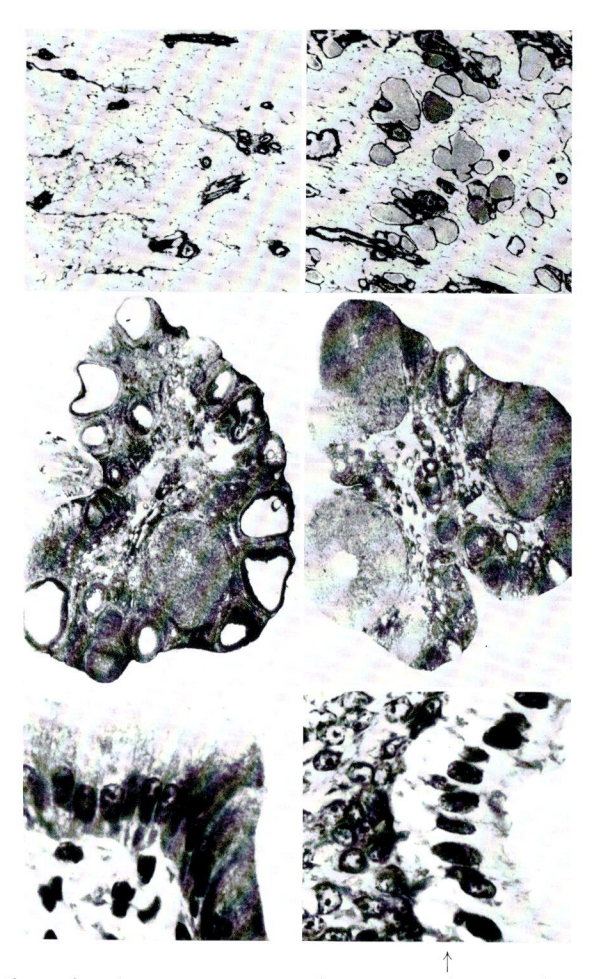

Fig. 42. Effect of 5 days' treatment with 5 mg perphenazine/kg/day on the mammary glands, ovaries and uteri of rats weighing 95—105 g. Treatment was started at early diestrus: l: control; r: treated rat. Note mammary development, luteotropic effect on the ovary and "lucid zone" (arrow) in treated uterus

amounts of estrogen, either by physiological estrogen present in the intact animal, or by exogenous estradiol in ovariectomized rats. Estrogen seemingly acts as a permissibe factor in perphenazine-induced lactation.

Partial mammary growth in ovariectomized rats induced by perphenazine (MTI: 3.8) seems to result mainly from stimulation of the secretion of mammotropic hormones from the anterior pituitary. This assumption is supported by the results of Talwalker and Meites (1961) who showed that anterior pituitary hormones alone, if given in sufficient amounts, induce mammary growth in the ovariectomized-hypophysectomized rat.

That the presence of the ovaries is not essential for perphenazine to exercise its full mammogenic effects is evident from the fact that a dose of 10 mg/kg/day for 5 days was fully effective in ovariectomized rats provided they had been primed with estradiol. Moreover, full lactation has also been obtained in adult non-primed male rats treated with perphenazine (Ben-David, 1968 b). The state of the ovaries provided no evidence that the mammotropic effect of perphenazine was due to release of FSH evoking an increase in estrogen production by the ovaries. Eckstein (1962) had reported that the ovaries of juvenile rats were very sensitive to exogenous FSH and that their weight greatly increased after administration of PMS. In the study by Ben-David (1968 b) perphenazine treatment of juvenile rats caused no increase in ovarian weight, in neither estradiol-primed rats nor in non-primed juvenile rats. Similarly, no increase in uterine weight was found, indicating that FSH secretion was not stimulated by perphenazine. In fact, it has been reported that phenothiazine derivatives, e. g. chlorpromazine and also reserpine, block rather than stimulate FSH-LH secretion (Sulman and Winnik, 1956; Winnik and Sulman, 1956; Nagata, 1957; Barraclough and Sawyer, 1957). Barraclough (1957) has also reported the induction of pseudopregnancy in rats by chlorpromazine and reserpine (cf. als Khazan, Sulman and Winnik, 1960).

When perphenazine was given at an appropriate stage of the cycle (starting at early diestrus), a typical "lucid zone" was obtained in the endometrium (Fig. 42). The "lucid zone" was first described in rats by Vokaer in 1950 as indicating the existence of active corpora lutea (pseudopregnancy). However, it is well recognized that, in rats and mice, the corpora lutea are not truly active during the normal estrous cycle, and a state of pseudopregnancy can be achieved only after release of adenohypophysial luteotropin which changes the cycling corpora lutea into active endocrine organs secreting mainly progestational compounds (VELARDO, DAWSON, OLSEN and HISAW, 1953). VELARDO (1958) also reported that a pseudopregnant state in rats could be achieved by perphenazine administration, and he confirmed the existence of this state by the induction of deciduomata in the rats.

We confirmed VELARDO's finding by reproducing the occurrence of the "lucid zone" in the endometrium of perphenazine-treated rats (Fig. 42). These experiments show that in virgin adult female rats mammary gland development and milk secretion were induced by 5 days' perphenazine (Trilafon) treatment. The drug, however, failed to induce similar lactation in either ovariectomized adult animals or in juvenile rats unless the animals were pretreated with small quantities of estradiol, indicating that, when the ovaries are absent or almost inactive, as in juvenile rats, estradiol-priming is a prerequisite for perphenazine-induced lactation. When perphenazine was given to adult rats starting at early diestrus, the following changes occurred: 1) a continuous stage of diestrus which lasted until about 7 days after the end of treatment; 2) the ovaries were predominantly occupied by corpora lutea; and 3) the "lucid zone", typical for pseudopregnancy, appeared in the uterine endometrium.

It is concluded that perphenazine treatment stimulates endogenous secretion of LTH, which in rats is identical with prolactin, and thus promotes lactogenesis.

Chapter 21

Pituitary-Adrenal Axis in Hypothalamic Lactation

The role of the adrenocortical hormones in mammary differentiation has not yet been fully investigated. Some results indicate, however, that these hormones influence ductal as well as alveolar development. It has, for instance, been reported that desoxycorticosterone stimulates ductal growth in the ovariectomized (SMITH and BRAVERMAN, 1953), and that injections of cortisol can enhance alveolar development in intact female rats (JOHNSON and MEITES, 1955). These observations confirm the widely held concept that ovarian steroids (LYONS, LI and JOHNSON, 1958) and adrenocorticosteroids (JACOB-

SOHN, 1962 a, b) are essential but not indispensable for mammary lobulo-alveolar growth in rats. There remains the decisive fact that lobulo-alveolar development is possible in ovariectomized-adrenalectomized-hypophysectomized rats injected with large doses of prolactin and somatotropin (TALWALKER and MEITES, 1961; MEITES, 1965).

While ovarian steroids seem to act as permissive factors in hypothalamic lactation (BEN-DAVID, 1968 b), the role of the adrenal corticosteroids in perphenazine-induced lactation is still unknown. This question is of special importance because of the above controversial reports concerning the role of adrenal corticosteroids in normal growth and secretion of the mammary gland. Whe have therefore undertaken a series of experiments in adrenalectomized, ovariectomized and adrenalectomized-ovariectomized adult female rats in order to elucidate the role of the pertinent steroids in perphenazine-induced hypothalamic lactation (BEN-DAVID and CHRAMBACH, 1969).

Intact, adrenalectomized, ovariectomized and adrenalectomized-ovariectomized adult female rats received 5 daily injections of 5 mg/kg/d perphenazine or control solution. All operations were performed under ether anesthesia on the first day of treatment prior to the first injection. Normal saline and 5% glucose solutions were given freely to all animals, which were kept at 24° C. The animals were killed on the sixth day. Five days after adrenalectomy alone had been performed, the mammotropic index was high (MTI: 3.5) and the mammary glands underwent no involution, whereas the mammary glands of intact control animals had an MTI of 2.4 only. Removal of both ovaries and adrenals resulted in involution of the mammary glands (MTI: 1.9). On the other hand, perphenazine treatment, which stimulated endogenous secretion of prolactin, resulted not only in prevention of mammary involution 5 days after adrenalectomy and ovariectomy, but also in mammary growth (MTI: 3.5); it also brought about an identical degree of mammary development (MTI: 3.5) in adrenalectomized rats. Still, in adrenalectomized or adrenalectomized-ovariectomized rats, the mammotropic effect of 5 mg/kg/d perphenazine was less than that obtained in intact rats (MTI: 4). It, thus, seems that lack of adrenocortical steroids in female rats has no effect on the degree of mammary growth in untreated animals, while in perphenazine-treated animals it slightly impairs the degree of mammary development. This is best shown by measuring the milk yield, as described in Fig. 43.

Absence of both ovarian and adrenocortical steroids resulted in involution of the mammary tissue of the untreated rats as well as impairment of the mammotropic effect of perphenazine. It seems most likely that, in the rat, the presence of minimal amounts of adrenocortical hormones is necessary for full development of the mammary

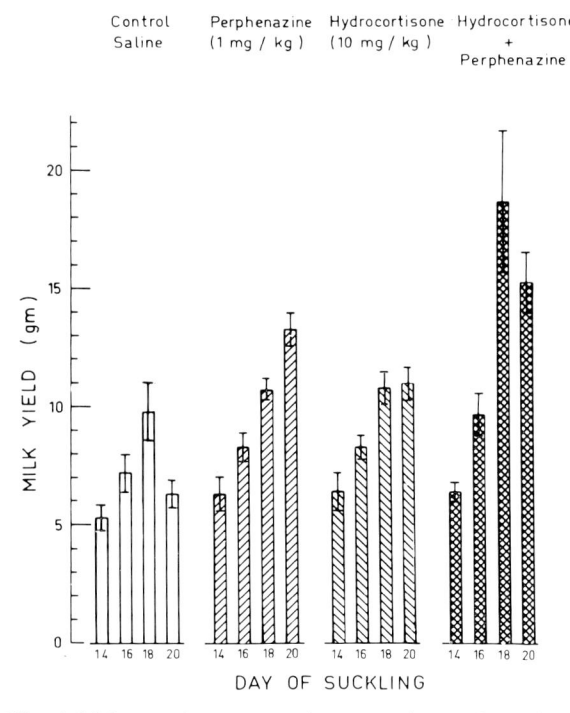

Fig. 43. Milk yield in nursing rat mothers: Hydrocortisone improves effect of perphenazine

tissue and initiation of milk secretion. In other words, the adrenal cortex is less important and possibly not essential at all for the first stages of mammary growth, i. e. mammogenesis, as evidenced by development of the ductal and lobulo-alveolar systems, while the cortical steroids seem to be necessary for the promotion of advanced development of the lobulo-alveolar system (colostrogenesis) and milk secretion (lactogenesis). Similar results were obtained in other experiments by NICOLL and MEITES (1964), HAMBERGER and AHRÉN (1964) and MEITES (1965).

These results were further confirmed when a corticoid was added to the above set-up. In adrenalectomized or in adrenalectomized-ovariectomized rats, the addition of hydrocortisone to perphenazine treatment resulted in maximal development of the mammary tissue with copious milk secretion (MTI:5). Thus, although prolactin is a powerful stimulus to milk secretion, its maximal effect cannot be achieved unless glucocorticoids are present. While hydrocortisone alone does not produce a secretory response (BARNAWELL, 1967), it potentiates the galactogenic property of prolactin.

Results of the present study might explain the many discrepancies in literature concerning the role of the adrenocortical hormones in lactation. It seems that, in any discussion of this problem, one must differentiate between the primary process of the actual mammary growth (mammogenesis and colostrogenesis) and the process of conversion of the developed gland into secreting structures (lactogenesis and galactopoiesis). With regard to the former, we found that, in the absence of steroids, the mammary glands underwent considerable growth in response to endogenous prolactin produced by perphenazine, but they never approached maximal levels. These results confirm the *in vivo* observation of MEITES (1965) that mammary growth could be achieved in male rats in the absence of gonadal and adrenocortical hormones by treatment with prolactin and growth hormone. The *in vitro* studies reported recently by DILLEY and NANDI (1968) also showed that mammary growth could be maintained in the absence of steroids. On the other hand, our findings in the rat are also in agreement with those reported by NELSON, GAUNT and SCHWEIZER (1943), FOLLEY (1952a), and by MEITES and NICOLL (1966) who claimed that adrenocortical hormones were important for the second stage of milk secretion. Different species, however, have different requirements for milk secretion. While both mouse and rat require LTH and/or STH plus corticoids as a minimum, the rabbit may need only either one (MEITES, HOPKINS and TALWALKER, 1963).

Summarizing we must bear in mind that *in vivo* adrenocortical hormones seem to be essential for milk secretion but that they are not necessary for the first stages of mammary growth. This holds true also in perphenazine-induced hypothalamic lactation—at least for the rat; other species may have different requirements. The requirements *in vitro* are described in Chapter 6.

Chapter 22

Pituitary-Thyroid Axis in Hypothalamic Lactation

a) Role of Thyroid in vivo and in vitro

A deficiency in thyroid hormone may result in reduced mammary growth, although the rat appears to be an exception in this respect (KRAGT and MEITES, 1965). Since hypothyroidism has been found to be associated with a decrease in secretion of pituitary LTH (McQUEEN-WILLIAMS, 1935; MEITES and TURNER, 1947 a), of STH, and ACTH (CONTOPOULOS, SIMPSON and KONEFF, 1958) as well as with irregularities in FSH-LH secretion (LEATHEM, 1959), the retardation of mammary development may be partly secondary to changes in secretion of these hormones. There is evidence that thyroid hormones can stimulate LTH and perhaps STH secretion by the rat anterior pituitary.

The *in vitro* ability of thyroxine and triiodothyronine to stimulate pituitary prolactin release is in agreement with *in vivo* experiments in which thyroid hormones were reported to stimulate pituitary prolactin secretion and milk production (GROSVENOR, 1961; MEITES, 1959 a, 1963). Thyroidectomy (McQUEEN-WILLIAMS, 1935) or thiouracil feeding (MEITES and TURNER, 1947 a) result in diminished pituitary prolactin content in rats. Direct stimulation of pituitary prolactin release by thyroid hormones *in vitro* does not exclude the possibility that *in vivo* these hormones may also promote prolactin release through the hypothalamus.

Thyroxine alone, and in combination with other hormones, has been shown transiently to increase milk yield in lactating rats (GROSVENOR and TURNER, 1959 a, b; DJOJOSOEBAGIO and TURNER, 1964) and in cows (BARTLETT, BURT, FOLLEY and ROWLAND, 1954), and to cause mammary gland development in mice (ANDERSON and TURNER, 1963). In the goat, removal of the thyroid results in a fall in milk yield of various degrees (SMITH, STOTT and WALKER, 1957). That is not due to the absence of the thyroid becomes evident from the fact that lactation is not affected when the thyroid is destroyed by radioiodine, leaving the parathyroid essentially normal (BRUCE and SLOVITER, 1957). These experiments suggest that the thyroid is not essential for milk secretion. It seems, therefore, preferable to approach the problem by avoiding operations and using "pharmacologically thyroidectomized animals", which can be achieved by thiouracil treatment.

b) Thyroid and Hypothalamic Mammary Growth

As it was not known whether or not the thyroid hormones were involved in perphenazine-induced hypothalamic lactation, this problem was studied by assessing the effect of propylthiouracil (PTU) and triiodothyronine (T-3) on physiological as well as on hypothalamic lactation (BEN-DAVID, DIKSTEIN and SULMAN, 1965). The mammotropic effect was evaluated by the mammotropic index (MTI). In the case of propylthiouracil treatment, 0.3% of the powder was added to the diet and given for 2 weeks prior to, and during, the 5 days' treatment with perphenazine. For estimation of milk yield, lactating rats, each nursing a litter of 6, were used. Beginning on the 7th day post-partum, daily s.c. injections of different drugs were given until the 20th day. Daily weights of dams and litters were recorded from the 7th to the 20th day post-partum and milk yield was estimated from the litters' weight gain during one hour of nursing. Ten hours before suckling, litters were isolated from their mothers without any food supply. One minute prior to suckling, 1 IU of synthetic oxytocin (Syntocinon) was given to all dams to obtain complete milk ejection.

A dose of 5 mg/kg/d perphenazine for 5 days caused considerable mammary gland development in rats primed with subthreshold doses of estradiol (MTI: 4.7), while triiodothyronine sodium (T-3) in a dose of 0.03 mg/kg produced no better mammotropic effect than saline (MTI: 1.2). Moreover, no dose of T-3 (0.003—0.03—0.3 mg/kg) significantly improved the mammotropic effect achieved by perphenazine (MTI: 4.2, 4.4, 4.6). On the other hand, propylthiouracil, which inhibits endogenous thyroxine production, caused mammary gland development in estrogen-primed rats when given together with saline and even improved the mammotropic effect of perphenazine (MTI: 5). This mammotropic effect obtained by propylthiouracil was abolished by additional treatment with 0.03 mg/kg/d T-3 in both experiments: propylthiouracil alone yielded MTI: 1.0, and propylthiouracil + perphenazine MTI: 4.4.

These results clearly indicate that a difference exists between the roles of endogenous and exogenous thyroxine in mammary development. It was shown that in perphenazine-lactating rats the mammotropic effect of propylthiouracil is due to a decrease in endogenous circulating thyroid hormones, since the addition of exogenous T-3 prevented mammary gland development altogether.

c) Thyroid and post-partum Galactopoiesis

A low dose of perphenazine (1 mg/kg/d), given to lactating rats, slightly increased milk yield during the whole period of suckling, and a sharp increase was obtained on the 20th day when the drug significantly prevented the natural decrease in milk yield observed in the control animals (Fig. 44). Propylthiouracil (which suppresses endo-

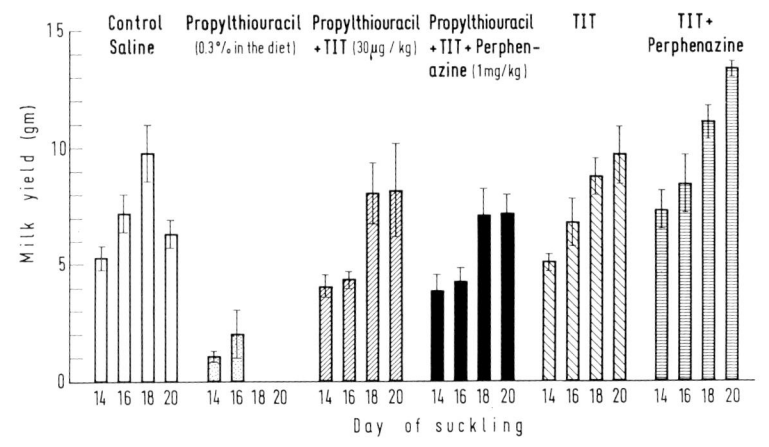

Fig. 44. Milk yield in nursing rat mothers. Propylthiouracil depresses milk yield; perphenazine and triiodothyronine (TIT) increases it

genous production of thyroxine) strongly suppressed milk production, so that the mothers failed to secrete milk on the 18th day post-partum, and the litters died on the 20th day. This suppression was prevented when combined treatment with T-3 (0.03 mg/kg/d) and propylthiouracil (0.3% in the diet) was given. Hence, T-3 did not significantly alter the milk yield either of normal lactating rats or of perphenazine-treated lactating animals. It prevented, however, the suppression of milk yield obtained by propylthiouracil.

d) Conclusion

While, physiologically, thyroid hormones inhibit mammary gland development in the first stage of lactation, they are essential for milk production in the second or third stage. On the other hand, exogenous T-3 did not show significant mammotropic effect, either on mammogenesis or on colostrogenesis, and a slight interference could even be observed. The mechanism by which thyroid hormones inhibit the first stages of mammary gland development is not clear. However, other authors (LEONARD and REECE, 1941; TRENTIN, HURST and TURNER, 1948; MEITES and KRAGT, 1964) also observed a depression of mammary gland development after administration of thyroid hormones. Propylthiouracil which stimulated mammary gland development and inhibited milk secretion, by decreasing circulating thyroid hormones, was active in artificial perphenazine-induced hypothalamic lactation as well as in normal post-partum lactation. This fact again indicates that thyroid hormones inhibit mammogenesis and colostrogenesis, but promote lactogenesis and galactopoiesis.

Chapter 23

Prolactin Release by Hypothalamus of Nursing Rat Mothers

It is well known that the hypothalamus governs the synthesis and release of tropins from the pituitary gland in different ways: hypothalamic stimulation is needed for release of most hypophysiotropins, with the exception of prolactin which is inhibited by the hypothalamus. The prolactin-inhibiting activity of hypothalami taken from cycling animals was proved and confirmed in many species (MEITES, NICOLL and TALWALKER, 1963), but, as prolactin secretion was believed to have no peripheral feedback mechanism, many workers looked for a competitive or complementary mechanism to the inhibitory one exerted by the hypothalamus.

The amount of milk produced by lactating rats increases steadily throughout lactation (MAYER, 1935; COX and MUELLER, 1937). GROSVENOR and TURNER (1958 a) found that pituitary prolactin levels in lactating rats increased from the second to the sixth day post-partum, while REECE and TURNER (1937) reported a 2-fold increase in anterior pituitary lactogen content in rats following parturition. WELLER showed in 1963 that hypothalami from milking cows had a high mammotropic effect when injected into estrogen-primed rats. The view that rat hypothalamic tissue can induce mammary secretion in rats was advanced by MEITES, NICOLL and TALWALKER (1963). RATNER and MEITES (1964) found that extracts from hypothalami of postpartum rats yielded *in vitro* more prolactin when incubated with normal pituitary glands than extracts prepared from hypothalami of virgin rats. KRAGT and MEITES reported in 1965 that the hypothalamus of a parent pigeon, on the day the young are hatched, stimulates prolactin release by the pigeon pituitary gland *in vitro*.

Indications that the CNA exerts a stimulating effect on prolactin secretion came from observations in humans and in farm animals. MEITES, TALWALKER and NICOLL (1960 a) were the first to report that injections of rat hypothalamic extracts initiated mammary secretion in estrogen-primed rats. In 1962, TURNER and GRIFFITH postulated that, under normal conditions, a "lactogen-releasing factor" builds up in the hypothalamus and lactogen stores in the anterior pituitary gland, and that, following a suckling stimulus (oxytocin release), the hypothalamic factor "is discharged into its portal blood system and stimulates the discharge of lactogen from the anterior pituitary." They also found that the low prolactin levels in the anterior pituitary gland of most species during pregnancy increase markedly following parturition (TURNER and GRIFFITH, 1962). These findings have been confirmed by us and recently also in *in vitro* work by our group (unpublished results). Thus, it seems that the prolactin-releasing activity of the post-partum rat hypothalamus is enhanced, possibly due to decreased production of the prolactin-inhibiting factor and increased production of a prolactin-releasing factor. Inasmuch as the hypothalamic homogenates contain numerous mediators, e. g., serotonin, catecholamines, acetylcholine and histamine, the possibility must be considered that these agents influence the *in vivo* release of prolactin in the test rats. It is generally believed (MEITES and NICOLL, 1966) that the effectiveness of a hypothalamic extract *in vivo* is due to its content of substances which can stimulate prolactin release indirectly, and MEITES, NICOLL and TALWALKER (1963) suggested that hypothalamic tissue induces adrenocortical stimulation which in turn elicits release of prolactin from the pituitary gland.

MISHKINSKY, KHAZEN and SULMAN (1968) studied the problem in adult virgin rats which had been primed with estradiol 8 μg/rat/day for 10 days and injected for a further 5 days with different hypothalamic and pituitarian extracts. The homogenates of hypothalami and of pituitaries of the donor rats were prepared in 3 ml of saline with a cone-glass electric homogenizer. The mother rats were sacrificed by a blow on the head on the specified day of lactation, and their hypothalami and pituitaries were removed within 1 min of death; hypothalami and pituitaries of virgin rats of corresponding ages and weights served as controls. Only fresh homogenates were used, in a proportion of 0.6 hypothalamus or pituitary gland per experimental rat per day.

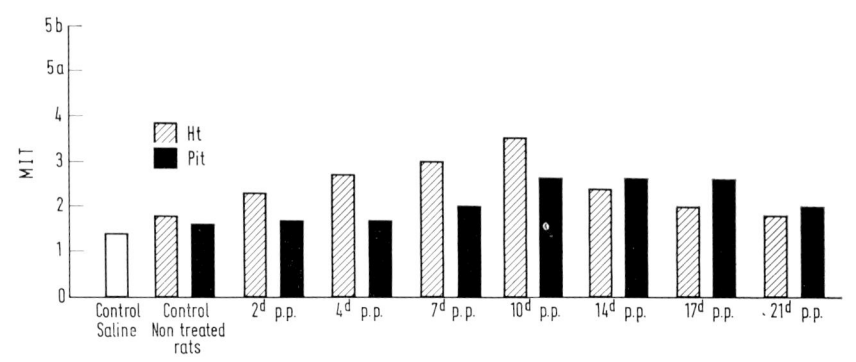

Fig. 45. Natural development of PIF suppression and PRF stimulation in nursing rat mothers: Average Mammotropic Index (MTI) of rats after 10 days' priming with estradiol (8 μg/d), followed by 5 days' injections of homogenates from hypothalami (Ht) or pituitaries (Pit) taken from nursing rat mothers 2—21 days post partum (p. p.). On the 10th day of nursing the hypothalamic PRF arrives at its peak, and is followed by a protracted peak of pituitary prolactin on days 10—17

A day-dependent curve of mammotropic index was obtained for the homogenates of hypothalami and pituitaries from post-partum rats injected into primed rats. The peak of these curves occurred on the 10th day post-partum in regard to the hypothalamic homogenates, and around the 14th day for the pituitary glands, i. e., maximal prolactin releasing activity appears in the hypothalamus on the 10th day post-partum, stimulating the pituitary gland to peak secretion of prolactin at the 10th—17th day (Fig. 45).

In conclusion it can be stated that these findings point towards impairment of the balance between PIF and a tentative PRF, indicating that the latter may be strongly stimulated during natural lactation. As to the part that oxytocin plays in this complex, it was believed for many years that oxytocin is the milk "let down" hormone by both central and peripheral mechanisms. From our results it seems that oxytocin, even in high doses (0.1—10 IU), does not release LTH from the pituitary, and therefore has no effect on mammary gland development. The final answer to this complex problem can only be established by localization of the hypothalamic site responsible for active prolactin release, using oxytocin retard preparations, able to supply a continuous release of small doses of oxytocin.

Chapter 24

Prolactin as Luteotropic Hormone

The ability of pituitary extracts to prolong the normal estrus cycle in rats and mice had been demonstrated (LONG and EVANS, 1922; TEEL, 1926) before the recognition of prolactin as a pituitary hormone (STRICKER and GRUETER, 1928, 1929). Later, prolactin was found to be responsible for prolongation of the duration of diestrus in the mouse (DRESEL, 1935) and rat (LAHR and RIDDLE, 1936). ASTWOOD (1941) found that, in hypophysectomized rats, injections of prolactin could produce functional corpora lutea. Prolactin induces pseudopregnancy in hypophysectomized rats and leads to the formation of deciduomata in a damaged uterine horn (EVANS, SIMPSON, LYONS and TURPEINEN, 1941). CLEMENS, SAR and MEITES (1968) have shown that the luteotropic effect of pituitary prolactin is necessary, during the first 6 days of pregnancy in the rat, to maintain progesterone secretion by the corpora lutea.

Homologous grafts of the anterior pituitary beneath the kidney capsule, which have been shown to secrete large amounts of prolactin, will also maintain the corporea lutea of rats in a functional state for long periods (EVERETT, 1954, 1956; NIKITOVITCH-WINER and EVERETT, 1958; DESCLIN, 1956). Reserpine has been shown to produce lactation in rats and at the same time to maintain the corpora lutea in a functional state (DE FEO, 1957; DESCLIN, 1958; BARRACLOUGH and SAWYER, 1959). The same effect has been shown for chlorpromazine (BARRACLOUGH and SAWYER, 1959) and for perphenazine (BEN-DAVID, 1968 b).

Luteotropic activity can, thus, be observed directly on the corpora lutea, either by gross examination (KOVACIC, 1968) or by histology, where the cells of the corpus luteum increase in size under the influence of prolactin (EVERETT, 1944; MANDL, 1959; WOLTHUIS, 1963). Another criterion could be the indirect result of lutein stimulation, as evidenced by prolongation of diestrus, induction of pseudopregnancy and deciduoma formation and the histological appearance of the uterus and vagina—all being the result of progesterone secretion. In fact, a luteotropic hormone has been defined as one capable of evoking progesterone secretion (ASTWOOD, 1953; GREEP, 1961). This problem was studied by BEN-DAVID and SULMAN (1966) who found that, while prolactin was effective in priming the endometrium by promoting secretion of progesterone from functional corpora lutea, it was quite unable to bring the progestational endometrium to the characteristic decidual stage (Fig. 46). For deciduoma formation an additional stimulus is needed, probably one associated with histamine release, e. g. pyrathiazine (SHELESNYAK, 1959).

A proper balance in secretion of estrogen and progesterone is necessary for maintenance of induced uterine decidual reactions and vaginal mucification in pseudopregnant rats. A study was, therefore, undertaken by JOSIMOVICH (1968) to determine whether primate luteotropins, human placental lactogen (HPL) and human chorionic gonadotropin (HCG) could maintain this balance in the absence of the pituitary gland. He found that combined therapy with HCG and HPL supported a normal decidual reaction in the traumatized uterine horn and vaginal mucification for 7 or 8 days after uterine trauma. It was concluded that HPL acted in conjunction with HCG treatment of hypophysectomized pseudopregnant rats to provide normal duration of induced uterine decidual reactions and vaginal mucification.

It appears that, in rats and mice, some action by LH on the corpora lutea may be essential before prolactin can stimulate luteal function. In the rat, prolactin does not *in vitro* influence glucose utilization (ARMSTRONG, KILPATRICK and GREEP, 1963) nor does it stimulate progesterone synthesis (ARMSTRONG, O'BRIEN and GREEP, 1964) by corpora lutea from pseudopregnant rats. LH, however, stimulates glucose metabolism and progesterone synthesis by these ovaries. GOSPODAROWICZ and LEGAULT-DEMARE (1963) found that neither prolactin nor HCG had an appreciable effect on incorporation of labeled acetate by rat corpora lutea *in vitro,* but that both hormones given together significantly increased acetate incorporation. These studies suggest that both prolactin and LH may be necessary for stimulation of luteal function in rats. It is of interest in this connection that prolactin is apparently necessary for ovarian depletion of ascorbic acid in response to LH (GUILLEMIN and SAKIZ, 1963).

The question arises for how long can prolactin exert a luteotropic action in rats. EVERETT (1954, 1956) and DESCLIN (1956) observed that in rats hypophysectomized during estrus, when new corpora

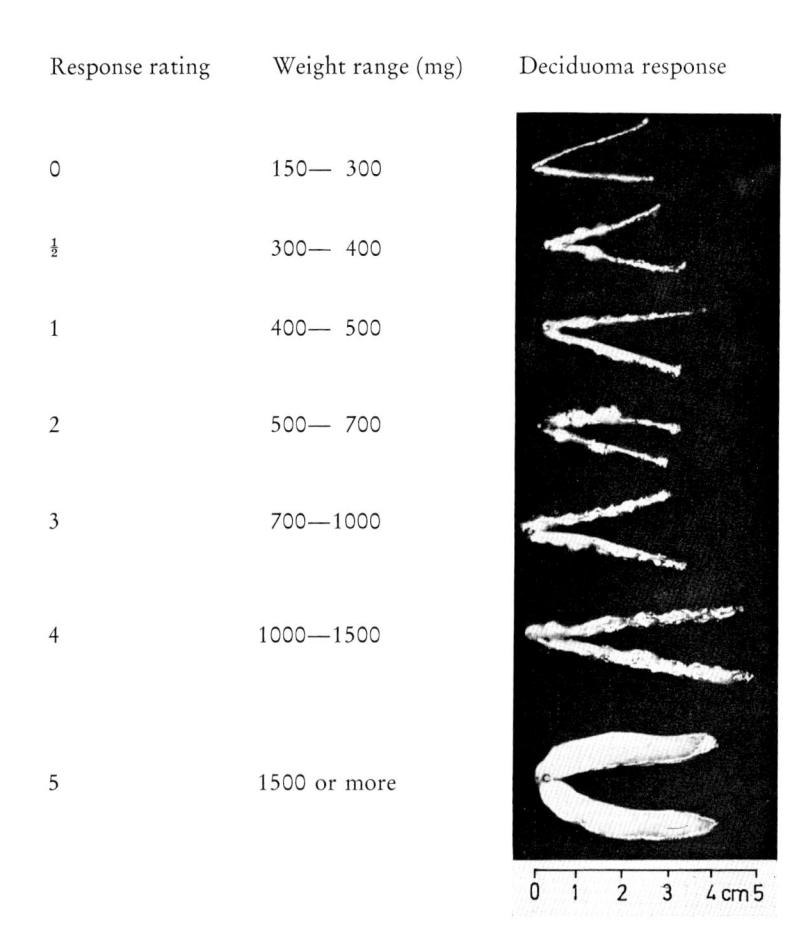

Response rating	Weight range (mg)	Deciduoma response
0	150— 300	
½	300— 400	
1	400— 500	
2	500— 700	
3	700—1000	
4	1000—1500	
5	1500 or more	

Fig. 46. Different degrees of deciduoma formation in the rat uterus after administration of prolactin (0.1—10 IU/rat), followed by pyrathiazine (1—20 mg/rat): 0, ¹/₂, 1—negative reaction; 2, 3, 4, 5—positive reaction. The degree of the reaction was not proportionate to the doses of prolactin injected. It corresponded, however, to the doses of pyrathiazine given. This shows that the luteotropic effect of prolactin is negligible for the decidual responses

lutea are present, these are maintained in a functional state more or less indefinitely when a pituitary is grafted into the kidney capsule. On the other hand, MALVEN, HANSEL and SAWYER (1967) found that they could not prolong pseudopregnancy in rats by either prolactin or LH. MEITES and SHELESNYAK (1957) observed that pregnancy in rats could be prolonged by injection of prolactin. It is possible that the estrogen present during the latter phase of pseudopregnancy in the rat is insufficient for prolactin to prolong it.

In addition to their luteotropic action, prolactin injections have recently been reported to exert a luteolytic effect in rats (MALVEN and SAWYER, 1966). The authors noted that injections of prolactin, first given several weeks after hypophysectomy, hastened morphological regression of corpora lutea in adult rats. They postulated that the corpora lutea lose their capacity to be functionally maintained by prolactin within a few days after hypophysectomy. Earlier, ROTH-CHILD (1965) had presented evidence that LH can exert a luteolytic effect in rats. Thus prolactin and LH may exert either luteotropic or luteolytic effects in rats, depending on the particular state of the corpora lutea.

In vitro studies (HUANG and PEARLMAN, 1962) failed to demonstrate any influence of prolactin on progesterone synthesis in the rat corpus luteum, whereas *in vivo* the luteotropic activity of this hormone is considered proven beyond doubt. Under similar experimental conditions, LH induced progesterone synthesis, thus fitting into the definition of luteotropic hormone.

In the ferret, pituitary stalk section resulted in the persistence of luteal function, suggesting that prolactin is luteotropic in the ferret as well as in the rat and mouse (DONOVAN, 1963b).

In the ewe, in which formation of corpora lutea had been artificially induced, prolactin caused progestational activity (MOORE and NALBANDOV, 1955). Similarly, intact ewes, infused with prolactin into the ovarian arteries, showed enhanced progesterone secretion into the ovarian veins (DOMANSKI and DOBROWOLSKI, 1966).

As for the cow, controversial reports have been published on the luteotropic activity of prolactin. MASON, MARSH and SAVARD (1962) reported that LH, when added to bovine corpus luteum slices incubated *in vitro*, induced an enhanced synthesis of progesterone. Prolactin, however, under similar conditions failed to enhance progesterone synthesis, i. e., did not prove luteotropic. A similar observation was reported by ARMSTRONG and BLACK (1966) and by others. In a more recent work, however, prolactin as well as LH were shown to augment progesterone synthesis and secretion rate in bovine luteal ovaries perfused *in vitro* (BARTOSIK, ROMANOFF, WATSON and SCRICCO, 1967).

In the rabbit, prolactin failed to maintain the corpora lutea after hypophysectomy (MAYER, 1951), but LH did (KILPATRICK, ARMSTRONG and GREEP, 1964). On the other hand, HILLIARD and SAWYER (1966) reported that, after continuous infusion of LH into rabbits *in vivo*, the initial enhanced progesterone secretion rate was not maintained. On this background of continuous LH infusion prolactin did enhance the progesterone secretion rate. In a subsequent study the same investigators demonstrated that, while LH accelerated the synthesis of progesterone in the rabbit ovarian interstitial tissue, cholesterol stores were depleted. Thus, steroidogenesis was subsequently depressed. The addition of prolactin at this stage promoted cholesterol storage and restored the sensitivity of the ovary to LH (HILLIARD, SPIES, LUCAS and SAWYER, 1968). In addition prolactin maintained the morphological structure of the interstitial tissue, the significance and function of which remain to be clarified.

In the hamster, GREENWALD (1967) demonstrated that functional luteal maintenance in the pregnant animal involves a hormone complex rather than a single factor. Synergism of FSH and prolactin was found to be a prerequisite for luteal maintenance in this species. Thus, in hypophysectomized hamsters with ectopic pituitary grafts secreting prolactin, luteal function was not maintained without additional administration of FSH (CHOUDARY and GREENWALD, 1967). Conversely, the addition of large doses of LH along with prolactin and FSH terminated pregnancy in pregnant hypophysectomized hamsters (GREENWALD, 1967).

Recent observations may eventually induce us to change our concept of a single luteotropic hormone and think in terms of a "luteotropic-hormone-complex"—the result of synergism between two or more pituitary tropins, and possibly estrogen as well (SPIES, HILLIARD and SAWYER, 1968). Such a synergism would explain the failure to demonstrate luteal maintenance and progesterone secretion from the corpus luteum by prolactin in many species, and the discrepancies between different research works. Animals with intact pituitaries react differently from hypophysectomized animals, and these, in turn, unlike isolated ovaries, e. g. *in vitro*.

There may be other reasons for failure to demonstrate the luteotropic activity of prolactin. (i) Faulty technique: BARTOSIK et al. (1967) claim that plain incubation *in vitro* might be unsuitable for the induction of progesterone secretion from luteal tissue and that in all the studies, in which such an effect has been reported, the ovaries were perfused, either *in vivo* or *in vitro*. They further suggest that "the stimulatory effect of prolactin depends on an interaction between the luteal cells and surrounding ovarian tissue." This last statement is further supported by the discovery that prolactin may exert a "tropic" effect on the ovarian interstitial tissue and not directly on the corpora lutea (HILLIARD et al., 1968). (ii) The luteotropic effect of prolactin may be subject to species specificity. (iii) The life span of the corpus luteum has been shown to be limited (ASTWOOD, 1941), hence its responsiveness to prolactin would be limited to a short period after ovulation, and any experiment carried out at a different time is doomed to failure. (iv) There always remains the possibility that prolactin is in fact not luteotropic in one species or another. In the rat luteotropic effects can be obtained by local depletion of monoamines in the median eminence of the hypothalamus (VAN MAANEN and SMELIK, 1968).

To conclude, the luteotropic activity of prolactin has been clearly demonstrated in the rat and mouse only. For other species, some evidence of the luteotropic effect of prolactin has been brought forward in the ferret (DONOVAN, 1963 a, b), the ewe (MOORE and NALBANDOV, 1955), the cow (BARTOSIK et al., 1967) and the rabbit (HILLIARD et al., 1968). Other studies have, however, failed to show any effect of prolactin on the luteal phase of the menstrual cycle in either the monkey (HISAW, 1944) or women (KUPPERMAN, FRIED and HAIR, 1944).

Part III

Prolactin Assay

Chapter 25

In vitro Methods

a) Effect of Hypothalamus

Earlier experiments on *in vitro* culture of mammalian pituitary glands have shown that these gradually lose their ability to secrete tropins into the nutrient medium. This was first proved for ACTH and subsequently for TSH and the gonadotropins (GUILLEMIN and ROSENBERG, 1955). Hormonal elaboration *in vitro* could, however, be restored after 3—4 days by the addition of hypothalamic extracts (GUILLEMIN, HEARN, CHEEK and HOUSHOLDER, 1957). These findings reconfirmed the stimulatory hypothalamic control of the anterior pituitary hormones. Most of the modern experiments were carried out in an artificial medium M-199 (Difco). The technique is not complicated (Figs. 47, 48).

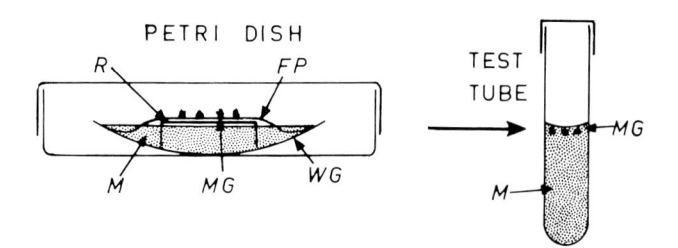

Fig. 47. Co-culture of hypothalamus, hypophysis and crop gland or mammary gland in medium 199: Petri dish; watch glass with plastic raft, filter paper, organ explants in medium 199. The Petri dish can be replaced by a test tube, since fatty organs float high enough for adequate oxygen supply.—*M* Medium (3 ml); *MG* Mammary Gland Fragments; *WG* Watch Glass; *R* Raft; *FP* Filter Paper

Effect of Hypothalamus 163

Fig. 48. MTI scoring *in vitro*: *In vitro* mammotropic index of mammary glands of female rats, weighing 270 ± 10 g, after fixation with Bouin's solution and staining with hematoxylin (24×).—MTI 0 represents normal gland pattern; MTI 1 shows stimulation by estrogen—"budding"; MTI 2 represents alveolar formation due to prolactin

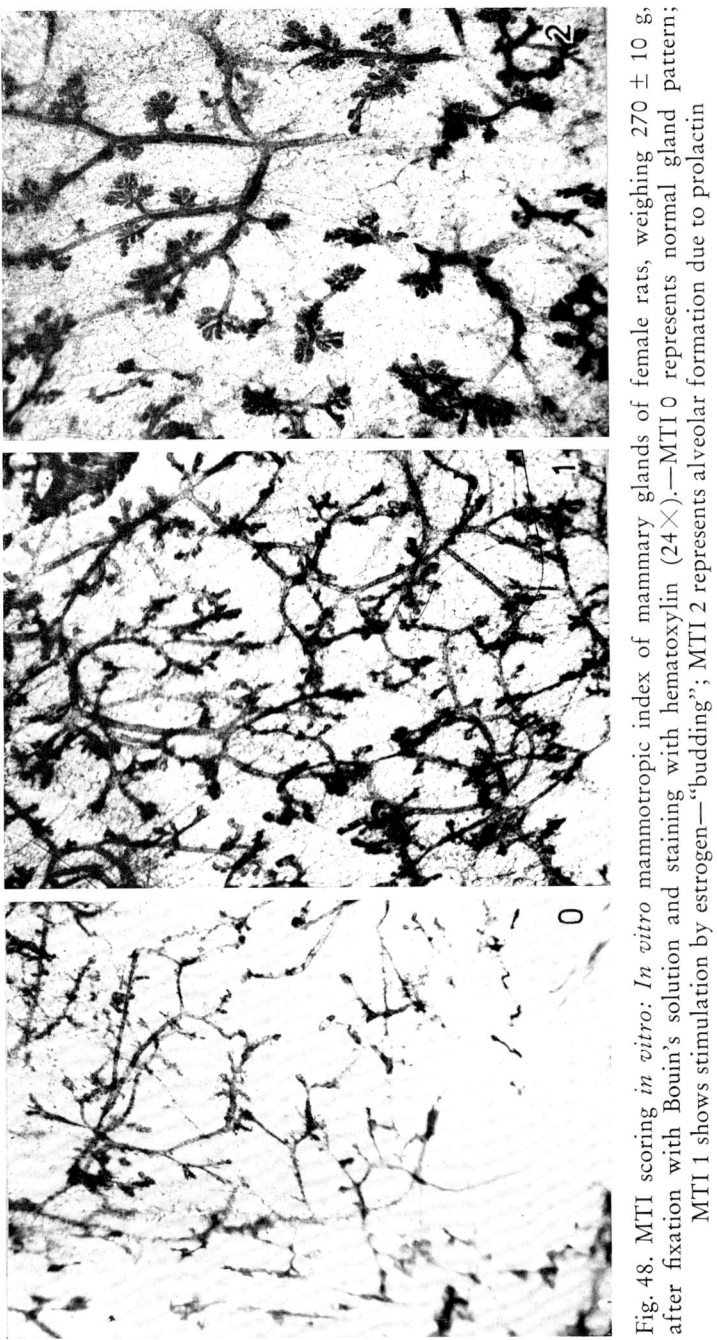

b) Effect of Pituitary

Many *in vitro* experiments, such as pituitary transplantation, stalk section or the introduction of specific hypothalamic lesions, have shown that under normal conditions prolactin release is permanently inhibited by the hypothalamus. Consequently, the pituitary could be expected to secrete prolactin *in vitro* even in the absence of hypothalamic influence. This assumption was confirmed by the results of PASTEELS (1961 b) and of MEITES, KAHN and NICOLL (1961), who concurrently, though under different experimental conditions, cultured rat pituitary explants *in vitro*. The explants may preserve their structure well for 4 weeks. In these organ cultures the anterior pituitary secreted large amounts of prolactin for up to 33 days after explantation (PASTEELS, H. BRAUMAN and J. BRAUMAN, 1963). This secretory activity was increased, rather than decreased, in the course of cultivation, and was shown to amount to several times the quantity of prolactin present in the pituitary prior to explantation. Thus the autonomous ability of the anterior pituitary to secrete prolactin *in vitro* was established (ROBBOY and KAHN, 1966).

In contradistinction to other pituitary hormones, the addition of hypothalamic fragments to pituitary cultures *in vitro* significantly decreased the amount of prolactin secreted into the medium (PASTEELS, 1961 a; DANON, DIKSTEIN and SULMAN, 1963), while other nervous tissue was inactive. This finding suggested the presence of an inhibitory factor in the hypothalamus. A further step was the demonstration that this hypothalamic inhibitory effect was mediated by a humoral—non-nervous—factor, i. e. it is independent of the presence of intact hypothalamic tissue *in vitro*. Thus PASTEELS (1962) and TALWALKER, RATNER and MEITES (1963) showed that hypothalamic homogenates or acid extracts could elicit the same inhibitory effect on prolactin secretion as the hypothalamic tissue itself.

c) Effect of Steroids

Anterior pituitary cultures were also shown to respond to stimulation by hormones added *in vitro* to the culture medium. Thus, NICOLL and MEITES (1962 a, 1963, 1964) demonstrated that estradiol stimulates the anterior pituitary directly when added to an *in vitro* culture, resulting in an increased rate of prolactin release. Other hormones, however, did not display any direct effect on the *in vitro* cultured pituitary.

Both GALA and REECE (1964 b), using organ culture techniques, and PASTEELS (1963), using tissue culture methods, were unable to demonstrate any effect of estrogen on prolactin secretion by rat adenohypophysis *in vitro*. The reasons for the discrepancy between the negative *in vitro* findings of these investigators and the positive ones *in vivo* (KANEMATSU and SAWYER, 1963 b; RAMIREZ and McCANN, 1964) and *in vitro* (RATNER, TALWALKER and MEITES, 1962; NICOLL and MEITES, 1962 a, b, c; BEN-DAVID, DIKSTEIN and SULMAN, 1964; NICOLL and MEITES, 1964; RATNER and MEITES, 1964) are not clear, but may be found in different procedures employed.

BEN-DAVID, DIKSTEIN and SULMAN (1964) explored the effects of other steroid hormones on the hypothalamus-pituitary axis *in vitro,* using the technique described above. They found cortisone and cortisol to be strong stimulants of prolactin release in organ co-culture, apparently acting directly on the pituitary. High doses of cortisol, applied by NICOLL and MEITES (1962 a, 1963, 1964), depressed prolactin. Progesterone inhibited prolactin release in our set-up, again acting directly on the pituitary, while testosterone at the dose level used did not affect prolactin release. SAR and MEITES (1968) found that progesterone, testosterone propionate and cortisol acetate decreased hypothalamic PIF content, increased pituitary prolactin concentration and induced mammary growth. Testosterone propionate and cortisol acetate also elicited mammary secretion.

Norethinodrel (Enovid) also reduces the PIF, thus yielding increased prolactin release. This effect is apparently due to its being metabolized into an estrogenic compound (MINAGUCHI and MEITES, 1967).

These results indicate that, of all steroids employed, only estradiol and cortisol were capable of directly stimulating the pituitary to increase prolactin release. This is in complete agreement with the many *in vivo* studies in which administration of moderate but not large doses of estrogen or cortisol increased pituitary prolactin content and initiated milk secretion. Since estradiol and cortisol can also deplete the hypothalamus of PIF, the probability exists that under *in vivo* conditions they act on both the pituitary and the hypothalamus to promote prolactin release. Our observation of a direct stimulation by estrogen of prolactin release from the rat anterior pituitary *in vitro* is in agreement with the results of NICOLL and MEITES (1962 a, 1963, 1964) but not with those of GALA and REECE (1964 b). Estradiol has been found to increase also prolactin release by the hamster pituitary *in vitro* (NICOLL, 1965). When incubated *in vitro,* the pituitaries of rats receiving estradiol injections release greater amounts of prolactin than pituitaries of control rats not injected with estradiol (RATNER and MEITES, 1964). Pituitary prolactin content in cycling rats (REECE and TURNER, 1937) and guinea pigs (REECE, 1939) is highest during proestrus and estrus and lowest during metestrus and diestrus (MEITES, 1966).

d) Effect of Neurohormones

Trying to assess the effects of drugs and hormones on the hypothalamus-pituitary axis, MEITES (1966) and his collaborators published a series of papers in which the drug under investigation was injected into the donor rats *in vivo* and the degree of inhibitory or stimulatory effect of the hypothalamic extract from treated animals on the pituitary *in vitro* was measured. They further introduced a short-term incubation technique, in which pituitary fragments were incubated and shaken for 2—4 hours in a Dubnoff shaker at 37° C under continuous O_2—CO_2 mixture gasing. By these techniques they found that adrenaline and acetylcholine (MITTLER and MEITES, 1967) reduced the hypothalamic prolactin-inhibitory factor, thus yielding a higher prolactin output from anterior pituitary cultures subjected to hypothalamic extracts from pre-treated animals. GALA and REECE (1965 a, b) found that adrenaline (and crude oxytocin) increased prolactin release. These findings require re-examination.

e) Effect of Phenothiazines

A different approach was taken by DANON, DIKSTEIN and SULMAN (1963). They tested the effect of the phenothiazine derivative, perphenazine, on a combined hypothalamus-pituitary organ culture and showed that perphenazine, added *in vitro,* could reduce or offset the effect of the hypothalamic prolactin-inhibiting factor. Perphenazine did not have any significant effect on the anterior pituitary itself, but the perphenazine-treated hypothalamus did not inhibit prolactin release. Thus, a depressant effect of perphenazine on the hypothalamic prolactin-inhibiting factor (PIF) was demonstrated *in vitro*. For other experiments, in which similar pharmacological effects of drugs on nervous tissue were studied *in vitro,* the reader is referred to a review by M. R. MURRAY (1965).

f) Effect of Reserpine

RATNER, TALWALKER and MEITES (1965) compared the effect of reserpine on the hypothalamus after its injection into rats and its addition *in vitro* to explants of the hypothalamus and pituitary incubated for 2 hours in a Dubnoff shaker. In both cases they found the same degree of depletion of hypothalamic PIF. This clearly indicates that the effect of the drug on the hypothalamus is similar *in vivo* and *in vitro*. NICOLL (1970) has, however, shown that MEITES' method of short incubation may yield false results. During the first 4 hours the hypothalamus produces a PIF, but later it produces a PRF which may give opposite results.

g) Effect of Thyroid

The profound effect of thyroid hormones on adenohypophysial function has long been recognized by endocrinologists. Although injections of thyroid hormones in rats can increase prolactin secretion by the adenohypophysis *in vivo* (MEITES, NICOLL and TALWALKER, 1963), the locus of action of the thyroid hormones was not established by these *in vivo* studies. NICOLL and MEITES (1963) observed that addition of either thyroxine or triiodothyronine (both at 0.1 µg per ml) to the medium of organ cultures of rat adenohypophysis resulted in a pronounced stimulation of prolactin secretion. A direct action of the thyroid hormones on the adenohypophysial prolactin cells was thus demonstrated. Evidently, of all the hormones tested, only estrogens and thyroid hormones were capable of stimulating prolactin secretion in the rat by a direct action on the adenohypophysial prolactin cells. In this context, it is pertinent to mention that COHERE, BOUSQUET and MEUNIER (1964) found that both thyroxine and estrogen stimulated development of the endoplasmic reticulum of the prolactin-producing cells of explants of rat adenohypophysis in organ culture.

h) Effect of Oxytocin

The hypothesis of BENSON and FOLLEY (1956) that prolactin release is stimulated by oxytocin has aroused considerable interest and controversy since it was first advanced. *In vitro* evidence has since been offered which contradicts the view of BENSON and FOLLEY (1956, 1957). NICOLL and MEITES (1962 b) studied the action of oxytocin (Pitocin) and vasopressin (Pitressin) on prolactin secretion by the adenohypophysis of rats in organ culture. No effect was observed with either neurohypophysial preparation in three days of culturing, a finding confirmed by PASTEELS (1963). TALWALKER, RATNER and MEITES (1963) also found that neither oxytocin nor vasopressin nor any of several other neuropharmacological agents were able to influence prolactin release. Our results agree with these findings.

The content of the prolactin-inhibiting factor in the hypothalamus has been reported to vary in oxytocin-mediated situations which may alter prolactin release. It drops during lactation (RATNER and MEITES, 1964), although it is not altered detectably following acute suckling (GROSVENOR, 1965 c).

i) Conclusion

The hypothalamus—pituitary organ co-culture is an excellent tool for prolactin assay, if its effect is evaluated either in the crop gland test or by addition of mammary tissue to the culture medium. Addition of hypothalamus of normal animals impairs prolactin secretion from the pituitary into the medium. Hypothalamus from nursing animals, however, increases prolactin production. Prolactin output is increased by addition of estrogens, norethinodrel, hydrocortisone and derivatives of phenothiazine or reserpine; thyroxine and triiodothyronine have a similar effect. Prolactin secretion is suppressed by the addition of progesterone and of most of its derivatives. The effect of adrenaline and acetylcholine is not clear. Testosterone, oxytocin and vasopressin are not active on prolactin release in this set-up. Unfortunately, the quantitative application of this test encounters a gamut of difficulties represented by the factors mentioned in this Chapter. It begins with the problem of creating a sensitive target organ by priming with estrogen or other steroids followed by the interaction of thyroxine, insulin and oxytocin. Thus, the combined organ co-culture could well serve for the determination of the prolactin-releasing or inhibiting activity of drugs if a standardization of all factors required could be achieved.

Chapter 26

In vivo Methods

This synopsis contains only the few methods adopted by us.

a) Rabbit Tests

α) *Rabbit Intraductal Method* (Lyons' Nipple Test, 1942, Fig. 49)

Priming: (a) 100 µg estradiol injected s.c. daily for 10 days into female rabbits (3.0 ± 0.2 kg B.W.), or: (b) 100 IU of HCG in one single i.v. injection in female rabbits (2.5 ± 0.2 kg B.W.).

Injections: Prolactin test starts on 11th day: one single injection of 0.25 ml mixed with 0.25 ml of gelatin (5% solution 40° C), injected through a 26-gauge hypodermic needle into one duct of a nipple. Controls of 0.25 or 1 IU standard prolactin prepared similarly are injected into another duct of the same nipple.

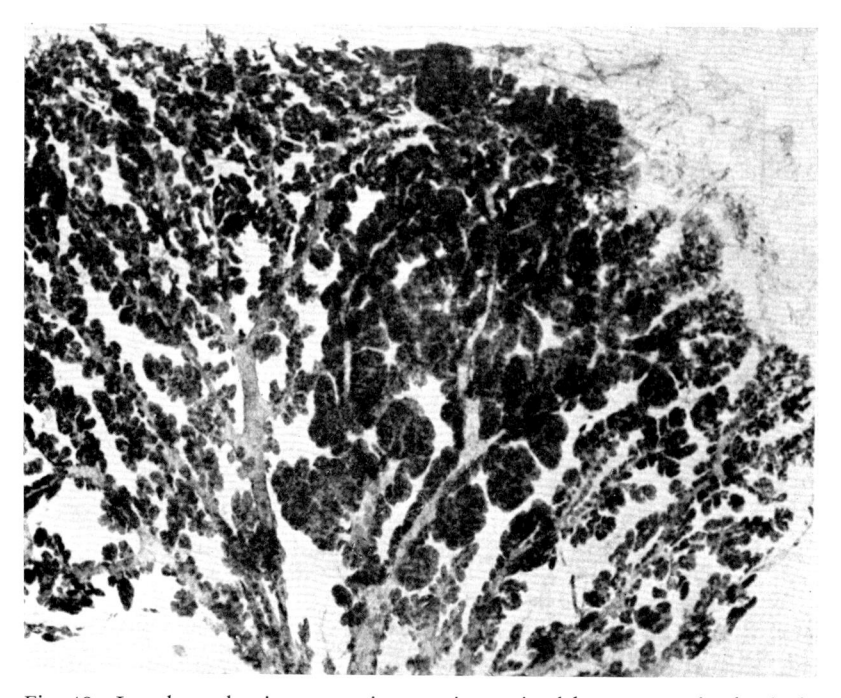

Fig. 49. Local prolactin assay in ovariectomized-hysterectomized virgin rabbits primed with estrogen and progesterone (by courtesy of W. R. Lyons, 1942, University of California, San Francisco Medical Center)

Assessment: After 3—4 days, comparison of the regions which contain milk, of both unknown and standard prolactin.

Evaluation: Inexact but convenient method, since 6—8 determinations can be performed on the same rabbit.

Sensitivity: 0.25—1 IU.

β) Rabbit Intracerebral Test (Stereotaxic Method) (MISHKINSKY, LAJTOS and SULMAN, 1966)

Priming: 100 µg estradiol injected s.c. daily for 10 days into female rabbits (3.0 ± 0.2 kg B.W.).

Technique: Implantation of microcrystals of prolactin-releasing substances in different sites of brain, especially hypothalamus.

Assessment: Copious lactation after 5 days.

Evaluation: Not quantitative, but convenient for localization of prolactin-inhibiting or releasing areas and for testing of lactogenic activity of unknown substance (Fig. 50).

b) Rat Tests

α) Mammotropic Index in Female Rats (TURNER, 1939)

Priming: Female rats (200 ± 10 g B.W.) receive not more than 8 µg/d estradiol s.c. daily for 10 days.

Injections: Prolactin test starts on 11th day: one s.c. injection of 0.2 ml daily for 5 days.

Autopsy: 16th day—rats killed, mammary glands fixed, mounted and stained (Fig. 51).

Evaluation: Test is highly specific if rats are correctly primed. Mammotropic Index (MTI) cf. Fig. 52. 1—2: neg., 3: ±, 4—5 pos.

Sensitivity: 2 IU exogenous prolactin, but very sensitive to endogenous prolactin release induced by drugs.

β) Milk Yield Test in post-partum Rats (GROSVENOR and TURNER, 1959 a)

Animals: Rat mothers (250 ± 20 g B.W.) with 6—10 litter mates each, housed in individual cages. No treatment before the 6th day post-partum.

Injections: Lactogenic preparations injected s.c. once daily during the 7th-20th day of suckling, saline controls required.

Suckling: Ten hours before suckling, litters are isolated from mothers which get 1 IU synthetic oxytocin s.c. immediately before suckling.

Milk yield: Estimated from pups' weight gain and mothers' weight loss during 1 hour or test suckling on days 14, 16, 18 and 20 postpartum.

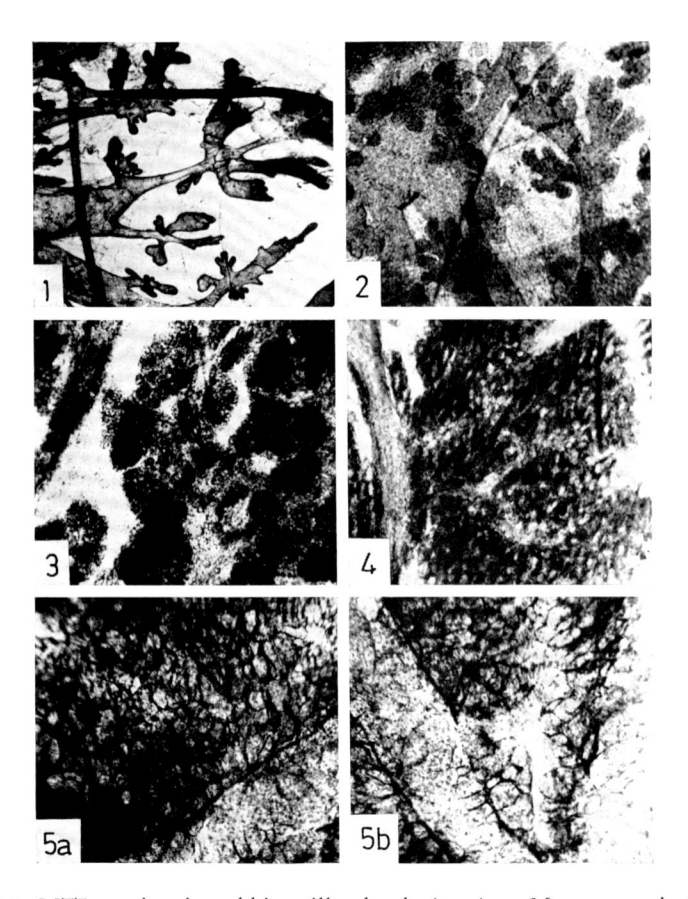

Fig. 50. MTI scoring in rabbit milk glands *in vivo:* Mammotropic index (MTI) of mammary glands of adult female rabbits shows different stages of development when examined in toto after fixation with Bouin's solution and staining with hematoxylin (12×). — MTI 1—2 represent the effect of ineffective drugs: normal gland pattern; MTI 3 represents gland after priming with chorionic gonadotropin; MTI 4—5 a—5 b represent increasing degrees of lactogenic effect (in 5 b the milk has been ejected)

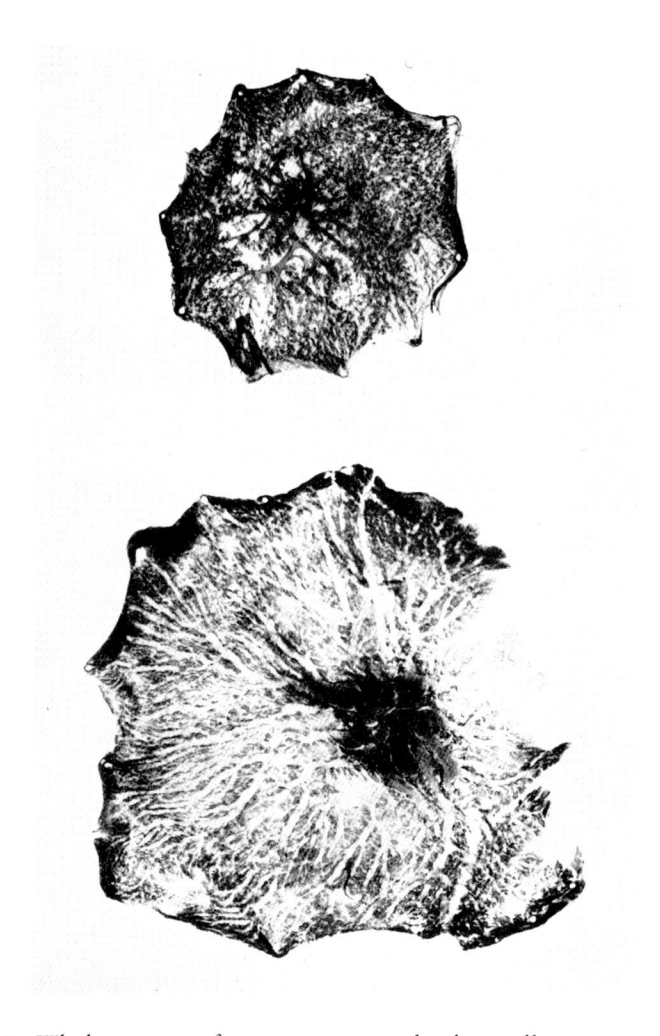

Fig. 51. Whole mount of rat mammary gland: small mount: normally developed gland of adult rat; large mount: same gland after induction of hypothalamic lactation by perphenazine

Evaluation: Test is not specific to exogenous prolactin because of surplus of endogenous prolactin. It allows a superior evaluation of the effect of inhibitory or auxiliary lactogenic factors.

Sensitivity: 20 IU prolactin.

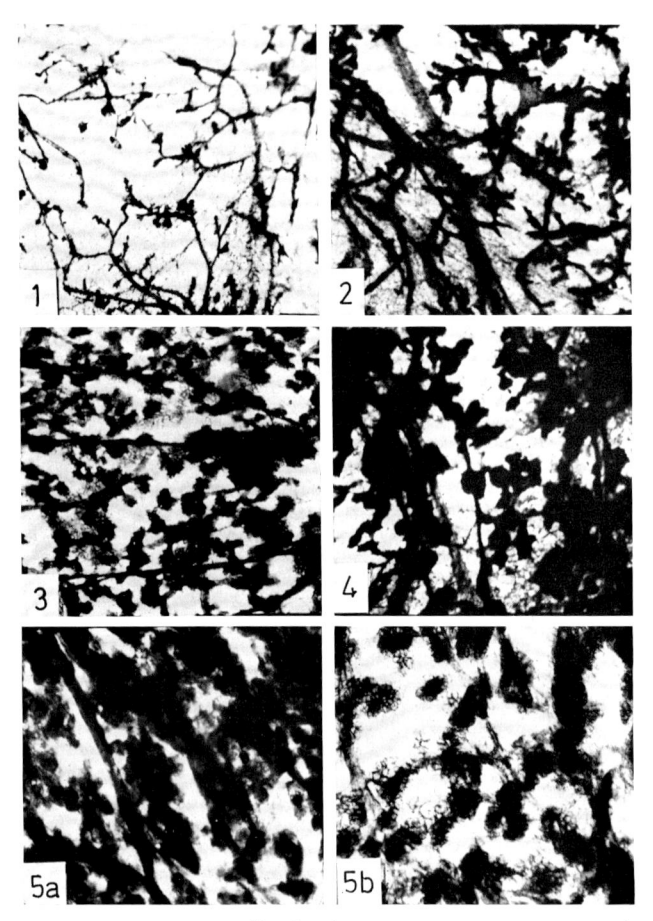

Fig. 52. MTI scoring in rat milk glands *in vivo:* Mammotropic index (MTI) of mammary glands of adult female rats in different stages of development, examined in toto after fixation with Bouin's solution and staining with hematoxylin (12×). MTI 1—2 represent the effect of estrogen and progesterone: normal gland pattern. MTI 3—5 represent increasing degrees of lactogenic effect. In 5 a the alveoles contain milk. In 5 b the milk has been ejected

c) Pigeon Tests

α) *Systemic Pigeon Crop Gland Weight Method* (RIDDLE, BATES and DYKSHORN, 1933)

Injections: 1 ml of each specimen injected daily s.c. into 4 pigeons for 4 days.

Autopsy: 5th day—birds killed and weighed, crop gland removed, spread, cleaned, blotted dry and weighed.

Evaluation: No practical use.

Sensitivity: 5 IU. Only high doses of prolactin can be estimated. High variability in response, depending on time, age and weight of the birds.

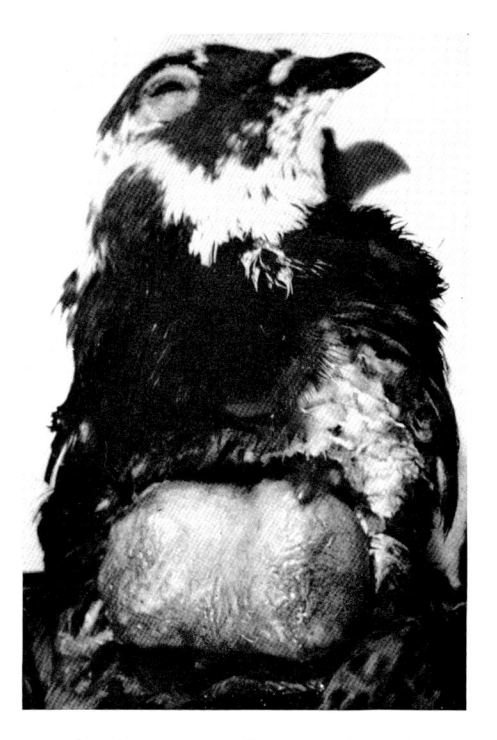

Fig. 53. Local crop gland response after intradermal injection: r: positive reaction; l: control reaction

β) Intradermal Pigeon Crop Gland Method (REECE and TURNER, 1937)

Injections: Minimal effective dose (0.05 IU prolactin in 0.05 ml normal saline) injected as standard in four equal parts during two days intracutaneously above one side of the crop gland. On the other side, dilutions of the material to be tested are injected in the same fashion (0.05 ml). The unknown material is diluted until the response is equal to that of the standard (Fig. 53). Addition of hydrocortisol (V. BEERSWORDT-WALLRABE, HERLYN and JANTZEN, 1965) or prednisolone (HERLYN, JANTZEN, FLASKAMP, HOFFMANN and V. BEERSWORDT-WALLRABE, 1969) reduces unspecific inflammatory reactions. Blood prolactin can be determined by the method of BEN-DAVID and CHRAMBACH (1970).

Autopsy: 5th day. Assessment based on: ridging, wrinkling and measuring of the proliferative area (Fig. 54). The dry mucosa weight method is most reliable (NICOLL, 1967, 1969).

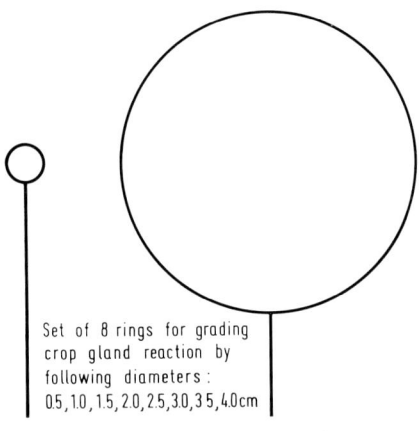

Fig. 54. Standardizing of crop gland reaction by 8 graduated wire rings, according to C. E. GROSVENOR and C. W. TURNER (1958): 0.5, 1, 1.5, 2, 2.5, 3, 3.5, 4 cm diameter

Requirements: 16 pigeons and 10 hours' working time for one titration.

Sensitivity: 0.05 IU, using 4 pigeons for each assay, differences of 50% are detectable. The H^3-methyl-thymidine uptake developed by BEN-DAVID is more sensitive but expensive (Fig. 55).

d) Concentration Methods

A major problem in the local assay methods is the occurrence of non-specific inflammatory or other reactions, resulting from the injection of tissue extracts or body fluids. These reactions may be massive in the serosa and other submucosal layers and thicken the crop sac, thus imitating a prolactin response. The experienced investigator can, in most cases, distinguish between a true prolactin response and a purely inflammatory reaction. However, the problem remains where there are minimal responses or overlapping prolactin and inflammatory reactions. This difficulty is minimized by the use of mucosal weight according to NICOLL (1967), but some mucosal involvement also occurs and may be misleading (NICOLL and BERN, 1964). This shortcoming has been overcome by the addition of hydrocortisone or prednisolone, as mentioned above. BAHN and BATES (1956) claimed

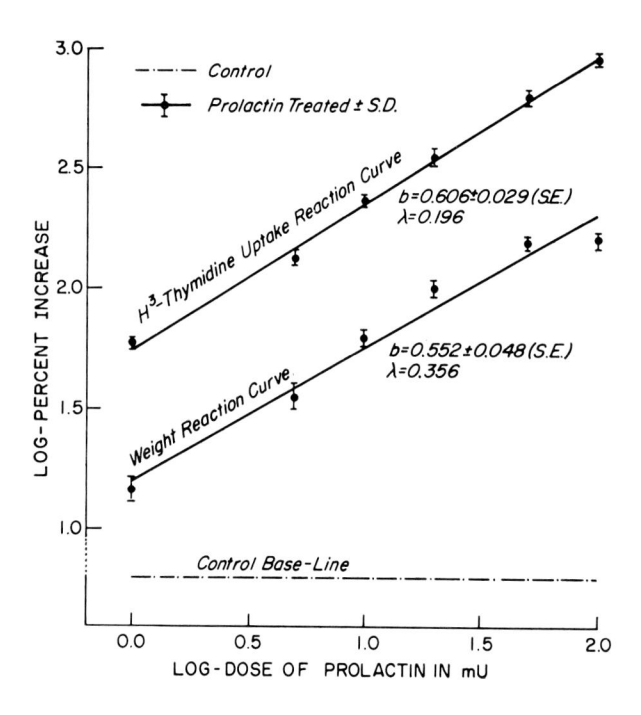

Fig. 55. Testing prolactin with tritiated thymidine. Relation between weight reaction curve of pigeon crop gland test and H^3-methyl-thymidine uptake

they could distinguish genuine prolactin reaction from non-specific responses by the cytoplasmic basophilia characteristic of the former and absent in inflammatory mucosal proliferation. Using this criterion, these authors could not support former reports (LYONS and PAGE, 1935; MEITES and TURNER, 1941; COPPEDGE and SEGALOFF, 1951) on the existence of measurable amounts of prolactin in blood and urine of human subjects.

SIMKIN and GOODART (1960) and SIMKIN and ARCE (1963) proposed using SULMAN's acid-acetone extraction method of chromatophorotropic activity (1952) for the extraction of prolactin from human blood, thus minimizing non-specific responses associated with the presence of foreign proteins. This procedure has been tested in our laboratory (unpublished data) and found unsatisfactory for the local crop-sac prolactin assay. A careful study of SIMKIN and GOODART's original paper reveals its non-specificity: in a recovery test, upon addition of 0.05 IU prolactin, they recovered ten times that activity as calculated on their own dose-response curve. The overlapping non-specific activity in this procedure is obvious.

Neither MEITES (1966) nor our group were able to confirm any of the results with rat or human blood claimed by SIMKIN and GOODART (1960). MOON (1966) also attempted to confirm their findings in rats, but without success. Inasmuch as the blood extracts employed elicit inflammatory reactions, it seems likely that the response measured by this technique is merely inflammation. Such non-specific inflammatory reactions may involve both the mucosal and submucosal tissues of the crop-sac (NICOLL and BERN, 1964), and can be elicited by diverse substances such as raw urine, mustard, and tannic acid (BAHN and BATES, 1956). We, therefore, tried to make use of the basophilia of the crop-sac cells in order to increase the specificity of the test by using basophil stains, as e. g. hematoxylin or methylenblue—unfortunately with negative results.

We presently use the following method for assaying prolactin in blood (BEN-DAVID and CHRAMBACH, 1970) (Fig. 56): The prolactin activity is extracted into the supernatant of serum brought to a concentration of 76% ethanol and 2% NaCl, according to CANFIELD and BATES (1965), at 0° C. Aliquots of serum are pipetted into 30 ml round bottom Corex centrifuges tubes. The appropriate amount of NaCl is dissolved in the serum by magnetic stirring. Cold absolute ethanol is added to a final concentration of 76%, the tube is covered with Parafilm, inverted several times, and the suspension stirred magnetically for 10 to 20 min. It is then centrifuged at 37,000 g for 20 min. Supernatants of aliquots of each pool or serum are combined in a 125 ml round bottom flask and subjected to flash evaporation, using a heating bath not exceeding 35° C and a "Rotovapor" connected to an oil pump with dry ice trap. Flash evaporation is terminated when the solution becomes cloudy. The cloudy solution is then transferred to

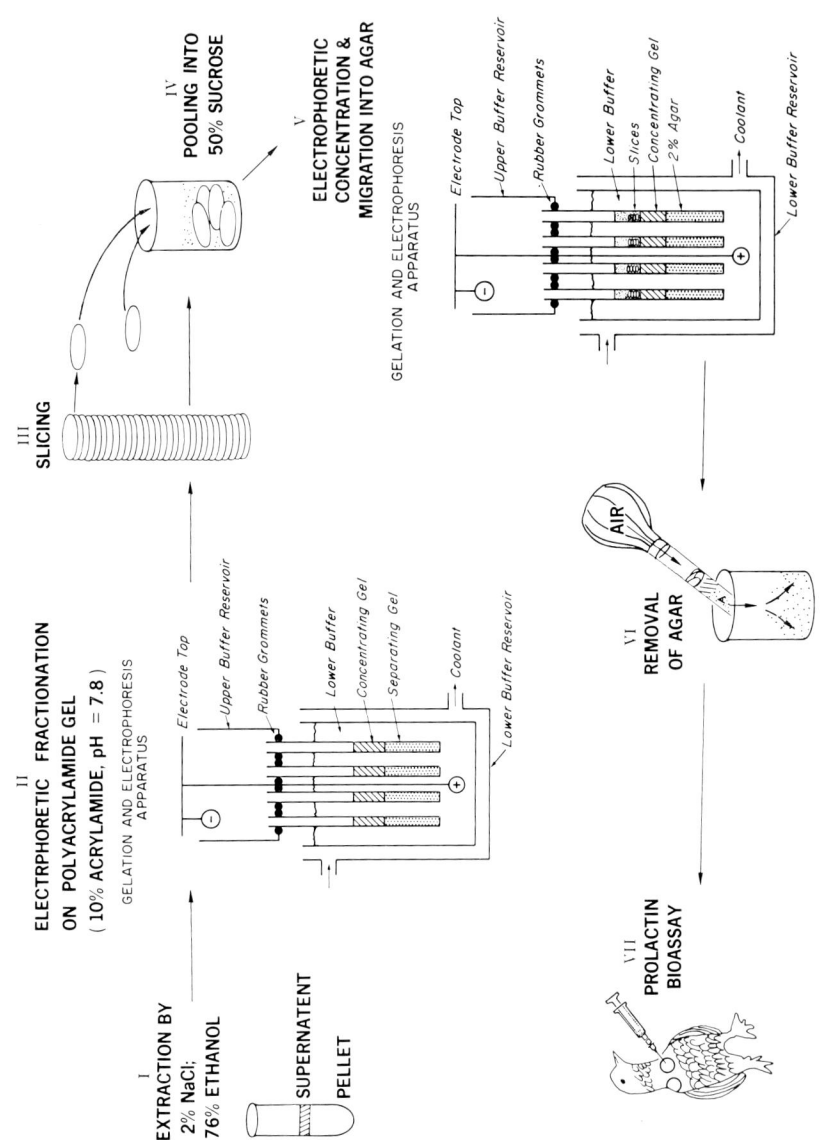

Fig. 56. Prolactin identification and assay in human blood: Stage I: extraction with 2% NaCl and 76% ethanol; Stage II: the lyophilized supernatant is subjected to electrophoretic fractionation on polyacrylamide gel; it can be injected directly into pigeons by the intradermal crop gland test. If prolactin identification and separation from STH is required—the following stages III—VII should be applied: Stage III: identification of active slices (Rf); Stage IV: pooling of active slices in 50% sucrose; Stage V: electrophoretic transport of prolactin into 0.2% agar; Stage VI: separation of agar from polyacrylamide gel by air blow; Stage VII: injection of agar prolactin concentrate into pigeons

dialysis tubing of 6.5 mm diameter [1]. The material is dialyzed at 4 °C against distilled water until the specific conductance of the dialyzate has reached the value obtained for distilled water. The dialyzed serum remnant within the tubing is lyophilized overnight and the product is stored at $-20°$ C. Prior to use the lyophilized serum extract is dissolved in 0.1 M Na_2HPO_4 with magnetic stirring. It is then neutralized by an equal volume of 0.1 M NaH_2PO_4. The total volume of phosphate buffer used is 0.2 ml per ml of serum. The equivalent of 1 ml serum is assayed in one pigeon and 5 pigeons are used for each assay point.

Chapter 27

Radioimmunoassay

Radioimmunoassays for rat, ovine and bovine prolactins were recently described (Fig. 57). Kwa and Verhofstad (1967) developed a radioimmunoassay for rat prolactin, using rat prolactin prepared from transplants of pituitary tumors as a reference antigen. These authors were able to detect changes in prolactin levels in blood of individual female rats through the estrus cycle and through pseudopregnancy by taking a 1 ml blood sample daily from the tail vein. High concentration of prolactin was found during the onset of lactation and unexpectedly low values during pregnancy (Kwa, 1968). A similar method for rat prolactin has been described by Niswender, Chen, Midgley, Meites and Ellis (1969). Bovine and ovine prolactin do not appear to crossreact in this radioimmunoassay. Quantities of NIH-P-B2 and NIH-P-S8 as high as 1 µg/assay tube did not inhibit the reaction between the antirat prolactin and the rat prolactin-[131]I. It appears that a radioimmunoassay specific for rat prolactin has been developed which is capable of measuring the prolactin in as little as 100 µl of serum from an immature rat and in 10 µl obtained from most females on the afternoon of proestrus. This radioimmunoassay

---

[1] The dialysis tubing is washed as follows: The tubing is boiled in 10% Na_2CO_3 until no more yellow material appears in the wash; it is then washed to neutrality with distilled water and stored at 4° C in 0.01 M NaEDTA of pH 7.0. Just prior to use it is washed thoroughly with distilled water to free it from EDTA.

method is estimated to be about 1,000—2,000 times more sensitive than the local pigeon crop bioassay. It has been used to measure serum levels of prolactin in rats throughout their cycle, and when bearing prolactin secreting tumors. The purification of rat prolactin for these studies has been described by Ellis, Grindeland, Nuenk and Callahan (1969).

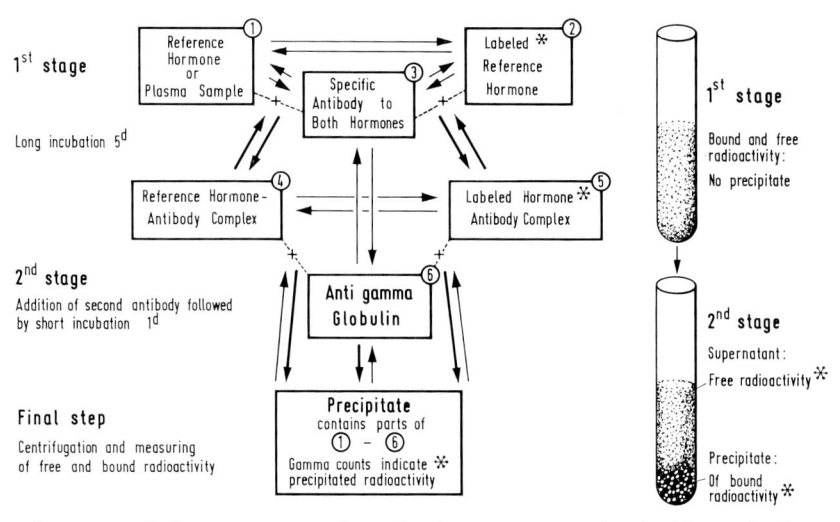

Fig. 57. Radioimmunoassay of prolactin. Diagram of "double-antibody" radioimmunoassay in three stages

Arai and Lee (1967) reported use of a double-antibody radioimmunoassay for ovine pituitary prolactin. The ovine prolactin antiserum showed crossreaction with bovine prolactin but not with human preparation. Using the radioimmunoassay procedure, Arai and Lee (1967) also found a gradual decline in prolactin levels during pregnancy from 100 mμg/ml in the normal non-pregnant sheep down to about 4 mμg/ml in the 4½ months of pregnancy. Subsequently they found a sharp rise post-partum and during lactation, up to 307 mμg/ml in the fourth month post-partum.

Plasma prolactin levels, detected in the goat by radioimmunoassay, were reported by Bryant, Connan and Greenwood (1968) and Bryant and Greenwood (1968). They found that delivery, milking and i.v. injections of oxytocin resulted in a pronounced increase in the plasma prolactin levels. Bryant and Greenwood also measured the rate of disappearance of ovine prolactin from the plasma after its i.v. injection into a goat over a 2-hr period. The half-time of disappearance from arterial, jugular and mammary blood was found to be 19 min. Cows seem to be able to lactate without significant increase of prolactin levels (SCHAMS and KARG, 1969). The only peak discovered by them was 24 hours before parturition, when estrogen levels are high.

Radioimmunoassay is the most sensitive test. When radioimmunoassays are used for the determination of protein hormones, it is important to remember that they are based on immunological specificity rather than on biological activity. It is clear that the molecular structural requirements for biological and immunological activities are not identical. For example, it has been demonstrated that the N-terminal part of ACTH is needed for biological activity, whereas the C-terminal section is needed for immunological activity (IMURA, SPARKS, GRODSKY and FORSHAM, 1965). Moreover, we found that the three forms of ovine prolactin detected through radioimmunoassay show four bands at polyacrylamide gel electrophoresis. But only a single main band exerts the biological activity of prolactin (BEN-DAVID and CHRAMBACH, unpublished data). This could simulate cross reactions with other tropic hormones or hormones derived from other species or even unspecific cross-reactions and thus provide false positive results. BRYANT, TECH, GREENWOOD and LINZEL (1968) actually found cross-reactions between ovine, bovine and caprine prolactin which allows the ovine prolactin radioimmunoassay to be applied to all these three species.

The amply discussed question of whether or not "human prolactin" does exist as a molecule separate from human growth hormone (LARON, ASSA and APOSTOLAKIS, 1965) has not yet been resolved because, radioimmunologically, "human prolactin" is not separable from human growth hormone. Using the pigeon crop bioassay and polyacrylamide gel electrophoresis, we were able to demonstrate different bands for human growth hormone and most active human prolactin preparations, such as preparation HSI-087-E of WILHELMI or preparation 36-R8 of LARON, ASSA and APOSTOLAKIS (1965). We could not, however, differentiate between these bands when using human growth hormone radioimmunoassay since they cross-react with each other (CHRAMBACH and BEN-DAVID, 1970). It is essential, therefore, to run from time to time a bioassay alongside with the radioimmunoassay to provide a permanent check of the latter.

Part IV

Problems of Lactation

Chapter 28

Lactation in Humans

a) Lactation and post-partum Amenorrhea

For successful lactation it is essential that the patient be endowed with breast tissue that is capable of normal response to estrogens, progesterone, and lactogenic hormones (Fig. 58). The rising titer of the steroid hormones during pregnancy causes maximum development of the lobulo-alveolar system and presumably sensitizes the glandular tissue for the action of lactogenic hormones. Prolactin is secreted in increased amounts during pregnancy, but its release from the pituitary and the placenta is inhibited by the high estrogen titer. It should be pointed out that the colostrum, that may exude or be expressed from the breasts during pregnancy, contains little, if any, milk and consists chiefly of desquamated epithelium and transudate. It is not until about the third day post-partum, when the estrogens have been nearly completely cleared, that the breasts become engorged by the active secretion of milk. It is presumed that this is caused by the release of LTH and STH following the removal of the estrogen "brake." For lactation to continue actively, it is necessary that there should be regular suckling or artificial pumping of the breasts. This suggests the importance of the neural pathway from the breasts to the hypothalamus mediating the continued secretion of oxytocin and lactogenic hormones.

b) Inhibition of Lactation and Breast Engorgement

Hormonal therapy with estrogens or androgens has been found helpful in preventing painful engorgement of the breasts in those patients who do not wish to nurse, and in alleviating the engorge-

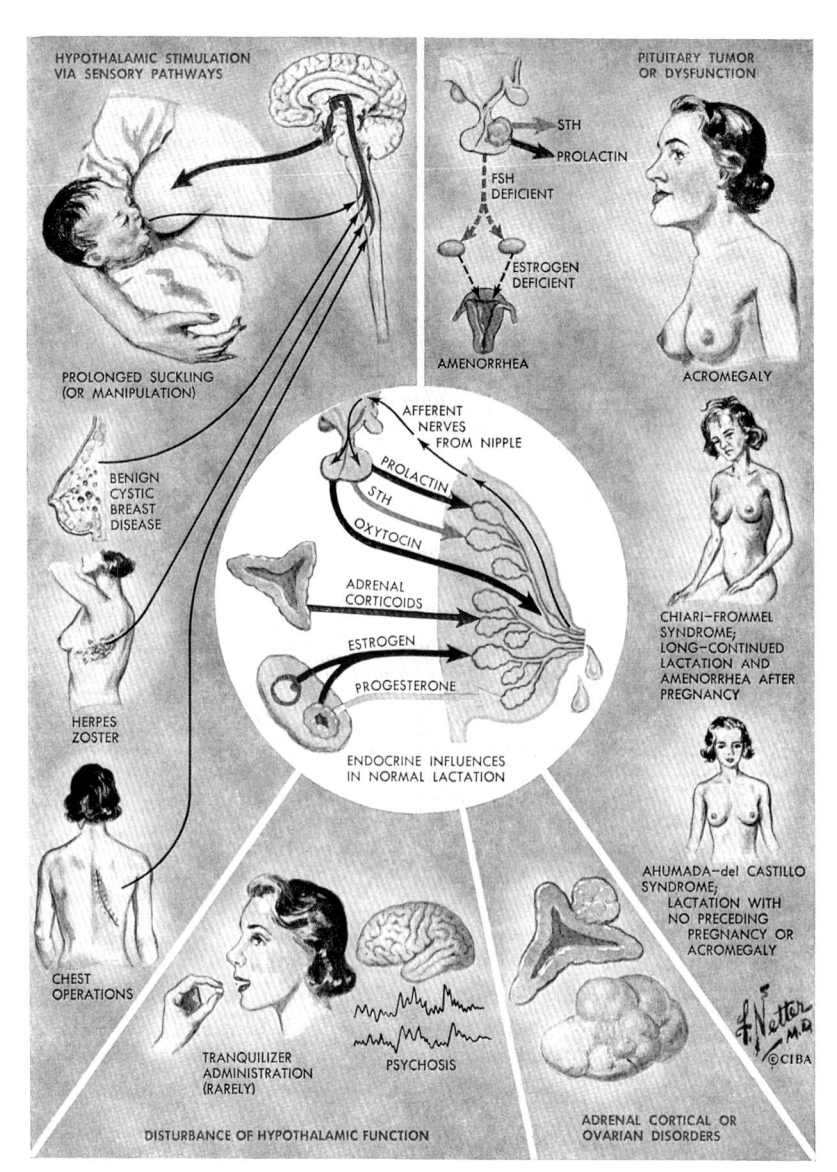

Fig. 58. Presentation of the various factors which may produce lactation or galactorrhea: Center: Afferent nerves from nipple, prolactin, STH, oxytocin, adrenal corticosteroids, estrogen, progesterone; Left side: Hypothalamic stimulation via sensory pathways operating possibly via the prolactin releasing factor (PRF), such as suckling, benign cystic breast disease, herpes zoster, chest operations. Lower part: Hypothalamic suppression by tranquilizers (including phenothiazines, reserpines, butyrophenones, etc.), in adrenal cortical or ovarian disorders, in Ahumada-del Castillo Syndrome and Chiari-Frommel Syndrome—all possibly operating via the prolactin-inhibiting factor (PIF). Right side: Pituitary tumor or dysfunction operating by direct release of STH and prolactin (acromegaly). (Copyright: The Ciba Collection of Medical Illustrations by FRANK H. NETTER, M. D.)

ment in patients who wish to wean their babies from breast feeding. Such therapy also helps in "drying up" the breasts if lactation has started, although some authorities are of the opinion that the decrease in milk flow is not influenced by hormonal therapy and is dependent entirely upon cessation of suckling.

Since 1937 it has been possible to inhibit the secretion of milk by the administration of different steroids. At first, injections of estradiol benzoate were used, but frequent ingestion of tablets containing small doses of estrone or estriol was also satisfactory. Lately, the synthetic estrogens—stilbestrol, dienestrol, and hexestrol, as well as ethinyl estradiol—have been more widely used owing to their low cost. Commonly, 5 mg stilbestrol six times a day for 4 to 10 days is given, but its effect is unpredictable according to HESSELTINE, BUSTAMANTE and NAVORI (1955). A combination of orally active, water-soluble, conjugated estrogens, obtained from the urine of pregnant mares (Premarin), and methyltestosterone is preferred by FISKIO (1955). The regimen is 3 tablets every 4 hours five times. While in about 40 per cent of the cases results are not good, there are no unpleasant side effects. It is claimed that similar combinations of ethinyl estradiol and methyltestosterone, having a synergistic suppressive action on lactation, also show less tendency to induce uterine bleeding on withdrawal. Tablets should be given at short intervals as these hormones are rapidly absorbed and excreted. The combination of 50 mg progesterone with long-acting estradiol benzoate (10 mg) by injection, given daily for 5 days, is more consistently effective.

Skepticism towards this treatment has been voiced recently by MACDONALD and O'DRISCOLL (1965). A double-blind trial of stilbestrol and placebo in 500 women was carried out to determine the effect of estrogens on discomfort from breast engorgement. Stilbestrol (10 mg/d) and placebo were almost equally effective, and it seems that the routine administration of small doses of estrogens to suppress lactation is not sufficient. Doses up to 100 mg/d may be required.

The use of a depot steroid mixture—Ablacton (Schering)—has recently been recommended for women. It contains estradiol benzoate (5 mg), estradiol valerianate (8 mg), ethinyl-testosterone acetate (20 mg) and testosterone enanthate (180 mg) and is given in one i.m. injection. This preparation is supposed to suppress hypothalamic PRF and pituitary LTH.

Alpha-ergocryptin suppresses lactation in rats, it has, however, not yet been tried in women (FLUECKIGER and WAGNER, 1968).

c) Persistence of Lactation and Menstrual Cycle

Persistence of lactation and spontaneous lactogenesis are rare conditions in modern civilized society. In western countries in the past, and even now in primitive races, it was economic policy for mothers and grandmothers to prolong lactation for many years by repeated suckling, so as to assist in feeding their large numbers of children and grandchildren. Some women secrete a little milk for a year or more after weaning, some do at the menopause; GESCHICKTER (1960) quotes a figure of 1 to 4 per cent of women in England and the United States with persistent lactation.

Spontaneous lactation (witch milk) often occurs in the breasts of newborn children and is no doubt due to the action of the maternal hormones. Many babies have slight breast development, and secretion of milk may develop after about a week. Their small breasts may be congested and hard. This usually subsides without trouble, but abscess formation sometimes appears.

In the normal lactating mother there is some degree of ovarian suppression, due probably to inhibition of pituitary FSH and LH. So, too, in cases of spontaneous galactorrhea, there is usually some associated estrogen deficiency manifesting itself in menstrual irregularity, hypomenorrhea, or incomplete development of the normal pattern of vaginal smear. There may, in addition, be varying degrees of atrophy of the genital tract and even of the ovaries themselves. Numerous reports have also confirmed a deficiency or absence of urinary FSH.

Evidence of overactivity of growth hormone exists in some of these cases, and acromegalic women sometimes have galactorrhea (Fig. 58). In 1954 FORBES et al. gave a full account of galactorrhea in 15 non-acromegalic women. The interesting finding was that in 8 of these patients there was evidence of a pituitary tumor, and in 3 biopsy confirmed the presence of a chromophobe adenoma. It is not known whether lactation is induced by overproduction of growth hormone and prolactin from the eosinophile cell tumor in the acromegalic or from the pre-eosinophilic cells of the chromophobe adenoma, or whether it is just a matter of pressure from a tumor or other stimulus affecting the hypothalamus. In all these cases, galactorrhea and amenorrhea were accompanied by absence of follicle-stimulating hormone excretion without any other signs of pituitary failure; in fact, the finding of hirsuties and elevated 17-ketosteroids in some cases would be consistent with hyperpituitarism.

d) Sterility during Lactation

The belief is current that lactation is associated with physiologic sterility, and also that it inhibits the return of uterine bleeding. The hormonal theory usually advanced is that these effects are due to pituitary prolactin suppressing the secretion of pituitary gonadotropins, especially LH.

Lactation is regarded generally to retard the return of uterine bleeding. DE LEE (1924) observed that in women who do not nurse uterine bleeding usually returned within six weeks of delivery and recurred quite regularly. However, in nursing mothers, bleeding was said to return within three months in 50% of the women, in 71% within six months, and in 81% before nursing was discontinued. MAZER and GOLDSTEIN (1932) cite a study of 1,200 lactating women according to which 55% experienced a return of uterine bleeding while still nursing. These data indicate that the "amenorrhea of lactation" is but a relative affair and becomes uncertain after 3 months. During lactation, episodes of uterine bleeding occur at irregular intervals, and the flow is variable in amount.

The question whether ovulation occurs during "lactational amenorrhea" arises because of the occurrence of conception prior to resumption of the menses after delivery. A study by GRIFFITH and MCBRIDGE (1939) of endometrial biopsy specimens, obtained from 21 lactating women during the 3rd to 24th week of puerperium, indicated that only one ovulated prior to the resumption of bleeding. In another study (TOPKINS, 1943) of 145 endometrial biopsy specimens, secured from 28 lactating women during the period of "lactational amenorrhea", only nine of these specimens (6%) showed progestational alterations, all of which were associated with the first episodes of flowing.

Other studies (LASS, SMELSER and KURZROK, 1938) indicate, however, that the early episodes of bleeding during lactation do not necessarily constitute true menstrual bleeding, i. e. they were not preceded by ovulation, corpus luteum formation or progestational proliferation of the endometrium. Forty-seven women, who were studied throughout their puerperia, had 194 regular episodes of bleeding, and 106 of these episodes (55%) were judged to have been preceded by sterile cycles since bleeding occurred from estrogenic endometria. Yet 82 episodes of bleeding (45%) occurred from progestational endometria and were thought to have been preceded by fertile cycles.

These different studies suggest 1) that ovulation does not generally occur during "lactational amenorrhea," and 2) that uterine bleeding during lactation may or may not be preceded by ovulation. Accordingly, sterility during lactation is relative and not absolute.

The advent of the contraceptive pills has made a more exact study imperative. CRONIN (1968) estimated that the non-fertile period in women who do not lactate lasts for 6 weeks after childbirth and in nursing mothers for 3 months. The validity of these estimates was checked by daily recordings of body-temperature in relation to menstruation in 93 non-lactators and in 81 women who did lactate. The mean time to first ovulation was 12 weeks in the 93 non-lactators, and a fertile ovulation preceded the first menstruation in 21 of them; there seems to be a 1-in-20 chance that a fertile ovulation will occur before the first menstruation and before the 7th week post-partum. Lactation extended the time lapse until the return of ovulation. Thus, provided breast-feeding is in progress and menstruation has not returned, ovulation before the end of week 10 post-partum will be a rare event.

Appreciation of the influence of lactation upon the resumption of ovulation after childbirth is important from the practical as well as the theoretical aspect. Definite knowledge of the earliest date at which ovulation can occur would make conception avoidance unnecessary before that date. The use of contraceptive methods—which may inhibit lactation—could thus be postponed to permit lactation to become well established—a consideration of great importance in countries in which the child depends on breast-milk for its intake of protein.

e) Galactorrhea as a General Syndrome

Galactorrhea (abnormal lactation) refers to mammary secretion of a milky fluid, which is nonphysiological in that it may be persistent, excessive, and unrelated to a recent pregnancy. The onset may date back to a normal post-partum lactation which did not cease (Fig. 58).

Amenorrhea which may, or may not, precede the lactation, genital atrophy, and other evidences of estrogen deficiency are frequent but not invariable phenomena. Pathological lactation occurs very seldom in females and is extremely rare in males.

The factors operative in abnormal lactation are not yet clear, especially because of the large variety of clinical conditions with which it is associated. In most instances, galactorrhea can be traced to an organic functional intracranial disorder, involving particularly the pituitary-hypothalamic structures. The final and effective link in the various pathogenetic mechanisms appears to be an unphysiological production of one or more adenohypophysial tropic hormones, i. e. LTH or STH. By further analogy with normal lactation, mechanical and psychic as well as humoral components seem to play a role. On the other hand, abnormal lactation is relatively rare in the various diseases occasionally associated with lactation.

Among the better known and more frequent causes of galactorrhea are pituitary tumors, with or without acromegaly. Eosinophilic, chromophobe, or basophilic neoplasms may be present. Galactorrhea and amenorrhea may be the first symptoms, so that, even when no tumor can be demonstrated, it is advisable to suspect such lesions in an incipient form. Other organic brain lesions include encephalitis, meningitis, and trauma.

Dysfunctional hypothalamus-pituitary disorders include the Chiari-Frommel syndrome which is characterized by galactorrhea, following pregnancy and nursing, in some cases due to unrecognized pituitary tumors, in others to primary ovarian failure; the Ahumada-del Castillo syndrome which is characterized by spontaneous galactorrhea, not related to pregnancy or acromegaly, sometimes due to pituitary neoplasm or to selective hyperpituitarism. The latter has recently been studied by SPELLACY, CARLSON and SCHADE (1968).

f) Forbes-Albright-Castillo Syndrome

This clinical entity which originates in the pituitary and is associated with lactation, has been described by FORBES, HENNEMAN, GRISWOLD and ALBRIGHT (1951, 1954) and by ARGONZ and DEL CASTILLO (1953). The clinical picture is characterized by the presence of amenorrhea (usually primary), a low FSH, and an acromegalic-like facies. Patients with this complaint usually have an enlarged sella turcica but do not show evidence of visual disturbances. They appear to be akin to frank acromegalics (who may also lactate), except that they do not show the organomegaly, increase in soft tissue of the hands, or marked encroachment on the visual fields usually associated with progressive enlargement of the sella. If menses are not desired in the patients with Forbes-Albright-Castillo syndrome, control of lactation may be accomplished sometimes by the use of 10.0 mg, three times daily, for 30 days, of norethindrone, norethandrolone, or norethynodrel. If lactation does not subside, the course of any of the 19-norsteroids should be continued beyond that period. The same schedule, using chlorotrianisene or clomiphene, may also be used with an equal degree of effectiveness where catamenia is desirable. Once lactation subsides, it does not usually reappear, but should it do so, another course of therapy may be instituted. Occasionally there are patients in whom lactation cannot be completely suppressed by this type of treatment and there is little that can be accomplished for these women since even twice the recommended dose of the 19-norsteroids would probably not be more effective either. In addition, when one increases the doses of the steroids to twice the initially recommended regimen, an increase in undesirable side effects may be noted. It should, however, be mentioned that the prospects for fertility of the lactating patients have become fairly hopeful since the introduction of clomiphene, which, however, does not unfortunately completely abolish lactation.

g) Chiari-Frommel Syndrome

In rare instances, amenorrhea and lactation may persist for many years in women who have had an otherwise normal post-partum period and have not developed any complicating organic disease. This syndrome was first described by CHIARI, BRAUN and SPAETH in 1852, and later elaborated by FROMMEL (1852, 1855) (Fig. 58). It occurs most frequently in primiparas, who had, as a rule, uncomplicated deliveries. The patients include both nursing and non-nursing mothers. Menstruation fails to reappear within the usual period of time and, subsequently, amenorrhea persists indefinitely. The breasts continue to lactate, and in many instances a copious, milky discharge lasts for many years in spite of efforts to dry up the breasts. In other instances, the milky discharge is scant and demonstrable only by manual expression from the breasts.

On examination the breasts are found to be well developed and to contain abundant glandular tissue, but in a few instances milky secretion is present in patients with small flaccid breasts. The other striking finding is a marked degree of hypoplasia of the uterus. The endometrium is usually atrophic. The ovaries are small. Galactorrhea usually precedes amenorrhea, though the reverse condition may occur as well. It may persist for years together with amenorrhea, or there may be temporary intermissions in which presumably follicle-stimulating hormone is resecreted so that menstrual rhythm returns and pregnancy may ensue. The uterus may be of normal size or may show varying degrees of atrophy.

Endocrine studies have consistently shown absence of urinary gonadotropins (FSH), markedly diminished excretion of estrogens, and normal urinary 17-ketosteroids. In some cases there was diminished basal metabolic rate. Many of the patients complain of marked tiredness and increased fatigue. As a rule, no gross nervous or psychic disturbance seems to be involved.

LIGGINS and IBBERTSON (1966) reported a successful quintuplet pregnancy following treatment with human pituitary gonadotropin in a patient with post-partum Chiari-Frommel syndrome. Preliminary investigation of the mother indicated that failure of ovulation resulted from a defect in pituitary-hypothalamic function. The response to clomiphene, which is assumed to cause a discharge of gonadotropin (GREENBLATT and MAHESH, 1964—1965), indicated normal gonadotropin production. Although gonadotropin release was not measured, a failure of release was inferred from the monophasic temperature chart and the low urinary estrogen levels. Excessive prolactin secretion resulting from loss of the normal hypothalamic control (PIF) was obviously the reason for the persistent lactation in this case. The human-growth-hormone immunoassay technique used does not distinguish between growth hormone and prolactin, which are immunologically very similar (BERSON and YALOW, 1964). Although STH levels were normal, the failure to suppress growth hormone to zero during a glucose infusion pointed to the possibility of an autonomous prolactin production.

The Chiari-Frommel syndrome is characterized by a marked resistance to therapy. It is usually difficult to induce bleeding even with large doses of estrogen and progesterone, but, once the first bleeding episode is induced, others can often be brought on more easily by cyclic administration of estrogen and progesterone. Subsequent resumption of spontaneous ovarian function is rare. Most patients also fail to react to gonadotropic stimulation. The addition of thyroid extract appears to be helpful in combating the fatigue but dose not influence the ovarian hypofunction. Clomiphene therapy is at present the best treatment, it reduces the milk flow, without, however, abolishing it entirely.

h) Sheehan's Syndrome

Selective hypopituitarism is a rare cause of galactorrhea, as is juvenile hypothyroidism with precocious menstruation. The possibility of compensatory pituitary adenoma cannot be excluded in these cases (MAAS, 1967).

i) Hypothalamic Lactation

Galactorrhea also rarely occurs in psychoses but often in patients receiving large doses of tranquilizers such as phenothiazine derivatives (cf. Chapters 14 and 15), rauwolfia alkaloids (cf. Chapter 16), butyrophenones (cf. Chapter 17) or methyldopa (cf. Chapter 18) (PETTINGER, HORWITZ and SJOERDSMA, 1963). As even imipramine (Tofranil) can evoke galactorrhea (KLEIN, SEGAL and PICHEL-WARNER, 1964), hypothalamic involvement is obvious. The pathogenetic mechanisms in prolonged suckling, benign cystic breast disease, dorsal herpes zoster or chest operations (SALKIN and DAVIS, 1949; GRIMM, 1955; QUINLAN, 1968) presumably involve stimulation of the sensory reflex arc from the chest region to the hypothalamus. The latter condition is especially resistant to steroid treatment. Other associated conditions include adrenocortical hyperfunction (tumor, hyperplasia), menopause, castration and ovarian tumors (SEBAOUN, DELZANT and PEILLON, 1967). The fact that medroxyprogesterone acetate (Provera) may induce galactorrhea in 9% of cases after administration of a total of more than 1 g given in 10—20 injections, points to the possibility that it is partly metabolized to an estrogenic compound which affects the hypothalamus and the pituitary (BOLOGNESE, PIVER and FELDMAN, 1967).

j) Promotion of post-partum Lactation

LYONS, LI, AHMAD and RICE-WRAY (1968) injected 22 lactating women, most of whom complained of insufficient milk secretion, subcutaneously for one week with daily 10 mg doses of a human pituitary preparation which had about 5 IU/mg of pigeon crop-stimulating activity; about 1 IU/mg of somatotropin; and lactogenic activity detectable at the 20 µg dose in hypophysectomized rats. Placebo treatment for 1 week preceded or followed hormone injections. The suckling babies were weighed daily, and average daily weight changes were compared during the 2 regimens. In 16 cases the average weight gain during the hormone period was over 100% greater than that during the placebo treatment.

k) Gynecomastia

α) Morphology

Enlargement of the male breast, owing to an increase in the glandular component, is known as gynecomastia. The degree of enlargement is variable, ranging from a barely visible, small, central, subareolar disc of mammary tissue to an extension the size of a normal female adolescent breast. It may be unilateral or bilateral, frequently painful and tender, and, occasionally, emitting a secretion. Its presence is sometimes difficult to ascertain in obese males, since their breast enlargement may be due entirely, or in large part, to fat deposition (pseudo-gynecomastia).

The histopathology is characterized by stimulation of ducts and proliferation of stroma. The ducts undergo lengthening and branching, with budding and formation of new ducts without alveoli. Simultaneously, there is an increase in the bulk of the stromal tissue, which is often hyalinized. Stromal replacement of adipose tissue and periductal infiltrations with round cells are common. Since such changes may be induced by the administration of estrogen to males or to unstimulated females, it is logical to seek a similar explanation in patients with gynecomastia, but efforts in this direction have been only partially successful, especially because the condition is encountered in a great variety of apparently unrelated pathological disorders. Easier to explain is its occurrence in certain physiologic states.

β) Physiologic States

Neonatally, slight transitory breast enlargement is common in both sexes (witch milk). This is presumably due primarily to high levels of maternal estrogens. Gynecomastia, often slight, bilateral, and frequently painful, occurs in about two-thirds of boys during puberty (Froehlich Syndrome). It subsides spontaneously within 2 to 4 years in over 90 per cent of cases. Involutional enlargement occurs in some men later in life.

γ) Pathological Conditions

Superficially, these appear to be divisible into endocrine and non-endocrine disorders. It seems, however, that their origin is ultimately endocrine-conditioned. Chronic hyperestrogenism, possibly synergized by the action of one or more anterior pituitary hormones (i. e. prolactin or growth hormone), appears to be the principal etiologic factor. Estrogenic preponderance (without overt feminization) may arise through increased estrogen production, impaired estrogen inactivation in liver cirrhosis, and reduced androgen elaboration, or even increased androgens converted into estrogens. Augmented mammary sensitivity to circulating estrogens may explain certain innocuous familial cases, among others. B-avitaminosis may be a contributing factor, as vitamin B components are required for steroid metabolism in the liver. In most pathological conditions involving the endocrine glands absolute or relative hyperestrogenism is involved.

In general, gynecomastia appearing in, or after, the third decade of life is most often a manifestation of a serious underlying disease operating via hypothalamic release of prolactin. Examples are disorders of the testes, e. g. tumors, especially chorionepithelioma (HUNT and BUDD, 1939), hypogonadism (KLINEFELTER, REIFENSTEIN and ALBRIGHT, 1942), male pseudo-hermaphroditism, adrenal cortex tumors or hyperplasia (WILKINS, 1948), hypothyroidism (BERGSTRAND, 1955; SILVER, 1958) and pituitary tumors, e. g. acromegaly or chromophobe adenoma (DAVIDOFF, 1926). Other disorders, which possibly suppress the hypothalamic PIF, include hematochromatosis (S. C. FINCH and C. A. FINCH, 1955), liver cirrhosis due to alcohol abuse with impaired steroid metabolism and increased estrogen levels (EMERSON, 1966), malnutrition with vitamin B deficiency which also impairs estrogen cleavage (KLATSKIN, SALTER and HUMM, 1947), refeeding after starvation (PLATT, SCHULTZ and KUNSTADTER, 1946; JACOBS, 1948), bronchogenic carcinoma (DEL CASTILLO, DE LA BALZE and REFORZO MEMBRIVES, 1941), and traumatic paraplegia (KUPPERMAN, 1963 a, b). Obviously, gynecomastia readily develops after prolonged administration of estrogens for the treatment of prostate carcinoma, or of estrogen-contaminated drugs. It also occurs infrequently after treatment with chorionic gonadotropin, and steroid derivatives, as e. g. androgens (Lewin, 1941), digitalis (Lewinn, 1953), spironolactone (Mann, 1963; Sussman, 1963; Clark, 1965). Phenothiazines (Winnik and Tennenbaum, 1955; Margolis and Gross, 1967), isoniazid (INH) (Koang, Hu, Tchén and Chu, 1957; Molina, Aberkane and Conquydouard, 1957), ethionamide (Trecator) (Gernez-Rieux, Taquet and Maquet, 1963) and reserpine (Meyler, 1966) may also provoke gynecomastia via suppression of the hypothalamic PIF. Diabetes mellitus, generalized exfoliative dermatitis and many other miscellaneous conditions may sometimes be accompanied by gynecomastia. Here the hypothalamic relation is not clear.

Klinefelter's syndrome is characterized by gynecomastia with semineferous tubular hyalinization, small atrophic testes, elevated urinary FSH excretion, eunuchoidism and sometimes mental deficiency. It is due to the presence of a supernumerary chromosome—these patients having 47 chromosomes instead of the normal 46.

A prerequisite for treatment is thorough diagnostic investigation. Hormonal therapy is of no avail (norethandrolone up to 60 mg/d was proposed by Kupperman, 1963 a, b). Mammoplasty is indicated exclusively for cosmetic reasons and only when spontaneous regression has not taken place. This surgical procedure usually improves the patient's ability to carry on a normal social life as a male, preventing recurrent embarrassment due to the feminine breast development. There are, however, cases who end up as transvestites (Foss, 1958).

l) Conclusion

In lactating women endogenous prolactin suppresses FSH for 4 weeks and LH for 12 weeks. The knowledge of these terms is important for patients who wish to start the use of contraceptive pills without impairing lactation. If lactation is to be suppressed, stilbestrol should be given at doses of over 10 mg/d, combination with progestational or androgenic steroids being optional. Such treatment does not affect galactorrhea in the Chiari-Frommel syndrome or in the Ahumada-del Castillo syndrome. This fact points to our theory that normal lactation is produced by a prolactin-releasing factor sensitive to steroids, while pathological galactorrhea is produced by an impaired prolactin-inhibiting factor. These diseases do not react to steroid therapy; they are, however, slightly responsive to clomiphene.

Hypothalamic depressants, such as derivatives of phenothiazine, reserpine, butyrophenone, methyldopa, medroxyprogesterone acetate (in doses exceeding 1 g), can produce galactorrhea by suppressing the prolactin-inhibiting center. Recently, human growth hormone has been successfully applied in the promotion of post-partum lactation.

Gynecomastia in man is produced by a similar hypothalamic mechanism, mostly due to serious underlying disease, if not produced by drugs accidentally contaminated with estrogenic hormone or by stilbestrol phosphate used for the treatment of inoperable prostate carcinoma.

Chapter 29

Conclusions

The mammary gland represents a unique organ from a neuroendocrine point of view, since its development and secretory function are regulated by a harmonious concord of most of the endocrine glands and the central nervous system. Few other tissues in the body are influenced by so many endocrine, neural, and environmental factors. All of the anterior and posterior pituitary hormones may act on the mammary gland, either directly or through hormones of target organs, and all of them are controlled by the hypothalamus. Parathormone, insulin, placental hormones, and perhaps relaxin may also influence mammary function. Since the survival of most mammalian species would not be possible without milk, which requires an immense supply of protein, fats, sugar and calcium, it is perhaps not surprising that so many mechanisms have been evoked to permit the growth and secretory function of the mammary gland.

The hormones mainly responsible for lactation are prolactin and growth hormone. Estrogen, progesterone, corticosteroids, thyroidine and insulin have been shown to be of secondary importance in mammary development. Prolactin and growth hormone are considered to be principally involved since they alone can induce mammary lobulo-alveolar development in the absence of gonadal steroids, whereas estrogen and progesterone have but little effect in the absence of the pituitary hormones. These two steroids are believed to act mainly by increasing the release of prolactin and growth hormone from the anterior pituitary, and by synergizing with these hormones.

Chapter 30

Research Problems

The goal of producing hypothalamic lactation has become more accessible but still requires much further study from varying aspects. Substances more specific than phenothiazine and its sulfoxides should in particular be synthetized. Reserpine and haloperidol derivatives are at present being studied. The combination of such prolactin releasers with STH releasers becomes imperative in cows and sheep, and we have now begun a large-scale study of the possibilities of releasing growth hormone in laboratory animals. Paracetamol and some of its derivatives prepared by us may release growth hormone (DIKSTEIN, GROTTO, ZOR, TAMARI and SULMAN, 1966). Combinations will be studied in cows and sheep.

To test the hypothalamic prolactin releasers on cows does not seem very promising since apparently cows do not require prolactin for lactation. We, therefore, studied this method in sheep. The experiment was started in the 4th month of lactation when the daily milk yield had a tendency to decline. We tested the following eight compounds: piperacetazine (Quide), perphenazine sulfoxide (Schering), perphenazine (Trilafon), propericiazine (Neuleptil), reserpine (Serpasil), haloperidol (Halidol), trifluperidol (Trifludol), Sum-3170 (Wander). They were injected into 24 Awassi ewes either i. m. or i. v. for 6 days, in doses ranging between 0.02 and 0.5 mg daily. A slight increase in lactation was obtained only with propericiazine (0.5 mg i. m.), reserpine (0.05 mg i. m.) and haloperidol (0.1 mg i. v.). These experiments need further study and extension since the daily milk yield in sheep is subject to considerable fluctuations.

The major problem yet to be solved is elucidation of the structure of the hypothalamic factors responsible for lactation. It appears from our studies that phenothiazine derivatives and other psychopharmaca suppress the prolactin-inhibiting factor, whereas delivery and nursing release a prolactin-releasing factor. We feel, therefore, that, in addition to the prolactin-inhibiting factor of the hypothalamus, a prolactin-releasing factor should be studied. What facilitates this approach is that such a factor might possibly also stimulate other pituitary hormones, e. g. those required for preparation of lactation (FSH and LH), milk secretion (LTH and ACTH) and for maintenance of lactation (STH and TSH). Stimulation of a prolactin-releasing center may become feasible in farm animals through the stereotaxic approach. The methods we have begun using in laboratory animals are:

a) electrical stimulation of various centers that are responsible for lactation and maternal behavior;

b) removal of lactation-releasing factors from the hypothalamus of lactating animals through introducing capillaries which work by capillary attraction;

c) elution of lactation-releasing factors from suckling farm animals by intraventricular injections and perfusion of the cerebral ventricles, especially on the 10th day post-partum which has been found to be the peak period.

A technique not yet in use in domestic animals would be the injection of lactation-promoting drugs into the internal carotid artery, a site from which it would directly stimulate the hypothalamus and pituitary. This method could then be used for initiating lactation and increasing lactation to maximum levels for an extended period. The effect of oxytocin in this system requires further study.

Recently we found that perphenazine lactation in rats is impaired by adrenalectomy in spite of high blood prolactin levels. This would justify the re-examination of the amorphous steroid fraction of the adrenal cortex. It may contain a "corticolactin", differing from the hitherto identified corticosteroids.

It is most unlikely that hypothalamic milk releasers will be used per os in veterinary or in human medicine, since concentration of the releaser substance in the milk is just as high as in the blood. This has been confirmed by us by using radioactive phenothiazine derivatives in rabbits. Still, their use in farm animals may be possible by direct implantation into the hypothalamus, which would obviate any excretion either in the urine or the milk.

Modern methods of radioimmunoassay, which would permit measuring hormone levels at the different stages of pregnancy and lactation, could yield exact information on the gamut of the hormones participating in lactation at each stage. Thus, their practical use, either by injection or through release by chemical substances, would become feasible in the promotion of lactation.

References

Abood, L. G.: Effect of chlorpromazine on phosphorylation of brain mitochondria. Proc. Soc. exp. Biol. (N. Y.) **88**, 688 (1955).
Ahrén, K., Jacobsohn, D.: Mammary gland growth in hypophysectomized rats injected with ovarian hormones and insulin. Acta physiol. scand. **37**, 190 (1956).
— — The action of cortisone on the mammary glands of rats under various states of hormonal imbalance. Acta physiol. scand. **40**, 254 (1957).
Alloiteau, J. J.: Action inhibitrice de la chlorpromazine sur le déclenchement de deux types de pseudo-gestation chez la ratte. C. R. Soc. Biol. (Paris) **151**, 207 (1957).
Anderson, R. R., Turner, C. W.: Synergistic effect of various hormones on experimental mammary gland grown in mice. Proc. Soc. exp. Biol. (N. Y.) **113**, 308 (1963).
Andersson, B., McCann, S. M.: Drinking, antidiuresis and milk ejection from electrical stimulation within the hypothalamus of the goat. Acta physiol. scand. **35**, 191 (1955).
Apostolakis, M.: Lack of high growth hormone activity in highly active human pituitary prolactin extracts. In: Excerpta medica, international congress series no. 83, proceedings of the second international congress of endocrinology, London, August, 1964. Amsterdam: Excerpta medica foundation 1964, p. 1257.
Arai, Y., Lee, T. H.: A double-antibody radioimmunoassay procedure for ovine pituitary prolactin. Endocrinology **81**, 1040 (1967).
Argonz, J., Del Castillo, E. B.: A syndrome characterized by estrogenic insufficiency, galactorrhea and decreased urinary gonadotropin. J. clin. Endocr. **13**, 79 (1953).
Arimura, A., Saito, T., Müller, E. E., Bowers, C. Y., Sawano, S., Schally, A. V.: Absence of prolactin-release inhibiting activity in highly purified LH-releasing factor. Endocrinology **80**, 972 (1967).
Armstrong, D. T., Black, D. L.: Influence of luteinizing hormone on corpus luteum metabolism and progesterone biosynthesis throughout the bovine estrous cycle. Endocrinology **78**, 937 (1966).
— Kilpatrick, R., Greep, R. O.: *In vitro* and *in vivo* stimulation of glycolysis in prepubertal rat ovary by luteinizing hormone. Endocrinology **73**, 165 (1963).
— O'Brien, J., Greep, R. O.: Effect of luteinizing hormone on progestin biosynthesis in the luteinized rat ovary. Endocrinology **75**, 488 (1964).

ASSAEL, M., GABAI, F., WINNIK, H. Z., KHAZAN, N., SULMAN, F. G.: Recession of exophthalmos after massive doses of reserpine. Lancet 1960 I, 499.

ASTWOOD, E. B.: The regulation of corpus-luteum function by hypophysial luteotrophin. Endocrinology 28, 309 (1941).

— Tests for luteotrophins. Ciba Foundation Colloq. Endocrinology 5, 74 (1953).

AVERILL, R. L. W.: Restoration of lactation in rats with hypothalamic lesions which inhibit lactation. J. Endocr. 31, 191 (1965).

BAHN, R. C., BATES, R. W.: Histologic criteria for detection of prolactin: lack of prolactin in blood and urine of human subjects. J. clin. Endocr. 16, 1337 (1956).

BAILEY, G. L., BARTLETT, S., FOLLEY, S. J.: Use of L-thyroxine by mouth for stimulating milk secretion in lactating cows. Nature (Lond.) 163, 800 (1949).

BALDWIN, R. L., MARTIN, R. J.: Effects of hypophysectomy and several hormone replacement therapies upon patterns of nucleic acid and protein synthesis and enzyme levels in lactating rat mammary glands. J. Dairy Sci. 51, 748 (1968 a).

— — Protein and nucleic acid synthesis in rat mammary glands during early lactation. Endocrinology 82, 1209 (1968 b).

BAN, T., SHIMIZU, S., KUROTSU, J.: Experimental studies on the relationship between the hypothalamus including area preoptica and lactation in rabbits. Med. J. Osaka Univ. 8, 345 (1958).

BARANYAI, P., NAGY, I., KURCZ, M.: Kühlbare horizontale Stärke-Gel-Elektrophorese-Einrichtung für die Fraktionierung der Eiweißstoffe in der Adenohypophyse. Clin. chim. Acta 16, 449 (1967).

BARDIN, C. W., LIEBELT, A. G., LIEBELT, R. A.: The direct effect of pituitary isografts on mammary gland development in the mouse. Proc. Soc. exp. Biol. (N. Y.) 110, 716 (1962).

BARLET, J. P.: Effets de la thyrocalcitonine sur la calcémie et la phosphatémie de la vache laitière. C. R. Acad. Sci. (Paris) Ser. D 265, 1075 (1967).

BARNAWELL, E. B.: Analysis of the direct action of prolactin and steroids on mammary tissue of the dog in organ culture. Endocrinology 80, 1083 (1967).

BARRACLOUGH, C. A.: Induction of pseudopregnancy in the rat by reserpine and chlorpromazine. Anat. Rec. 127, 262 (1957).

— SAWYER, C. H.: Blockade of the release of pituitary ovulating hormone in the rat by chlorpromazine and reserpine: possible mechanisms of of action. Endocrinology 61, 341 (1957).

— — Induction of pseudopregnancy in the rat by reserpine and chlorpromazine. Endocrinology 65, 563 (1959).

BARRETT, R. J., FRIESEN, H., ASTWOOD, E. B.: Characterization of pituitary and peptide hormones by electrophoresis in starch gel. J. biol. Chem. 237, 432 (1962).

BARTLETT, S., BURT, A. W. A., FOLLEY, S. J., ROWLAND, S. J.: Relative galactopoietic effects of 3:5:3′-triiodo-L-thyronine and L-thyroxine in lactating cows. J. Endocr. **10**, 192 (1954).
BARTOSIK, D., ROMANOFF, E. B., WATSON, D. J., SCRICCO, E.: Luteotropic effects of prolactin in the bovine ovary. Endocrinology **81**, 186 (1967).
BATES, R. W., GARRISON, M. M., CORNFIELD, J.: An improved bio-assay for prolactin using adult pigeons. Endocrinology **73**, 217 (1963).
— LAHR, E. L., RIDDLE, O.: The gross action of prolactin and follicle-stimulating hormone on the mature ovary and sex accessories of fowl. Amer. J. Physiol. **111**, 361 (1935).
— RIDDLE, O.: Effect of volume used for injection in micro-assay of prolactin. Proc. Soc. exp. Biol. (N. Y.) **44**, 505 (1940).
— — Annual variation in the response of crop-sacs and viscera of pigeons to prolactin. Endocrinology **29**, 702 (1941).
BAYLE, J. D., ASSENMACHER, I.: Absence de stimulation du jabot du pigeon après autogreffe hypophysaire. C. R. Acad. Sci. (Paris) **261**, 5667 (1965).
BECK, J. C., GONDA, A., HAMID, M. A., MORGEN, R. O., RUBENSTEIN, D., MCGARRY, E. E.: Some metabolic changes induced by primate growth hormone and purified ovine prolactin. Metabolism **13**, Suppl. 1108 (1964 a).
— MCGARRY, E. E., DAWSON, K. G., GONDA, A., HAMID, M. A., RUBINSTEIN, D.: Comparative metabolic actions of primate growth hormone and non-primate prolactin in man. In: Excerpta medica, International congress series no. 83, proceedings of the second international congress of endocrinology, London, Aug., 1964. Amsterdam: Excerpta medica Foundation 1964 b, p. 1242.
BEN-DAVID, M.: A sensitive bioassay for prolactin based on H^3-methyl-thymidine uptake by the pigeon-crop mucous epithelium. Proc. Soc. exp. Biol. (N. Y.) **125**, 705 (1967).
— Induction of hypothalamic lactation in Fischer inbred rats by depot injection of fluphenazine enanthate. Neuroendocrinology **3**, 65 (1968 a).
— The role of the ovaries in perphenazine-induced lactation. J. Endocr. **41**, 377 (1968 b).
— Mechanism of induction of mammary differentiation in Sprague-Dawley female rats by perphenazine. Endocrinology **83**, 1217 (1968 c).
— CHRAMBACH, A.: Stimulation of pituitary release and production of prolactin by perphenazine. J. biol. Chem. (in press) (1970).
— DIKSTEIN, S., SULMAN, F. G.: Effects of different steroids on prolactin secretion in pituitary-hypothalamus organ co-culture. Proc. Soc. exp. Biol. (N. Y.) **117**, 511 (1964).
— — — Production of lactation by non-sedative phenothiazine derivatives. Proc. Soc. exp. Biol. (N. Y.) **118**, 265 (1965).
— — — Role of the thyroid in hypothalamic lactation. Proc. Soc. exp. Biol. (N. Y.) **121**, 873 (1966).
— KHAZEN, K., KHAZAN, N., SULMAN, F. G.: Correlation between depressant and mammotropic effects of fifteen reserpine analogues. Arch. int. Pharmacodyn. **171**, 274 (1968).

BEN-DAVID, M., RODERIG, H., KHAZEN, K., SULMAN, F. G.: Effect of different steroids on lactating rats. Proc. Soc. exp. Biol. (N. Y.) **120**, 620 (1965).
— SULMAN, G. F.: The mechanism of systemic deciduoma production. Arch. int. Pharmacodyn. **159**, 161 (1966).
— — The role of the adrenals in perphenazine-induced lactation. Acta endocr. (Kbh.) (in press) (1970).
BENSON, G. K.: Pituitary-adrenal relationships in the retardation of mammary gland involution by oxytocin. J. Endocr. **20**, 91 (1960).
— COWIE, A. T., COX, C. P., GOLDZWEIG, S. A.: Effects of estrone and progesterone on mammary development in the guinea-pig. J. Endocr. **15**, 126 (1957).
— — TINDAL, J. S.: The pituitary and the maintenance of milk secretion. Proc. roy. Soc. B. **149**, 330 (1958).
— FOLLEY, S. J.: Oxytocin as stimulator for the release of prolactin from the anterior pituitary. Nature (Lond.) **177**, 700 (1956).
— — The effect of oxytocin on mammary gland involution in the rat. J. Endocr. **16**, 189 (1957).
BERGENSTAL, D. M., LIPSETT, M. B.: The anabolic effect of sheep prolactin in man. J. clin. Invest. **37**, 877 (1958).
BERGSTRAND, C. G.: A case of hypothyroidism with signs of precocious sexual development. Acta endocr. (Kbh.) **20**, 338 (1955).
BERNTHSEN, A.: Zur Kenntnis des Methylenblau und verwandter Farbstoffe. Ber. dtsch. chem. Ges. **16**, 2897 (1883).
BERSON, S. A., YALOW, R. S.: Immunoassay of protein hormones. In: The Hormones, Vol. IV. Eds.: G. PINCUS, K. V. THIMANN, and E. B. ASTWOOD. New York-London: Academic Press 1964, p. 557.
BERSWORDT-WALLRABE, I. VON, HERLYN, U., JANTZEN, K.: A modification of the pigeon crop sac assay for determination of lactogenic hormone (prolactin) in human serum. Acta endocr. (Kbh). suppl. **100**, 169 (1965).
BHARGAVA, K. P., CHANDRA, O.: Tranquillizing and hypotensive activities of twelve phenothiazines. Brit. J. Pharmacol. **22**, 154 (1964).
BISSET, G. W.: The milk-ejection reflex and actions of oxytocin, vasopressin and synthetic analogues on the mammary gland. In: Handbuch der experimentellen Pharmakologie. Vol. XXIII; Neurohypophysial hormones and similar polypeptides. Berlin-Heidelberg-New York: Springer 1968, p. 475.
BLAXTER, K. L.: The preparation and biological effects of iodinated proteins. The effect of iodinated protein feeding on the lactating cow. I. The effects of preparations of low activity and of iodinated ardein. J. Endocr. **4**, 237 (1945).
BOLOGNESE, R. J., PIVER, M. S., FELDMAN, J. D.: Galactorrhea and abnormal menses associated with a long-acting progesterone. J. Amer. med. Ass. **199**, 42 (1967).
BOOT, L. M., ROEPCKE, G., MUEHLBOCK, O.: Prolacting-producing pituitary tumors arising in pituitary isografts in mice. In: Excerpta Medica, International congress series No. 33, Proceedings of the second inter-

national congress of endocrinology, London, August 1964, 1058. Amsterdam: Excerpta medica foundation 1964.
BRAUDE, R., MITCHELL, K. G.: Observations on the relationship between oxytocin and adrenaline in milk ejection in the sow. J. Endocr. 8, 238 (1952).
BROWN-GRANT, K., CROSS, B. A. (Eds.): Recent studies on the hypothalamus. Brit. med. Bull. 22, 195 (1966).
BRUCE, H. M.: Suckling stimulus and lactation. Proc. roy. Soc. B 149, 421 (1958).
— PARKES, A. S.: An olfactory block to nidation in mice. In: Advances in neuroendocrinology, p. 282. Ed.: A. V. NALBANDOV. Urbana (Ill.): Univ of Illinois Press 1963.
— SLOVITER, H. A.: Effect of destruction of thyroid tissue by radioactive iodine on reproduction in mice. J. Endocr. 15, 72 (1957).
BRUMBY, P. J., HANCOCK, J.: The galactopoietic rate of growth hormone in dairy cattle. N. Z. Jl. Sci. Technol. 36, 417 (1955).
BRYANT, G. D., CONNAN, R. H., GREENWOOD, F. C.: Changes in plasma prolactin induced by acepromazine in sheep. J. Endocr. 41, 613 (1968).
— GREENWOOD, F. C.: Plasma prolactin levels in the goat; physiological and experimental modification. J. Endocr. 40, IV (1968).
— TECH, B., GREENWOOD, F. C., LINZEL, J. L.: Radioimmunoassay of ovine, caprine and bovine prolactin in plasma; physiological and experimental changes in levels. Excerpta Medica, International congress series No. 157, Third international congress of endocrinology, Mexico, D. F., June 30—July 5, 1968. Amsterdam: Excerpta medica foundation 1968, p. 76.
BUTT, W. R.: In: Immunochemical studies on human growth hormone; a consideration of the human growth hormone-antihuman growth hormone system and its application to the assay of growth hormone. Eds.: M. M. and S. L. KAPLAN: Questions in Grumbach. Ciba Found. Colloq. Endocr. 14, 103 (1962).
CANFIELD, C. J., BATES, R. W.: Nonpuerperal galactorrhea. New Engl. J. Med. 273, 897 (1965).
CASTILLO, E. B. DEL, DE LA BALZE, F. A., REFORZO MEMBRIVES, J.: Ginecomastia y cancer de pulmón. Sem. méd. (B. Aires) 1, 1419 (1941).
CHADWICK, A., FOLLEY, S. J., GEMZELL, C. A.: Lactogenic activity of human pituitary growth hormone. Lancet 1961 I, 241.
CHEN, T. T., JOHNSON, R. E., LYONS, W. R., LI, C. H., COLE, R. D.: Hormonally induced mammary growth and lactation in the absence of the thyroid. Endocrinology 57, 153 (1955).
CHEN, C. L., MEITES, J.: Effect of thyroxine and thiouracil on hypothalamic PIF and pituitary prolactin levels. Proc. Soc. exp. Biol. (N. Y.) 131, 576 (1969).
CHEN, H. C., WILHELMI, A. G.: Purification of human growth hormone. Progr. 46th meet. Endocrine Soc. 46, 115 (1964).

CHIARI, J. B. V. L., BRAUN, C., SPAETH, J.: Klinik der Geburtshilfe und Gynaekologie. Erlangen: Enke 1852.

CHILD, K. J., SUTHERLAND, P., TOMICH, E. G.: Some effects of reserpine on barbitone anaesthesia in mice. Biochem. Pharmacol. **6**, 252 (1961).

— — — Barbitone induction time, body temperature and the actions of reserpine on mice. Biochem. Pharmacol. **11**, 475 (1962).

CHOUDARY, J. P., GREENWALD, G. S.: Effect of an ectopic pituitary gland on luteal maintenance in the hamster. Endocrinology **81**, 542 (1967).

CHRAMBACH, A., BEN-DAVID, M.: The molecular identity of rat prolactin derived from serum and pituitary under conditions of both natural and pharmacological-induced lactation. J. biol. Chem. (in press) (1970).

CLARK, E.: Spironolactone therapy and gynecomastia. J. Amer. med. Ass. **193**, 163 (1965).

CLEMENS, J. A., MEITES, J.: Prolactin implant into the median eminence inhibits pituitary prolactin secretion, mammary growth and luteal function. Physiologist **10**, 144 (1967).

— — Inhibition by hypothalamic prolactin implants of prolactin secretion, mammary growth and luteal function. Endocrinology **82**, 878 (1968).

— MINAGUCHI, H., STOREY, R., VOOGT, J. L., MEITES, J.: Induction of precocious puberty in female rats by prolactin. Neuroendocrinology **4**, 150 (1969).

— SAR, M., MEITES, J.: Termination of pregnancy by prolactin implants in the median eminence of the rat. Proc. Int. Un. Physiol. Sci. **7**, 1 (1968).

— — — Termination of pregnancy in rats by a prolactin implant in median eminence. Proc. Soc. exp. Biol. (N. Y.) **130**, 628 (1969).

CLEVERLEY, J. D.: The detection of oxytocin release in response to conditioned stimuli associated with machine milking in the cow. J. Endocr. **40**, II (1967).

COHERE, G., BOUSQUET, J., MEUNIER, J. M.: Ultrastructure d'explants de glande pituitaire de ratte adulte cultivés sur milieux artificiels; action de variations hormonales. C.R. Soc. Biol. (Paris) **158**, 1056 (1964).

CONTOPOULOS, A. N., SIMPSON, M. E., KONEFF, A.: Pituitary function in the thyroidectomized rat. Endocrinology **63**, 642 (1958).

COPPEDGE, R. L., SEGALOFF, A.: Urinary prolactin excretion in man. J. clin. Endocr. **11**, 465 (1951).

COPPOLA, J. A., LEONARDI, R. G., LIPPMANN, W., PERRINE, J. W., RINGLER, I.: Induction of pseudopregnancy in rats by depletors of endogenous catecholamines. Endocrinology **77**, 485 (1965).

CORBIN, A., COHEN, A. I.: Effect of median eminence implants of LH on pituitary LH of female rats. Endocrinology **78**, 41 (1966).

— STORY, J. C.: "Internal" feedback mechanism: Response of pituitary FSH and of stalk-median eminence follicle stimulating hormone-releasing factor to median eminence implants of FSH. Endocrinology **80**, 1006 (1967).

CORDELLI, F., CASTELLI, A., GORELLI, L.: Effect of hormones on lactation: I. Effect of Insulin. Lattante **31**, 546 (1960). (Title in Italian: Azioni ormoniche sulla lattazione Nota I. Azione dell'insulina.)

COTES, P. M., CRICHTON, J. A., FOLLEY, S. J., YOUNG, F. G.: Galactopoietic activity of purified anterior pituitary growth hormone. Nature (Lond.) **164**, 992 (1949).

COURVOISIER, S., DUCROT, R., JULOU, L.: Nouveaux aspects expérimentaux de l'activité centrale des dérivés de la phénothiazine. In: Psychotropic drugs, p. 373. Eds.: S. GARATTINI and V. GHETTI. Amsterdam: Elsevier 1957.

— FOURNEL, J., DUCROT, R., KOLSKY, M., KOETSCHET, P.: Propriétés pharmacodynamiques du chlorhydrate de chloro-3 (diméthyl-amino-3 propyl)-10-phénothiazine (4.560 R.P.); étude expérimentale d'un nouveau corps utilisé dans l'anesthésie potentialisée et dans l'hibernation artificielle. Arch. int. Pharmacodyn. **92**, 305 (1953).

COWIE, A. T.: Influence on the replacement value of some adrenal-cortex steroids of dietary sodium and synergism of steroids in lactating adrenalectomized rats. Endocrinology **51**, 217 (1952).

— Hormones and lactation. N.I.R.D. paper No. 1762. Nat. Agr. Adv. Connc. Quart. Rev. **2**, 32 (1956).

— The maintenance of lactation in the rat after hypophysectomy. J. Endocr. **16**, 135 (1957).

— The hormonal control of milk secretion. In: Milk; the mammary gland and its secretion. Vol. I, p. 163. Eds.: S. K. KON and A. T. COWIE. New York-London: Academic Press 1961.

— Anterior pituitary function in lactation. In: The Pituitary Gland. Vol. II, p. 412. Eds.: G. W. HARRIS and B. T. DONOVAN. London: Butterworths 1966.

— FOLLEY, S. J.: The role of the adrenal cortex in mammary development and its relation to the mammogenic action of the anterior pituitary. Endocrinology **40**, 274 (1947).

— — Neurophysial hormones and the mammary gland. In: The neurohypophysis, p. 183. Ed.: H. HELLER. London: Butterworths 1957.

— — The mammary gland and lactation. In: Sex and internal secretions. Vol. I, p. 590. Ed.: W. C. YOUNG, 3rd ed. Baltimore: Williams & Wilkins 1961.

— — MALPRESS, F. H., RICHARDSON, K. C.: Studies on the hormonal induction of mammary growth and lactation in the goat. J. Endocr. **8**, 64 (1952).

— HARTMANN, P. E., TURVEY, A.: The maintenance of lactation in the rabbit after hypophysectomy. J. Endocr. **43**, 651 (1969).

— KNAGGS, G. S., TINDAL, J. S., TURVEY, A.: The milking stimulus and mammary growth in the goat. J. Endocr. **40**, 243 (1968).

— TINDAL, J. S.: Maintenance of lactation in adrenalectomized rats with aldosterone and 9 halo derivatives of hydrocortisone. Endocrinology **56**, 612 (1955).

Cowie, A. T., Tindal, J. S.: Adrenalectomy in the goat; replacement therapy and the maintenance of lactation. J. Endocr. 16, 403 (1958).
— — Benson, G. K.: Pituitary grafts and milk secretion in hypophysectomized rats. J. Endocr. 21, 115 (1960).
Cox, W. M., Mueller, A. J.: The composition of milk from stock rats and an apparatus for milking small laboratory animals. J. Nutr. 13, 249 (1937).
Cronin, T. J.: Influence of lactation upon ovulation. Lancet 1968 II, 422.
Cross, B. A.: Sympathetico-adrenal inhibition of the neurohypophysical milk ejection mechanism. J. Endocr. 9, 7 (1953).
— Neurohormonal mechanisms in emotional inhibition of milk ejection. J. Endocr. 12, 29 (1955).
— Neural control of lactation. In: Milk, the mammary gland and its secretion, Vol. I. Eds.: S. K. Kon and A. T. Cowie. New York-London: Academic Press 1961, p. 229.
— Harris, G. W.: The role of the neurohypophysis in the milk-ejection reflex. J. Endocr. 8, 148 (1952).
Danon, A., Dikstein, S., Sulman, F. G.: Stimulation of prolactin secretion by perphenazine in pituitary-hypothalamus organ culture. Proc. Soc. exp. Biol. (N. Y.) 114, 366 (1963).
— — — In-vivo assay for hypothalamic prolactin-inhibiting factor. J. Endocr. 46, 237 (1970).
— Sulman, F. G.: Storage of prolactin-inhibiting factor in the hypothalamus of perphenazine-treated rats. Neuroendocrinology (in press) (1970).
Dao, T. L., Gawlak, D.: Direct mammotrophic effect of a pituitary homograft in rats. Endocrinology 72, 884 (1963).
Davidoff, L. M.: Studies in acromegaly. III. The anamnesis and symptomatology in one hundred cases. Endocrinology 10, 461 (1926).
De Feo, V. J.: Effect of large doses of reserpine on the deciduoma response. Anat. Rec. 127, 409 (1957).
— Reynolds, S. R. M.: Modification of the menstrual cycle in rhesus monkeys by reserpine. Science 124, 726 (1956).
Delay, J., Deniker, P.: Les hormones. Journées thérap. Paris 1, 97 (1953).
De Lee, J. B.: The principles and practice of obstetrics. 4th ed. Philadelphia: Saunders 1924.
Desclin, L.: A propos du mécanisme des oestrogènes sur la lobe antérieur de l'hypophyse. Ann. Endocr. (Paris) 11, 656 (1950).
— Hypothalamus et libération d'hormone lutéotrophique. Expériences de greffe hypophysaire chez le rat hypophysectomisé. Action lutéotrophique de l'ocytocine. Ann. Endocr. (Paris) 17, 586 (1956).
— Action de la réserpine sur la glande mammaire et sur l'ovaire du rat. Ann. Endocr. (Paris) 19, 1225 (1958).
— Flament-Durand, J.: Effect of reserpine on the morphology of pituitaires grafted into the hypothalamus or under the kidney capsule in the rat. J. Endocrinol. 43, LIX (1969).

DESJARDINS, C., PAAPE, N. J., TUCKER, H. A.: Contribution of pregnancy, fetal placentas and deciduomas to mammary gland and uterine development. Endocrinology **83**, 907 (1968).

DHARIWAL, A. P. S., GROSVENOR, C. E., ANTUNES-RODRIGUES, J., MCCANN, S. M.: Correspondence of prolactin-inhibition and LH-releasing activities after gel filtration of hypothalamic extracts on sephadex. Progr. 47th meet. Endocrine Soc. **47**, 127 (1965).

— — — — Studies on the purification of ovine prolactin-inhibiting factor. Endocrinology **82**, 1236 (1968).

DIKSTEIN, S., GROTTO, M., ZOR, U., TAMARI, M., SULMAN, F. G.: The stimulatory effect of paracetamol and its derivatives on growth and the rat tibia test. J. Endocr. **36**, 257 (1966).

— SULMAN, F. G.: Drugs acting on the hypothalamus-anterior pituitary axis. Obstet. gynec. Surv. **21**, 531 (1966).

DILLEY, W. G., NANDI, S.: Rat mammary gland differentiation *in vitro* in the absence of steroids. Science **161**, 59 (1968).

DIXON, J. S., LI, C. H.: Chemistry of prolactin. Metabolism **13**, 1093 (1964).

DJOJOSOEBAGIO, S., TURNER, C. W.: Effect of a combination of lactogenic, growth, thyroid, and parathyroid hormones on lactation in rats. Proc. Soc. exp. Biol. (N. Y.) **116**, 213 (1964).

DOMANSKI, E., DOBROWOLSKI, W.: Perfusion of an organ in situ as a method in endocrinological investigations. In: Excerpta medica, international congress series no. 111, second international congress on hormonal steroids, Milan, Italy, May 23—28, 1966. Amsterdam: Excerpta medica foundation 1966, p. 259.

DONOVAN, B. T.: The effect of pituitary stalk section on luteal function in the ferret. J. Endocr. **27**, 201 (1963 a).

— Pituitary stalk section and corpus luteum function in the ferret. J. Physiol. (Lond.) **165**, 23P (1963 b).

— VAN DER WERFF TEN BOSCH, J. J.: The hypothalamus and lactation in the rabbit. J. Physiol. (Lond.) **137**, 410 (1957).

DRESEL, I.: The effect of prolactin on the estrus cycle of non-parous mice. Science **82**, 173 (1935).

DURLACH, J.: L'action mammotrope de la réserpine. Presse méd. **65**, 2060 (1957).

EBERT, A. G., HESS, S. M.: The distribution and metabolism of fluphenazine enanthate. J. Pharmacol. exp. Ther. **148**, 412 (1965).

ECKSTEIN, B.: The development of the ovary of the rat. II. The effect of chorionic gonadotrophin and serum gonadotrophin. Acta endocr. (Kbh.) **41**, 35 (1962).

EHRHARDT, K.: Über das Laktationshormon des Hypophysenvorderlappens. Münch. med. Wschr. **29**, 1163 (1936).

ELLIS, S., GRINDELAND, R. E., NUENKE, J. M., CALLAHAN, P. X.: Purification and properties of rat prolactin. Endocrinology **85**, 886 (1969).

ELY, F., PETERSON, W. E.: Factors involved in the ejection of milk. J. Dairy Sci. **24**, 211 (1941).

EMERSON, K., JR.: Diseases of the breast. In: Principles of internal medicine, p. 532. Eds.: T. R. HARRISON, R. D. ADAMS, I. L. BENNETT, W. H. RESNIK, G. W. THORN and M. M. WINTROBE. New York: McGraw Hill 1966.

EVANS, H. M., SIMPSON, M. E., LYONS, W. R., TURPEINER, K.: Anterior pituitary hormones which favor the production of traumatic uterine placentomata. Endocrinology **28**, 933 (1941).

EVERETT, J. W.: Evidence suggesting a role of the lactogenic hormone in the estrous cycle of the albino rat. Endocrinology **35**, 507 (1944).

— Luteotrophic function of autografts of the rat hypophysis. Endocrinology **54**, 685 (1954).

— Functional corpora lutea maintained for months by autografts of rat hypophyses. Endocrinology **58**, 786 (1956).

— Central neural control of reproductive functions of the adenohypophysis. Physiol. Rev. **44**, 373 (1964).

— QUINN, D. L.: Differential hypothalamic mechanisms inciting ovulation and pseudopregnancy in the rat. Endocrinology **78**, 141 (1966).

FAURET, E.: Vergleichende Untersuchungen über die Entwicklung und Funktion der Milchdrüsen. Experimentelle Untersuchungen über die Wirkung von Follikelhormonzufuhr auf die Milchdrüsen säugender Ratten. Arch. Gynäk. **171**, 342 (1941).

FELDBERG, W.: A new concept of temperature control in the hypothalamus. Proc. roy. Soc. Med. **58**, 395 (1965).

FERGUSON, K. A., WALLACE, A. L. C.: Starch-gel electrophoresis of anterior pituitary hormones. Nature (Lond.) **190**, 629 (1961).

FINCH, S. C., FINCH, C. A.: Idiopathic hemachromatosis, an iron storage disease. Medicine (Baltimore) **34**, 381 (1955).

FISKIO, P. W.: Combined steroids to suppress lactation. GP (Kansas) **11**, 70 (1955).

FLERKO, B.: Control of gonadotropin secretion in the female. In: Neuroendocrinology. Vol. I, p. 613. Eds.: L. MARTINI and W. F. GANONG. New York-London: Academic Press 1966.

— SZENTAGOTHAI, J.: Estrogen sensitive nervous structures in the hypothalamus. Acta endocr. (Kbh.) **26**, 121 (1957).

FLORINI, J. R., TONELLI, G., BREUER, C. B., COPPOLA, J., RINGLER, I., BELL, P. H.: Characterization and biological effects of purified placental protein (human). Endocrinology **79**, 692 (1966).

FLUECKIGER, E., WAGNER, H. R.: 2-Br-Alpha-Ergokryptin: Beeinflussung von Fertilität und Laktation bei der Ratte. Experientia (Basel) **24**, 1130 (1968).

FLUX, D. S.: The effect of adrenal steroids on the growth of the mammary glands, uteri, thymus and adrenal glands of intact, ovariectomized and estrone-treated ovariectomized mice. J. Endocr. **11**, 238 (1954).

— The value of some steroids in replacement therapy in adrenalectomized and adrenalectomized-ovariectomized lactating rats J. Endocr. **12**, 57 (1955).

FLUX, D. S., FOLLEY, S. J., ROWLAND, S. J.: The effect of adrenocorticotrophic hormone on the yield and composition of the milk of the cow. J. Endocr. **10**, 333 (1954).
— MUNFORD, R. E.: The effect of adrenocorticotrophin on the mammary glands of intact mature female mice of the CHI strain. J. Endocr. **14**, 343 (1957).
FOLLEY, S. J.: The adrenal cortex and the mammary gland. In: The Suprarenal Cortex. Ed.: J. M. YOFFEY. Proceedings of the Colston Research Society's Fifth Symposium. Bristol, April 1952. London: Butterworth 1952 a.
— Lactation. In: Marshall's physiology of reproduction. Vol. II, p. 525. Ed.: A. S. PARKES. London: Longmans, Green 1952 b.
— The physiology and biochemistry of lactation. Edinburgh-London: Oliver and Boyd 1956.
— COWIE, A. T.: Adrenalectomy and replacement therapy in lactating rats. Yale J. Biol. Med. **17**, 67 (1944).
— KON, S. K.: The effect of sex hormones on lactation in the rat. Proc. roy. Soc. B **124**, 476 (1938).
— MALPRESS, F. H.: Hormonal control of mammary growth. In: The Hormones. Vol. I, p. 695. Eds.: G. PINCUS and K. V. THIMANN. New York: Academic Press 1948 a.
— — Hormonal control of lactation. In: The Hormones. Vol. I, p. 745. Eds.: G. PINCUS and K. V. THIMANN. New York: Academic Press 1948 b.
— YOUNG, F. G.: Further experiments on the continued treatment of lactating cows with anterior pituitary extracts. J. Endocr. **2**, 226 (1940).
FORBES, A. P., HENNEMAN, P. H. GRISWOLD, G. C., ALBRIGHT, F.: A syndrome distinct from acromegaly, characterized by spontaneous lactation, amenorrhea and low follicle-stimulating hormone excretion. J. clin. Endocr. **11**, 749 (1951).
— — — — Syndrome characterized by galactorrhea, amenorrhea, and low urinary FSH: Comparison with acromegaly and normal lactation. J. clin. Endocr. **14**, 265 (1954).
FORREST, J. S., PIETTE, L. H.: New techniques of identification of individual oxidative metabolites derived from phenothiazine and related compounds. Neuropsychopharmacol. **3**, 397 (1964).
FORSYTH, I. A.: Lactogenic and pigeon crop-stimulating activities of a human placental lactogen preparation. J. Endocr. **37**, XXXV (1967).
— The lactogenic activity of human growth hormone *in vivo*. In: Excerpta media, international congress series no. 158, Growth hormone, Proceedings of the first international symposium, Milan, Italy, September, 1967. Amsterdam: Excerpta medica foundation 1968, p. 364.
— FOLLEY, S. J., CHADWICK, A.: Lactogenic and pigeon crop-stimulating activities of human pituitary growth hormone preparations. J. Endocr. **31**, 115 (1965).
FOSS, G. L.: Disturbances of Lactation. Clin. Obstet. Gynec. **1**, 245 (1958).

FRIESEN, H.: Purification of a placental factor with immunological and chemical similarity to human growth hormone. Endocrinology **76**, 369 (1965).
FROMMEL, R.: Über puerperale Atrophie des Uterus. Z. Geburtsh. Gynäk. **7**, 305 (1882).
— Bericht über Frauenkrankheiten der Städtischen Frauenklinik Wien. Z. Geburtsh. Gynäk. 1845—1855.
GALA, R. R., REECE, R. P.: Effect of a reserpine analogue, methyl-18-epi-o-methyl-reserpate hydrochloride, on lactogen release from rat anterior pituitary. Endocrinology **72**, 649 (1963).
— — Influence of hypothalamic fragments and extracts on lactogen production *in vitro*. Proc. Soc. exp. Biol. (N. Y.) **117**, 833 (1964 a).
— — Influence of estrogen on anterior pituitary lactogen production *in vitro*. Proc. Soc. exp. Biol. (N. Y.) **115**, 1030 (1964 a).
— — Influence of neurohumors on anterior pituitary lactogen production *in vitro*. Proc. Soc. exp. Biol. (N. Y.) **120**, 220 (1965 a).
— — *In vitro* lactogen production by anterior pituitaries from various species. Proc. Soc. exp. Biol. (N. Y.) **120**, 263 (1965 b).
GATSCHEW, E.: Reserpin, Prolactin-Freisetzung und Milchzuckersynthese. Arzneimittel-Forsch. **16**, 75 (1966).
GAUNT, R., CHART, J. J., RENZI, A. A.: Interactions of drugs with endocrines. Ann. Rev. Pharmacol. **3**, 109 (1963).
— EVERSOLE, W. J., KENDALL, E. C.: Influence of some steroid hormones on lactation in adrenalectomized rats. Endocrinology **31**, 84 (1942).
— RENZI, A. A., ANTONCHAK, N., MILLER, G. J., GILMAN, M.: Endocrine aspects of the pharmacology of reserpine. Ann. N. Y. Acad. Sci. **59**, 22 (1954).
GERNEZ-RIEUX, C., TACQUET, A., MAQUET, V.: Les perfusions d'éthionamide dans le traitement de la tuberculose pulmonaire. G. ital. Chemioter. **10**, 87 (1963).
GESCHICKTER, C. F.: Diseases of the breast. Philadelphia: Lippincott 1960.
GOSPODAROWICZ, D., LEGAULT-DEMARE, J.: Etude de l'activité biologique *in vitro* des hormones gonadotropes. IV. Action synergique de la prolactine et de l'hormone chorionique humaine sur le corps jaune de rat *in vitro*. Acta endocr. (Kbh.) **42**, 509 (1963).
GRAHAM, W. R., JR.: The action of thyroxine on the milk and milk-fat production of cows. Biochem. J. **28**, 1368 (1934).
GREENBLATT, R. B., MAHESH, V. B.: Induction of ovulation with clomiphene citrate. In: Yearbook of endocrinology, p. 248. Ed.: T. B. SCHWARTZ. Chicago: Yearbook medical publishers 1964—1965.
GREENWALD, G. S.: Luteotropic complex of the hamster. Endocrinology **80**, 118 (1967).
GREEP, R. O.: Physiology of the anterior hypophysis in relation to reproduction. In: Sex and internal secretions, Vol. I. 3rd ed. Ed.: W. C. YOUNG. Baltimore: Williams & Wilkins 1961, p. 240.

GREIG, M. E., MAYBERRY, T. C.: The relationship between cholinesterase activity and brain permeability. J. Pharmacol. exp. Ther. **102**, 1 (1951).
GRIFFITH, L. S., McBRIDE, W. P. L.: Clinical uses of endometrial biopsy. J. Mich. med. Soc. **38**, 1064 (1939).
GRIMM, E. G.: Non-puerperal galactorrhea. Quart. Bull. Northw. Univ. med. Sch. **29**, 350 (1955).
GRÖNROOS, M., KALLIOMÄKI, J. L., KEYRILÄINEN, T. O., MARJANEN, P.: Effects of reserpine and chlorpromazine on the mammary glands of the rat. Acta endocr. (Kbh.) **31**, 154 (1959).
GROSVENOR, C. E.: Effect of experimentally induced hypo- and hyper-thyroid states upon pituitary lactogenic hormone concentration in rats. Endocrinology **69**, 1092 (1961).
— Evidence that exteroceptive stimuli can release prolactin from the pituitary gland of the lactating rat. Endocrinology **76**, 340 (1965 a).
— Effect of nursing and stress upon prolactin-inhibiting activity of the rat hypothalamus. Endocrinology **77**, 1037 (1965 b).
— Effect of nursing and stress upon "prolactin-inhibiting" activity of rat hypothalamus. Fed. Proc. **24**, 190 (1965 c).
— KRULICH, L., McCANN, S. M.: Depletion of pituitary concentration of growth hormone as a result of suckling in the lactating rat. Endocrinology **82**, 617 (1968).
— McCANN, S. M., NALLAR, R.: Inhibition of nursing-induced and stress-induced fall in pituitary prolactin concentration in lactating rats by injection of acid extracts of bovine hypothalamus. Endocrinology **76**, 883 (1965).
— TURNER, C. W.: Pituitary lactogenic hormone concentration and milk secretion in lactating rats. Endocrinology **63**, 535 (1958 a).
— — Effect of lactation upon thyroid secretion rate in the rat. Proc. Soc. exp. Biol. (N. Y.) **99**, 517 (1958 b).
— — Effect of growth hormone and oxytocin upon milk yield in the lactating rat. Proc. Soc. exp. Biol. (N. Y.) **100**, 158 (1959 a).
— — Thyroid hormone and lactation in the rat. Proc. Soc. exp. Biol. (N. Y.) **100**, 162 (1959 b).
— — Pituitary lactogenic hormone concentration during pregnancy in the rat. Endocrinology **66**, 96 (1960).
GUILLEMIN, R.: Personal communication (1967).
— HEARN, W. R., CHEEK, W. R., HOUSHOLDER, D. E.: Control of cortico-trophin release: further studies with *in vitro* methods. Endocrinology **60**, 488 (1957).
— ROSENBERG, B.: Humoral hypothalamic control of anterior pituitary: A study with combined tissue cultures. Endocrinology **57**, 599 (1955).
— SAKIZ, E.: Quantitative study of the response to LH after hypophys-ectomy in the ovarian ascorbic acid depletion test. Endocrinology **72**, 813 (1963).
GUTH, P. S., SPIRTES, M. A.: The phenothiazine tranquilizers: biochemical and biophysical actions. Int. Rev. Neurobiol. **7**, 231 (1964).

HAASE, H. J., JANSSEN, P. A. J.: The action of neuroleptic drugs. Chicago: Yearbook Medical Publ. 1965.
HAHN, D. W., TURNER, C. W.: Effects of ovariectomy-hypophysectomy and adrenalectomy-ovariectomy-hypophysectomy on feed intake and mammary gland growth as measured by DNA. Proc. Soc. exp. Biol. (N. Y.) **126,** 476 (1967).
HAMBERGER, L., AHRÉN, K.: Influence of the adrenal cortex on growth processes in the rat mammary gland. J. Endocr. **30,** 171 (1964).
HARRIS, G. W.: Neural control of the pituitary gland. London: Arnold 1955.
— DONOVAN, B. T., Eds.: The pituitary gland. Vol. III. London: Butterworths 1966.
HEITZMAN, R. J.: The hormonal induction of glucose-6-phosphate metabolizing enzymes in the mammary gland of the rabbit. J. Endocr. **41,** XVI (1968).
HERLYN, U., GELLER, H. F., VON BERSWORDT-WALLRABE, I., VON BERSWORDT-WALLRABE, R.: Pituitary lactogenic hormone release during onset of pseudopregnancy in intact rats. Acta endocr. (Kbh.) **48,** 220 (1965).
— JANTZEN, K., FLASKAMP, D., HOFFMANN, H., VON BERSWORDT-WALLRABE, I.: A modification of the pigeon crop sac assay for lactotrophic hormone determinations by means of the addition of prednisolone. Acta endocr. (Kbh.) **60,** 555 (1969).
HERMAN, H. A., GRAHAM, W. R., JR., TURNER, C. W.: The effect of thyroid and thyroxine on milk secretion in dairy cattle. Res. Bull. Mo. agric. Exp. Stn. No. **275** (1938).
HESSELTINE, H. C., BUSTAMANTE, J., NAVORI, C. A.: Limited influence of diethylstilbestrol on lactation. Amer. J. Obstet. Gynec. **69,** 686 (1955).
HILLIARD, J., SAWYER, C. H.: Effect of prolactin on steroidogenesis and cholesterol storage in the rabbit ovary. In: Excerpta medica, international congress series no. 111, second international congress on hormonal steroids, Milan, Italy, May 23—28, 1966, p. 195. Amsterdam: Excerpta medica foundation 1966.
— SPIES, H. G., LUCAS, L., SAWYER, C. H.: Effect of prolactin on progestin release and cholesterol storage by rabbit ovarian interstitium. Endocrinology **82,** 122 (1968).
HISAW, F. L.: The placental gonadotrophin and luteal function in monkeys (Macaca mulatta). Yale J. Biol. Med. **17,** 119 (1944).
HOLLISTER, L. E.: Adverse reactions to phenothiazines. J. Amer. med. Ass. **189,** 311 (1964).
HOOPER, J. H., WELCH, V. C., POINT, P., SHACKELFORD, R. T.: Abnormal lactation associated with tranquilizing drug therapy. J. Amer. med. Ass. **178,** 506 (1961).
HORNYKIEWICZ, O.: Dopamine (3-hydroxytyramine) and brain function. Pharmacol. Rev. **18,** 925 (1966).
HOTOVY, R., KAPFF-WALTER, J.: Die pharmakologischen Eigenschaften des Perphenazinsulfoxyds. Arzneimittel-Forsch. **10,** 638 (1960).

HUANG, W. Y., PEARLMAN, J. W. H.: The corpus luteum and steroid hormone formation. Studies on luteinized rat ovarian tissue *in vitro*. J. biol. Chem. **237**, 1060 (1962).
HUNT, V. C., BUDD, J. W.: Gynecomastia associated with interstitial cell tumor of the testicle. J. Urol. (Baltimore) **42**, 1242 (1939).
IMURA, H., SPARKS, L. L., GRODSKY, G. M., FORSHAM, P. H.: Immunologic studies of adrenocorticotropic hormone (ACTH): Dissociation of biologic and immunologic activities. J. clin. Endocr. **25**, 1361 (1965).
JACOBS, E. C.: Gynecomastia following severe starvation. Ann. intern. Med. **28**, 792 (1948).
JACOBSOHN, D.: Mammary gland growth in relation to hormones with metabolic actions. Proc. roy. Soc. B. **149**, 325 (1958).
— Thyroxin and the reaction of the mammary glands to ovarian steroids in hypophysectomized rats. J. Physiol. (Lond.) **148**, 10 P (1959).
— Hormonal regulation of mammary gland growth. In: Milk, the mammary gland and its secretion. Vol. I, p. 127. Eds.: S. K. KON and A. T. COWIE. New York-London: Academic Press 1961.
— Modification by estrogens of the reaction of the rat's mammary gland to androgens. Acta endocr. (Kbh.) **41**, 88 (1962 a).
— Interactions of estrogens and androgens on the mammary gland and growth of other tissues in hypophysectomized rats treated with insulin, cortisone and thyroxine. Acta endocr. (Kbh.) **41**, 287 (1962 b).
JANSSEN, P. A. J.: Haloperidol and related butyrophenones. In: Psychopharmacological agents. Vol. II, p. 199. Ed.: M. GORDON. New York-London: Academic Press 1967 a.
— The pharmacology of haloperidol. Int. J. Neuropsychiat. **3**, Suppl. 1, 10 (1967 b).
— NIEMECEERS, C. J. E., SCHELLEKENS, K. H. L.: Is it possible to predict the clinical effects of neuroleptic drugs (major tranquilizers) from animal data? Arzneimittel-Forsch. **15**, 1196 (1965).
— VAN DER WESTERINGH, C., JAGENEAU, A. H. M., DEMOEN, P. J. A., HERMANS, B. K. F., VAN DAELE, G. H. P., SCHELLEKENS, K. H. L., VAN DER EYCKEN, C. A. M., NIEMEGEERS, C. J. E.: Chemistry and pharmacology of CNS depressants related to 4-(4-hydroxy-4-phenylpiperidino) butyrophenone. Part I Synthesis and screening data in mice. J. med. pharm. Chem. **1**, 281 (1959).
JOHNSON, R. M., MEITES, J.: Effects of cortisone, hydrocortisone, and ACTH on mammary growth and pituitary prolactin content of rats. Proc. Soc. exp. Biol. (N. Y.) **89**, 455 (1955).
— — Effects of suckling, pitocin and cervical stimulation on release of prolactin in the parturient rat. Amer. J. med. Sci. **16**, 1045 (1957).
— — Effects of cortisone acetate on milk production and mammary involution in parturient rats. Endocrinology **63**, 290 (1958).
JONGH, S. E. DE: Corpus luteum und Laktation. Acta brev. neerl. Physiol. **2**, 199 (1932).

JOSIMOVICH, J. B.: Maintenance of pseudopregnancy in the rat by synergism between human placental lactogen and chorionic gonadotrophin. Endocrinology **83**, 530 (1968).
— MACLAREN, J. A.: Presence in the human placenta and term serum of a highly lactogenic substance immunologically related to pituitary growth hormone. Endocrinology **71**, 209 (1962).
KANEMATSU, S., HILLIARD, J., SAWYER, C. H.: Effect of hypothalamic lesions on pituitary prolactin content in the rabbit. Endocrinology **73**, 345 (1963 a).
— — — Effects of reserpine on pituitary prolactin content and its hypothalamic site of action in the rabbit. Acta endocr. (Kbh.) **44**, 467 (1963 b).
— SAWYER, C. H.: Effects of intrahypothalamic implants of reserpine on lactation and pituitary prolactin content in the rabbit. Proc. Soc. exp. Biol. (N. Y.) **113**, 967 (1963 a).
— — Effects of intrahypothalamic and intrahypophysial estrogen implants on pituitary prolactin and lactation in the rabbit. Endocrinology **72**, 243 (1963 b).
KAPLAN, S. L., GRUMBACH, M. M.: Immunoassay of human chorionic "Growth Hormone-Prolactin" in serum and urine. Science **147**, 751 (1965 a).
— — Serum chorionic "Growth Hormone-Prolactin" and serum pituitary growth hormone in mother and fetus at term. J. clin. Endocr. **25**, 1370 (1965 b).
KATZ, S., MOLITCH, M., MCCANN, S. M.: Feedback of hypothalamic growth hormone (GH) implants upon the anterior pituitary (AP). Program of the 49th Meeting of the Endocrine Society, p. 23, 1967.
KHAN, M. Y., BERNSTORE, E. C.: Effect of chlorpromazine and reserpine upon pituitary function. Exp. Med. Surg. **22**, 363 (1964).
KHAZAN, N., BEN-DAVID, M., MISHKINSKY, J., KHAZEN, K., SULMAN, F. G.: Dissociation between mammotropic and sedative effects of non-hormonal hypothalamic tranquilizers. Arch. int. Pharmacodyn. **164**, 258 (1966).
— — SULMAN, F. G.: ADH-like effect of tranquilizers in amphibians. Proc. Soc. exp. Biol. (N. Y.) **112**, 490 (1963).
— PRIMO, C., DANON, A., ASSAEL, M., SULMAN, F. G., WINNIK, H. Z.: The mammotropic effect of tranquilizing drugs. Arch. int. Pharmacodyn. **136**, 291 (1962).
— SULMAN, F. G.: Melanophore-dispersing activity of reserpine in Rana frogs. Proc. Soc. exp. Biol. (N. Y.) **107**, 282 (1961).
— — WINNIK, H. Z.: Effect of reserpine on pituitary-gonadal axis. Proc. Soc. exp. Biol. (N. Y.) **105**, 201 (1960).
— — — Activity of pituitary-adrenal cortex axis during acute and chronic reserpine treatment. Proc. Soc. exp. Biol. (N. Y.) **106**, 579 (1961).
KHAZEN, K., MISHKINSKY, J., BEN-DAVID, M., SULMAN, F. G.: Lactogenic effect of phenothiazine-like drugs. Arch. int. Pharmacodyn. **174**, 428 (1968 a).

KHAZEN, K., MISHKINSKY, J., BEN-DAVID, M., SULMAN, F. G.: Structure-activity-relationship of mammotropic phenothiazine derivatives. Arch. int. Pharmacodyn. **171**, 251 (1968 b).

KILPATRICK, R., ARMSTRONG, D. T., GREEP, R. O.: Maintenance of the corpus luteum by gonadotrophins in the hypophysectomized rabbit. Endocrinology **74**, 453 (1964).

KINROSS-WRIGHT, J., CHARALAMPOUS, K. D.: A controlled study of a very long-acting phenothiazine preparation. Int. J. Neuropsychiat. **1**, 66 (1965).

KITAY, J. I., ALTSCHULE, M. D.: The pineal gland; a review of the physiologic literature. Cambridge (Mass.): Harvard University Press 1954.

KLATSKIN, G., SALTER, W. T., HUMM, F. D.: Gynecomastia due to malnutrition. 1. Clinical studies. Amer. J. med. Sci. **213**, 19 (1947).

KLEIN, J. J., SEGAL, R. L., PICHEL-WARNER, R. R.: Galactorrhea due to imipramine; Report of a case. New Engl. J. Med. **271**, 510 (1964).

KLINEFELTER, H. F., JR., REIFENSTEIN, E. C., JR., ALBRIGHT, F.: Syndrome characterized by gynecomastia, aspermatogenesis without A-leydigism, and increased excretion of follicle-stimulating hormone. J. clin. Endocr. **2**, 615 (1942).

KOANG, N. K., HU, T. C., TCHEN, C. L., CHU, T. H.: Endocrine function during treatment of pulmonary tuberculosis with INH. China med. J. **75**, 100 (1957).

KON, S. K., COWIE, A. T. (Eds.): Milk; the mammary gland and its secretion. I—II. New York-London: Academic Press 1961.

KOVACIC, N.: Luteotrophic activity of human growth hormone (Raben). Nature (Lond.) **195**, 1210 (1962 a).

— Prolongation of diestrus in the mouse as a quantitative assay of luteotrophic activity of prolactin. J. Endocr. **24**, 227 (1962 b).

— The deciduoma assay: a method for measuring prolactin. J. Endocr. **28**, 45 (1963).

— Mouse hyperaemic corpora lutea assay of prolactin. J. Reprod. Fertil. **15**, 259 (1968).

KRAGT, C. L., MEITES, J.: Stimulation of pigeon pituitary prolactin release by pigeon hypothalamic extract *in vitro*. Endocrinology **76**, 1169 (1965).

— — Dose-response relationships between hypothalamic PIF and prolactin release by rat pituitary tissue *in vitro*. Endocrinology **80**, 1170 (1967).

KRONFELD, D. S., MAYER, G. P., ROBERTSON, J. McD., RAGGI, F.: Depression of milk secretion during insulin administration. J. Dairy Sci. **46**, 559 (1963).

KUMARESAN, P., ANDERSON, R. R., TURNER, C. W.: Effect of corticosterone upon mammary gland growth of pregnant rats. Endocrinology **81**, 658 (1967).

KUPPERMAN, H. S.: Human Endocrinology. Vol. II, p. 453. Philadelphia: Davis 1963 a.

— Human Endocrinology. Vol. II, p. 604 and 644. Philadelphia: Davis 1963 b.

KUPPERMAN, H. S., FRIED, P., HAIR, L. Q.: The control of menorrhagia by prolactin. Amer. J. Obstet. Gynec. **48**, 228 (1944).

KURAMITSU, C., LOEB, L.: The effect of suckling and castration on the lactating mammary gland in rat and guinea pig. Amer. J. Physiol. **56**, 40 (1921).

KURCZ, M.: Studies on the neuro-humoral regulation of lactation. In: Symp. on Reproduction (Aug.—Sep. 1967, Pecs, Hungary), p. 175. Budapest: Akademai Kiado 1967.

KUROSHIMA, A., ARIMURA, A., BOWERS, C. Y., SCHALLY, A. V.: Inhibition by pig hypothalamic extracts of depletion of pituitary prolactin rats following cervical stimulation. Endocrinology **78**, 216 (1966).

KWA, H. G.: Personal communication (1968).

— VERHOFSTAD, F.: Radioimmunoassay of rat prolactin. Biochim. biophys. Acta (Amst.) **133**, 186 (1967).

LAHR, E. L., RIDDLE, O.: Temporary suppression of estrous cycles in the rat by prolactin. Proc. Soc. exp. Biol. (N. Y.) **34**, 880 (1936).

LARON, Z.: The immunological properties and biological activities of various growth hormones. Proc. Sec. Int. Congress Endocrin., p. 1195. Amsterdam: Excpt. Med. Publ. 1965.

— ASSA, S.: Immunological investigations of human prolactin preparation. In: Excerpta medica, international congress, Series no. 83, proceeding of the second international congress of endocrinology, London, August 1964, p. 1259. Amsterdam: Excerpta medica foundation 1964.

— — APOSTOLAKIS, M.: Immunological studies of a human prolactin preparation. Acta endocr. (Kbh.) Suppl. **100**, 164 (1965).

LASS, P. M., SMELSER, J., KURZROK, R.: Studies relating to time of human ovulation III. During lactation. Endocrinology **23**, 39 (1938).

LAUNCHBURY, A. P.: "Nuclear-modified" phenothiazines. J. Hosp. Pharm. **21**, 267 (1964).

LEATHEM, J. H.: Extragonadal factors in reproduction: In: Recent progress in the endocrinology of reproduction, p. 179. Ed.: C. W. LLOYD. New York: Academic Press 1959.

LEECH, F. B., BAILEY, G. L.: The effect on the health of lactating cows of treatment with galactopoietic doses of thyroxine or iodinated casein. J. agric. Sci. **43**, 236 (1953).

LEGORI, M., MOTTA, M., ZANISL, M., MARTINI, L.: "Short feedback loops" in ACTH control. Acta endocr. (Kbh.) Suppl. **100**, 155 (1965).

LEONARD, S. L., REECE, R. P.: The relation of the thyroid to mammary gland growth in the rat. Endocrinology **28**, 65 (1941).

LEVY, R. D., SAMPLINER, J.: Prolactin, immunologic evidence of species specificity. Proc. Soc. exp. Biol. (N. Y.) **109**, 672 (1962).

LEWIN, M. L.: Gynecomastia: The hypertrophy of the male breast. J. clin. Endocr. **1**, 511 (1941).

LEWINN, E. B.: Gynecomastia during digitalis therapy; report of eight additional cases with liver-function studies. New Engl. J. Med. **248**, 316 (1953).

References

Li, C. H.: Biochemistry of prolactin. In: Milk, the mammary gland and its secretion. Vol. I, p. 205. Eds.: S. K. Kon and A. T. Cowie. New York-London: Academic Press 1961.
— In: Discussion on prolactin (panel). Ciba Foundation Colloq. Endocr. **14,** 360 (1962).
Liggins, G. C., Ibbertson, H. K.: A successful quintuplet pregnancy following treatment with human pituitary gonadotrophin. Lancet **1966 I,** 114.
Lissen, A. W., Parkes, M. W.: The relation between sedation and body temperature in the mouse. Brit. J. Pharmacol. **12,** 245 (1957).
Long, J. A., Evans, H. M.: Estrous cycle in the rat and its associated phenomena. Mem. Univ. Calif. **6,** 1 (1922).
Lyons, W. R.: Lobulo-alveolar mammary growth in the rat. Colloq. int. center nat. rech. sci. **32,** 29 (1951).
— Hormonal synergism in mammary growth. Proc. roy. Soc. B. **149,** 303 (1958).
— Li, C. H., Ahmad, N., Rice-Wray, E.: Mammotrophic effects of human hypophysial growth hormone preparations in animals and man. In: Excerpta medica, international congress series no. 158, Growth Hormone, proceedings of the first international symposium, Milan, Italy, September, 1967, p. 349. Amsterdam: Excerpta medica foundation 1968.
— — Cole, R. D., Johnson, R. E.: Some of the hormones required by the mammary gland in its development and function. J. clin. Endocr. **13,** 836 (1953).
— Dixon, J. S.: The physiology and chemistry of the mammotrophic hormone. In: The Pituitary Gland. London: Butterworths 1966.
— — Johnson, R. G.: The enhancing effect of somatotropin on the mammary growth induced in rats with estrin, progestin and mammotropin. J. clin. Endocr. **12,** 937 (1952).
— — — The hormonal control of mammary growth and lactation. Recent Progr. Hormone Res. **14,** 219 (1958).
— — — Biologic activities of human hypophysial mammotrophin. Program of the Forty-third Meeting of the Endocrine Society. New York 1961, p. 4.
— Page, E.: Detection of mammotropin in the urine of lactating women. Proc. Soc. exp. Biol. (N. Y.) **32,** 1049 (1935).
Maanen, J. N. van, Smelik, P. G.: Induction of pseudopregnancy in rats following local depletion of monoamines in the median eminence of the hypothalamus. Neuroendocrinology **3,** 177 (1968).
Maas, J. M.: Amenorrhea-galactorrhea syndrome before, during, and after pregnancy. Fertil. and Steril. **18,** 857 (1967).
McCalister, A., Welbourn, R. B.: Stimulation of mammary cancer by prolactin and the clinical response to hypophysectomy. Brit. med. J. **1962 I,** 1669.
McCann, S. M., Mack, R., Gale, C.: The possible role of oxytocin in stimulating the release of prolactin. Endocrinology **64,** 870 (1959).

MacDonald, D., O'Driscoll, K.: Suppression of lactation, a double-blind trial. Lancet **1965 II**, 623.

McGarry, E. E., Beck, J. C.: The hyperglycaemic and hypoglycaemic effects of human growth hormone. 4ème Congrès de la Fédération internationale du Diabète, Geneva 1961. Comptes rendus. Vol. I. Genève: Editions médecine et hygiène 1961, p. 84.

— Rubinstein, D., Beck, J. C.: Growth hormones and prolactins; biochemical immunological and physiological similarities and differences. Ann. N. Y. Acad. Sci. **148**, 559 (1968).

McQueen-Williams, M.: Decreased mammotropin in pituitaries of thyroidectomized (maternalized) male rats. Proc. Soc. exp. Biol. (N. Y.) **33**, 406 (1935).

Maickel, R. P., Stern, D. N., Brodie, B. B.: In: Biochemical and neurophysiological correlation centrally acting drugs; Proceedings 2nd international pharmacological meeting, Prague, 1963. Vol. 2. Eds.: P. Paoleth and E. Trabucchi. London: Pergamon 1963, p. 225.

Malven, P. V., Hansel, W., Sawyer, C. H.: A mechanism antagonizing the luteotrophic action of exogenous prolactin in rats. J. Reprod. Fertil. **13**, 205 (1967).

— Sawyer, C. H.: A luteolytic action of prolactin in hypophysectomized rats. Endocrinology **79**, 268 (1966).

Mandl, A. M.: Corpora lutea in senile virgin laboratory rats. J. Endocr. **18**, 438 (1959).

Mann, N. M.: Gynecomastia during therapy with spironolactone. J. Amer. med. Ass. **184**, 778 (1963).

Margolis, I. B., Gross, C. G.: Gynecomastia during phenothiazine therapy. J. Amer. med. Ass. **199**, 942 (1967).

Marshall, W. K., Leiberman, D. M.: A rare complication of chlorpromazine treatment. Lancet **1956**, 270, 162.

Martini, L., Ganong, W. F.: Neuroendocrinology. Vol. I. New York-London: Academic Press 1966.

Mason, N. R., Marsh, J. M., Savard, K.: An action of gonadotropin *in vitro*. J. biol. Chem. **237**, 1801 (1962).

Matthies, D. L.: Studies of the luteotropic and mammotropic factor found in trophoblast and maternal peripheral blood of the rat at midpregnancy. Anat. Rec. **159**, 55 (1967).

Mayer, D. T.: Rat's milk and the stomach contents of suckling rats. J. Nutr. **10**, 343 (1935).

Mayer, G.: Le prolactine, facteur lutéotrophique. Arch. Sci. physiol. **5**, 247 (1951).

Mazer, C., Goldstein, L.: Clinical endocrinology of the female. Philadelphia: Saunders 1932.

Meites, J.: Le déclenchement de la lactation au moment de la parturition. Ann. Endocr. (Paris) **17**, 519 (1956).

— Induction of lactation in rabbits with reserpine. Proc. Soc. exp. Biol. (N. Y.) **96**, 728 (1957 a).

MEITES, J.: Effects of growth hormone on lactation and body growth of parturient rats. Proc. Soc. exp. Biol. (N. Y.) **96**, 730 (1957 b).
— Mammary growth and lactation. In: Reproduction in domestic animals, Vol. I. Eds.: H. H. COLE and P. T. CUPPS. New York-London: Academic Press 1959 a, p. 539.
— Hormonal prolongation of lactation for 75 days after litter removal in parturient rats. Fed. Proc. **18**, 103 (1959 b).
— Farm animals; hormonal induction of lactation and galactopoiesis. In: Milk; the mammary gland and its secretion. Vol. I, p. 321. Eds.: S. K. KON and A. T. COWIE. New York-London: Academic Press 1961.
— In: Pharmacological control of release of hormones including anti-diabetic drugs; Proceedings first international pharmacological meeting, Stockholm 1961. Vol. I, p. 151. Ed.: R. GUILLEMIN. London: Pergamon 1961 a.
— Pharmacological control of prolactin secretion and lactation. Biochem. Pharmacol. **8**, 27 (1961 b).
— Neurohumoral and pharmacological control of prolactin secretion. Scientific group on the physiology of lactation. Geneva: World Health Organization 1963.
— Maintenance of the mammary lobulo-alveolar system in rats after adreno-orchidectomy by prolactin and growth hormone. Endocrinology **76**, 1220 (1965).
— Control of mammary growth and lactation. In: Neuroendocrinology, Vol. I. Eds.: L. MARTINI and W. F. GANONG. New York-London: Academic Press 1966, p. 669.
— HOPKINS, T. E.: Mechanism of action oxytocin in retarding mammary involution: study in hypophysectomized rats. J. Endocr. **22**, 207 (1961).
— — TALWALKER, P. K.: Induction of lactation in pregnant rabbits with prolactin, cortisol acetate or both. Endocrinology **73**, 261 (1963).
— KAHN, R. H., NICOLL, C. S.: Prolactin production by rat pituitary *in vitro*. Proc. Soc. exp. Biol. (N. Y.) **108**, 440 (1961).
— KRAGT, C. L.: Effects of a pituitary homotransplant and thyroxine on body and mammary growth in immature hypophysectomized rats. Endocrinology **75**, 565 (1964).
— NICOLL, C. S.: Hormonal prolongation of lactation for 75 days after litter withdrawal in portpartum rats. Endocrinology **65**, 572 (1959).
— — Adenohypophysis: Prolactin. Ann. Rev. Physiol. **28**, 57 (1966).
— — TALWALKER, P. K.: Effects of reserpine and serotonin on milk secretion and mammary growth in the rat. Proc. Soc. exp. Biol. (N. Y.) **101**, 563 (1959).
— — — Local action of oxytocin on mammary glands of postpartum rats after litter removal. Proc. Soc. exp. Biol. (N. Y.) **103**, 118 (1960).
— — — The central nervous system and the secretion and release of prolactin. Advances in neuroendocrinology, p. 238. Ed.: A. V. NALBANDOV. Urbana (Ill.): Univ. of Illinois Press 1963.

Meites, J., Sgouris, J. T.: Can the ovarian hormones inhibit the mammary response to prolactin? Endocrinology **53**, 17 (1953).
— — Effects of altering the balance between prolactin and ovarian hormones on initiation of lactation in rabbits. Endocrinology **55**, 530 (1954).
— Shelesnyak, M. C.: Effects of prolactin on duration of pregnancy, viability of young and lactation in rats. Proc. Soc. exp. Biol. (N. Y.) **94**, 746 (1957).
— Talwalker, P. K., Nicoll, C. S.: Initiation of lactation in rats with hypothalamic or cerebral tissue. Proc. Soc. exp. Biol. (N. Y.) **103**, 298 (1960 a).
— — — Failure of oxytocin to initiate mammary secretion in rabbits or rats. Proc. Soc. exp. Biol. (N. Y.) **105**, 467 (1960 b).
— Turner, C. W.: Extraction and assay of lactogenic hormone in postpartum urine. J. clin. Endocr. **1**, 918 (1941).
— — Studies concerning the mechanism controlling the initiation of lactation at parturition. I. Can estrogen suppress the lactogenic hormone of the pituitary? Endocrinology **30**, 711 (1942 a).
— — Studies concerning the mechanism controlling the initiation of lactation at parturition. II. Why lactation is not initiated during pregnancy. Endocrinology **30**, 719 (1942 b).
— — Studies concerning the mechanism controlling the initiation of lactation at parturition. III. Can estrogen account for the precipitous increase in the lactogen content of the pituitary following parturition? Endocrinology **30**, 726 (1942 c).
— — Studies concerning mechanism controlling initiation of lactation at parturition. IV. Influence of suckling on lactogen content of pituitary of postpartum rabbits. Endocrinology **31**, 340 (1942 d).
— — Effect of thiouracil and estrogen on lactogenic hormone and weight of pituitaries of rats. Proc. Soc. exp. Biol. (N. Y.) **64**, 488 (1947).
— — Studies concerning the induction and maintenance of lactation. I. The mechanism controlling the initiation of lactation at parturition. Res. Bull. Mi. agric. Exp. Stn. No. **415** (1948).
— — Lactogenic hormone. In: Hormone assay, p. 237. Ed.: C. W. Emmens. New York: Academic Press 1950.
Mena, F., Beyer, C.: Induction of milk secretion in the rabbit by lesions in the temporal lobe. Endocrinology **83**, 618 (1968).
Mercier-Parot, L.: Troubles du développement post-natal du rat après administration de cortisone à la mère gestante ou allaitante. C. R. Acad. Sci. (Paris) **240**, 2259 (1955).
Mess, B., Martini, L.: The central nervous system and the secretion of anterior pituitary trophic hormones. Recent Advances in Endocrinology. VIIIth edition, p. 1. Ed.: V. H. T. James. London: Churchill 1968.
Meyler, L. (Ed.): Side effects of drugs; adverse reactions as reported in the medical literature of the world 1963—1965. Vol. V. Amsterdam: Excerpta medica foundation 1966.

MINAGUCHI, H., CLEMENS, J. A., MEITES, J.: Changes in pituitary prolactin levels in rats from weaning to adulthood. Endocrinology **82**, 555 (1968).
— MEITES, J.: Effects of norethynodrel-mestranol combination (Enovid) on hypothalamic and pituitary hormones in rats. Endocrinology **81**, 826 (1967).
MISHKINSKY, J., DIKSTEIN, S., BEN-DAVID, M., AZEROUAL, J., SULMAN, F. G.: A sensitive *in vitro* method for prolactin determination. Proc. Soc. exp. Biol. (N. Y.) **125**, 360 (1967).
— KHAZEN, K., GIVANT, Y., SULMAN, F. G.: Mammotropic and neuroleptic effects of butyrophenones in the rat. Arch. int. Pharmacodyn. (in press) (1969).
— — SULMAN, F. G.: Prolactin-releasing activity of the hypothalamus in post-partum rats. Endocrinology **82**, 611 (1968).
— — — Mammotropic effect of fluphenazine enanthate in the rat. Neuroendocrinology **4**, 321 (1969).
— LAJTOS, Z. K., SULMAN, F. G.: Initiation of lactation by hypothalamic implantation of perphenazine. Endocrinology **78**, 919 (1966).
— NIR, I., LAJTOS, Z. K., SULMAN, F. G.: Mammary development in pinealectomized rat. J. Endocr. **36**, 215 (1966).
— — SULMAN, F. G.: Internal feedback of prolactin in the rat. Neuroendocrinology **5**, 48 (1969).
— ZANBELMAN, L., KHAZEN, K., SULMAN, F. G.: Mammotropic and prolactin-like effects of placenta and amniotic fluid of rat and human origin. J. Endocr. **46**, 15 (1970).
MITTLER, J. C., MEITES, J.: Effects of epinephrine and acetylcholine on hypothalamic content of prolactin inhibiting factor. Proc. Soc. exp. Biol. (N. Y.) **124**, 310 (1967).
MIZUNO, H., TALWALKER, P. K., MEITES, J.: Influence of hormones on tumor growth and plasma prolactin levels in rats bearing pituitary "mammotropic" tumor. Cancer Res. **24**, 1433 (1964).
MOLINA, C., ABERKANE, B., CONQUY-DOUARD, T. A. M.: Les gynécomasties chez les tuberculeux pulmonaires; à propos de cinq observations. Maroc. méd. **36**, 635 (1957).
MOON, R. C.: Mammary growth in rats treated with somatotropin during pregnancy and/or lactation. Proc. Soc. exp. Biol. (N. Y.) **118**, 181 (1965).
— Personal communication (1966).
MOORE, W. W., NALBANDOV, A. V.: Maintenance of corpora lutea in sheep with lactogenic hormone. J. Endocr. **13**, 18 (1955).
MUNFORD, R. E.: The effect of cortisol acetate on estrone-induced mammary gland growth in immature ovariectomized albino mice. J. Endocr. **16**, 72 (1957).
MURRAY, M. R.: Nervous tissues *in vitro*. Cells and tissues in culture. Vol. II, p. 373. Ed.: E. N. WILLMER. New York-London: Academic Press 1965.

NAGATA, G.: Influences of chlorpromazine on the pituitary function especially corticotropin and gonadotrophin secretion. Rep. Shionogi Res. Lab. **7,** 71 (1957).

NANDI, S.: Endocrine control of mammary gland development and function in the C3H/Hc CRGL mouse. J. nat. Cancer Inst. **21,** 1039 (1958 a).

— Role of somatotropin in mammogenesis and lactogenesis in C3H/Hc CRGL mice. Science **128,** 772 (1958 b).

— Hormonal control of mammogenesis and lactogenesis in the C_3H/He CRGL mouse. Univ. Calif. Publ. Zool. **65,** 4 (1959).

— BERN, H. A.: The hormones responsible for lactogenesis in BALB/c CRGL mice. Gen. comp. Endocr. **1,** 195 (1961).

NELSON, W. O.: Endocrine control of the mammary gland. Physiol. Rev. **16,** 488 (1936).

— GAUNT, R., SCHWEIZER, M.: Effects of adrenal cortical compounds on lactation. Endocrinology **33,** 325 (1943).

NICOLL, C. S.: Neural regulation of adenohypophysial prolactin secretion in tetrapods: indications from *in vitro* studies. J. exp. Zool. **158,** 203 (1965).

— Bio-assay of prolactin. Analysis of the pigeon crop-sac response to local prolactin injection by an objective and quantitative method. Endocrinology **80,** 641 (1967).

— Electrophoretic comparison of the tetrapod prolactins. Excerpta medica, international congress series no. 157, third international congress of endocrinology, Mexiko, D. F., June 30—July 5, 1968, p. 77. Amsterdam: Excerpta medica foundation 1968.

— Bio-assay of prolactin. Acta endocr. (Kbh.) **60,** 91 (1969).

— Assay of hypothalamic factors which regulate prolactin secretion. In: Hypophysiotropic hormones of the hypothalamus: assay and chemistry, p. 115. Ed.: J. MEITES. Baltimore: Williams & Wilkins Publ. 1970.

— BERN, H. A.: "Prolactin" and the pituitary glands of fishes. Gen. comp. Endocr. **4,** 457 (1964).

— MEITES, J.: Prolongation of lactation in the rat by litter replacement. Proc. Soc. exp. Biol. (N. Y.) **101,** 81 (1959).

— — Estrogen stimulation of prolactin production by rat adenohypophysis *in vitro*. Endocrinology **70,** 272 (1962 a).

— — Failure of neurohypophysical hormones to influence prolactin secretion *in vitro*. Endocrinology **70,** 927 (1962 b).

— — Prolactin secretion *in vitro*: comparative aspects. Nature (Lond.) **195,** 606 (1962 c).

— — Prolactin secretion *in vitro*: effects of thyroid hormones and insulin. Endocrinology **72,** 544 (1963).

— — Prolactin secretion *in vitro*: effects of gonadal and adrenal cortical steroids. Proc. Soc. exp. Biol. (N. Y.) **117,** 579 (1964).

— TALWALKER, P. K., MEITES, J.: Initiation of lactation in rats by nonspecific stresses. Amer. J. Physiol. **198,** 1103 (1960).

николови́ч-Winer, M. B., Everett, J. W.: Comparative study of luteotropin secretion by hypophysial autotransplants in the rat. Effects of site and stages of the estrus cycle. Endocrinology **62**, 522 (1958).

Nir, I., Mishkinsky, J., Eshchar, N., Sulman, F. G.: Effect of pinealectomy on milk yield in rats. J. Endocr. **42**, 161 (1968).

Niswender, G. D., Chen, C. L., Midgley, A. R., Jr., Meites, J., Ellis, S.: Radioimmunoassay for rat prolactin. Proc. Soc. exp. Biol. (N. Y.) **130**, 793 (1969).

Palladino, A., Butcher, R. L., Fugo, N. W.: An additional effect of yohimbine-HCl. Proc. Soc. exp. Biol. (N. Y.) **127**, 408 (1968).

Paloyan, E., Kolar, J. B., Ernst, K., Paloyan, D., Harper, P. V.: The effect of glucagon on serum calcium. Fed. Proc. **24**, 323 (1965).

— Paloyan, D., Harper, P. V.: Glucagon-induced hypocalcemia. Metabolism **16**, 35 (1967).

Parkes, C. M., Brown, G. W., Monck, E. M.: The general practitioner and the schizophrenic patient. Brit. med. J. **1962 I**, 972.

Pasteels, J. L.: Premiers résultats de culture combinée in vitro d'hypophysis et d'hypothalamus, dans le but d'en apprécier la sécrétion de prolactin. C. R. Acad. Sci. (Paris) **253**, 3074 (1961 a).

— Sécrétion de prolactine par l'hypophyse en culture de tissus. C. R. Acad. Sci. (Paris) **253**, 2140 (1961 b).

— Administration d'extraits hypothalamiques à l'hypophyse de rat in vitro, dans le but d'en contrôler la sécrétion de prolactine. C. R. Acad. Sci. (Paris) **254**, 2664 (1962).

— Recherches morphologiques et expérimentales sur la sécrétion de prolactine. Arch. Biol. (Liège) **74**, 440 (1963).

— Hormone de croissance et prolactine dans l'hypophyse humaine. Ann. Endocr. (Paris) **28**, 117 (1967).

— Brauman, H., Brauman, J.: Étude comparé de la sécretion d'hormone somatotrope par l'hypophyse humaine in vitro, et son activité lactogénique. C. R. Acad. Sc. (Paris) **256**, 2031 (1963).

Pavel, S., Petrescu, S.: Inhibition of gonadotrophin by a highly purified pineal peptide and by synthetic arginine vasotocin. Nature (Lond.) **212**, 1054 (1966).

Peake, G. T., McKeel, D. W., Jarett, L., Daughaday, W. H.: Ultrastructural histologic and hormonal characterization of a prolactin-rich human pituitary tumor. Endocrinology **29**, 1383 (1969).

Petersen, W. E.: Lactation. Physiol. Rev. **24**, 340 (1944).

Peterson, P. V., Nielson, I. M.: Thioxanthene derivatives. Psychopharmacological agents. Vol. I, p. 301. Ed.: M. Gordon. New York-London: Academic Press 1964.

Pettinger, W. A., Horwitz, D., Sjoerdsma, A.: Lactation due to methyldopa. Brit. med. J. **1963 I**, 1460.

Platt, S. S., Schulz, R. Z., Kunstadter, R. H.: Hypertrophy of male breast associated with recovery from starvation. Bull. U.S. Army med. Dep. **7**, 403 (1946).

PLATT, R., SEARS, H. T. N.: Reserpine in severe hypertension. Lancet **1956 I,** 401.
PLUMER, A. J., EARL, A., SCHNEIDER, J. A., TRAPOLD, J., BARRETT, W.: Pharmacology of *Rauwolfia* alkaloids, including reserpine. Ann. N. Y. Acad. Sci. **59,** 8 (1954).
POLISHUK, W. Z., KULCSAR, S.: Effects of chlorpromazine on pituitary function. J. clin. Endocr. **16,** 292 (1956).
POPOVICH, S.: Prolactin releasing factor. Dokl. Akad. Nauk SSSR, Dtd. Biol. **121,** 186 (1958).
PROP, F. J. A.: Effects of hormones on mouse mammary glands *in vitro*. Analysis of the factors that cause lobulo-alveolar development. Path. et Biol. **9,** 640 (1961 a).
— Hormones and mammary tissue *in vitro*. Acta physiol. pharmacol. neerl. **10,** 305 (1961 b).
— Sensitivity to prolactin of mouse mammary glands *in vitro*. Exp. Cell Res. **24,** 629 (1961 c).
PURSHOTTAM, N.: Effects of tranquilizers on induced ovulation in mice. Amer. J. Obstet. Gynec. **83,** 1405 (1962).
QUINN, D. L., EVERETT, J. W.: Delayed pseudopregnancy induced by selective hypothalamic stimulation. Endocrinology **80,** 155 (1967).
QUINLAN, J. J.: Lactation following thoracotomy. Canad. J. Surg. **11,** 60 (1968).
RACADOT, J.: Neurosécrétion et activité thyroidienne chez la chatte au cours de la gestation et de l'allaitement. Ann. Endocr. (Paris) **18,** 628 (1957).
RAJKOVITZ, K.: Localization of pituitary function in the diencephalon. In: Hypothalamic control of the anterior pituitary. Eds.: J. STENTAGOTHAI, B. FLERKO, B. MESS and B. HALASZ. Budapest: Akademiai Kaido 1962.
RALSTON, N. P., COWSERT, W. C., RAGSDALE, A. C., HERMAN, H. A., TURNER, C. W.: The yield and composition of the milk of dairy cows and goats as influenced by thyroxine. Res. Bull. Mo. agric. Exp. Stn. No. **317,** 1940.
RAMIREZ, V. D., MCCANN, S. M.: Induction of prolactin secretion by implants of estrogen into the hypothalamo-hypophysial region of female rats. Endocrinology **75,** 206 (1964).
RATNER, A., MEITES, J.: Depletion of prolactin-inhibiting activity of rat hypothalamus by estradiol or suckling stimulus. Endocrinology **75,** 377 (1964).
— TALWALKER, P. K., MEITES, J.: Effect of estrogen administration *in vivo* on prolactin release by rat pituitary *in vitro*. Proc. Soc. exp. Biol. (N. Y.) **112,** 12 (1962).
— — — Effect of reserpine on prolactin-inhibiting activity of rat hypothalamus. Endocrinology **77,** 315 (1965).
RAY, E. W., AVERILL, S. C., LYONS, W. R., JOHNSON, R. E.: Rat placental hormonal activities corresponding to those of pituitary mammotropin. Endocrinology **56,** 359 (1955).

REECE, R. P.: Lactogen content of female guinea pig pituitary. Proc. Soc. exp. Biol. (N. Y.) **42**, 54 (1939).
— TURNER, C. W.: The lactogenic and thyrotropic hormone content of the anterior lobe of the pituitary gland. Mo. agric. Exp. Sta. Res. Bull. **266** (1937).
REICHLIN, S.: Functions of the median-eminence gland. New Engl. J. Med. **275**, 600 (1966).
REINEKE, E. P.: Unpublished observations (1942) quoted by MEITES, J.: Thyroactive substances. In: Milk, the mammary gland and its secretion. Vol. I, p. 353. Eds.: S. K. KON ond A. T. COWIE. New York-London: Academic Press 1961.
REISFELD, R. A., TONG, G. L., RICKES, E. L., BRINK, N. G., STEELMAN, S. L.: Purification and characterization of sheep prolactin. J. Amer. chem. Soc. **83**, 3717 (1961).
RIDDLE, O., BATES, R. W., DYKSHORN, S. W.: The preparation, identification and assay of prolactin — a hormone of the anterior pituitary. Amer. J. Physiol. **105**, 191 (1933).
RIMOIN, D. L., HOLZMAN, B., MERIMEE, J., RABINOWITZ, D., BARNES, C., TYSON, J. E. A., MCKUSICK, V. A.: Lactation in the absence of human growth hormone. J. clin. Endocr. **28**, 1183 (1968).
ROBBOY, S. J., KAHN, R. H.: Zone electrophoresis and prolactin activity of rat adenohypophysis cultivated *in vitro*. Endocrinology **78**, 440 (1966).
ROBINSON, B.: Breast changes in the male and female with chlorpromazine or reserpine therapy. Med. J. Aust. **44**, 239 (1957).
ROSSI, G. V.: The psychotherapeutic agents. Amer. J. Pharm. **136**, 6 (1964).
ROTH, J., GLICK, S. M., CUATRECASAS, P., HOLLANDER, C. S.: Acromegaly and other disorders of growth hormone secretion. Amer. intern. Med. **66**, 760 (1967).
ROTHCHILD, I.: The corpus luteum-hypophysis relationships; the luteolytic effect of luteinizing hormone (LH) in the rat. Acta endocr. (Kbh.) **49**, 107 (1965).
SALKIN, D., DAVIS, E. W.: Lactation following thoracoplasty and pneumonectomy. J. thorac. Surg. **18**, 580 (1949).
SAR, M., MEITES, J.: Changes in pituitary prolactin release and hypothalamic PIF content during the estrus cycle of rats. Proc. Soc. exp.
SAWYER, C. H.: Induction of lactation in the rabbit with reserpine. Anat. Rec. **127**, 362 (1957).
SCHALLY, A. V., KUROSHIMA, A., ISHIDA, Y., REDDING, T. W., BOWERS, C. Y.: The presence of prolactin inhibiting factor (PIF) in extracts of beef, sheep and pig hypothalami. Proc. Soc. exp. Biol. (N. Y.) **118**, 350 (1965). Biol. (N. Y.) **125**, 1018 (1967).
— — Effects of progesterone, testosterone, and cortisol on hypothalamic prolactin-inhibiting factor and pituitary prolactin content. Proc. Soc. exp. Biol. (N. Y.) **127**, 426 (1968).

Schally, A. V., Meites, J., Bowers, C. Y., Ratner, A.: Identity of prolactin inhibiting factor (PIF) and luteinizing hormone-releasing factor (LRF). Proc. Soc. exp. Biol. (N. Y.) 117, 252 (1964).
— Steelman, S. L., Bowers, C. Y.: Effect of hypothalamic extracts on release of growth hormone *in vitro*. Proc. Soc. exp. Biol. (N. Y.) 119, 208 (1965).
Schams, D., Karg, H.: Radioimmunologische Bestimmung von Prolaktin im Blutserum vom Rind. Milchwissenschaft 24, 263 (1969).
Scharrer, E., Scharrer, B.: Neuroendocrinology. New York: Columbia Univ. Press 1963.
Schlittler, E., Plummer, A. J.: Tranquilizing drugs from Rauwolfia. In: Psychopharmacological agents. Vol. I, p. 9. Ed.: M. Gordon. New York-London: Academic Press 1964.
Sebaoun, J., Delzant, G., Peillon, F.: Les tumeurs à prolactines. Bull. Soc. med. Hôp. Paris 118, 525 (1967).
Segaloff, A., Steelman, S. L.: The human gonadotropins. Recent Progr. Hormone Res. 15, 127 (1959).
Selye, H.: Influence of the uterus on ovary and mammary gland. Proc. Soc. exp. Biol. (N. Y.) 31, 488 (1934 a).
— On the nervous control of lactation. Amer. J. Physiol. 107, 535 (1934 b).
— The effect of cortisol upon the mammary glands. Acta endocr. (Kbh.) 17, 394 (1954).
— Der derzeitige Stand der Stressforschung. In: Psyche und Hormon. Eds.: C. A. Joël, H. Meng, P. Parin, H. Selye, and F. G. Sulman. Bern: Huber 1960, p. 125.
— Collip, J. B., Thomson, D. L.: Nervous and hormonal factors in lactation. Endocrinology 18, 237 (1934).
Shaw, J. C.: The hypophysical growth hormone, nature and actions. New York: McGraw-Hill 1955, p. 486.
— Chung, A. C., Bunding, I.: The effect of pituitary growth hormone and adrenocorticotropic hormone on established lactation. Endocrinology 56, 327 (1955).
Shelesnyak, M. C.: Fall in uterine histamine associated with ovum implantation in pregnant rat. Proc. Soc. exp. Biol. (N. Y.) 100, 380 (1959).
Silver, H. K.: Juvenile hypothyroidism with precocious sexual development. J. clin. Endocr. 18, 886 (1958).
Simkin, B., Arce, R.: Prolactin activity in blood during the normal human menstrual cycle. Proc. Soc. exp. Biol. (N. Y.) 113, 485 (1963).
— Goodart, D.: Preliminary observations on prolactin activity in human blood. J. clin. Endocr. 20, 1095 (1960).
Sluyser, M., Li, C. H.: Studies of pituitary lactogenic hormone. XXIII. Isolation of the monomer of ovine prolactin. Arch. Biochem. 104, 50 (1964).
Smith, T. C., Braverman, L. E.: The action of desoxycorticosterone acetate on the mammary gland of the immature ovariectomized rat. Endocrinology 52, 311 (1953).

SMITH, V. R., STOTT, G. H., WALKER, C. W.: Observations on parathyroidectomized goats. J. Anim. Sci. **16**, 312 (1957).
SPELLACY, W. N., CARLSON, K. L., SCHADE, S. L.: Human growth hormone studies in patients with galactorrhea (Ahumada-del Castillo syndrome). Amer. J. Obstet. Gynec. **100**, 84 (1968).
SPIES, H. G., HILLIARD, Z., SAWYER, C. H.: Maintenance of corpora lutea and pregnancy in hypophysectomized rabbits. Endocrinology **83**, 354 (1968).
STEINER, W. G., HIMWICH, H. E.: Effects of antidepressant drugs on limbic structures of rabbit. J. nerv. ment. Dis. **137**, 277 (1963).
— — Electroencephalographic changes following administration of N-dimethylacetamide and other antitumor agents to rabbits. Int. J. Neuropharmacol. **3**, 327 (1964).
STRICKER, P., GRUETER, F.: Action du lobe antérieur de l'hypophyse sur la montée laiteuse. C.R. Soc. Biol. (Paris) **99**, 1978 (1928).
— — Recherches expérimentelles sur les fonctions du lobe antérieur de l'hypophyse; influence des extraits du lobe antérieur sur l'appareil génital de la lapine, et sur la montée laiteuse. Presse méd. **37**, 1268 (1929).
SULMAN, F. G.: Simple test for blood-ACTH. Lancet **1952 I**, 1161.
— Chromatophorotropic activity of human blood: review of 1,200 cases. J. clin. Endocr. **16**, 755 (1956 a).
— Experiments on the mechanism on the "push and pull" principle. J. Endocr. **14**, XXVII (1956 b).
— The mechanism of the "Push and Pull" principle. II. Endocrine effects of hypothalamus depressants of the phenothiazine group. Arch. int. Pharmacodyn. **118**, 298 (1959).
— TWERSKY, J.: Lack of effect of prolactin on milk production of cows. Refuah Veterinarith (J. Isr. Vet. Med. Ass.) **5**, 53, 73 (1948).
— WINNIK, H. Z.: Hormonal effects of chlorpromazine. Lancet **1956**, 270, 161.
SUSSMAN, R. M.: Sprironolactones and gynecomastia. Lancet **1963 I**, 58.
SWANSON, E. W.: The effect upon milk production and body weight of varying withdrawal periods after thyroactive supplement feeding. J. Dairy Sci. **37**, 1212 (1954).
TALEISNIK, S., ORIAS, R., OLMOS, J.: Topographic distribution of the melanocyte-stimulating hormone-releasing factor in rat hypothalamus. Proc. Soc. exp. Biol. (N. Y.) **122**, 325 (1966).
— TOMATIS, M. E.: Melanocyte-stimulating hormone-releasing and inhibiting factors in two hypothalamic extracts. Endocrinology **81**, 819 (1967).
TALWALKER, P. K., MEITES, J.: Mammary lobulo-alveolar growth induced by anterior pituitary hormones in adreno-ovariectomized and adreno-ovariectomized-hypophysectomized rats. Proc. Soc. exp. Biol. (N. Y.) **107**, 880 (1961).
— NICOLL, C. S., MEITES, J.: Induction of mammary secretion in pregnant rats and rabbits by hydrocortisone acetate. Endocrinology **69**, 802 (1961).

TALWALKER, P. K., RATNER, A., MEITES, J.: In vitro inhibition of pituitary prolactin synthesis and release by hypothalamic extract. Amer. J. Physiol. **205**, 213 (1963).

TEEL, H. M.: The effect of injecting anterior hypophysical fluid on the course of gestation in the rat. Amer. J. Physiol. **79**, 170 (1926).

THIÉBLOT, L.: Le rôle functionnel de l'épiphyse. Rev. Canad. Biol. **13**, 189 (1954).

THOMAS, J. W.: Hormonal relationships and applications in the production of meat, milk and eggs. Agricultural Board, Natl. Res. Council Publ. No. 266, p. 47. Washington, D.C. 1953.

— MOORE, L. A.: Thyroprotein feeding to dairy cows during successive lactations. J. Dairy Sci. **36**, 657 (1953).

TINDAL, J. S.: The lactogenic response of the rabbit to reserpine. J. Endocr. **18**, XXXVI (1959).

— KNAGGS, G. S., TURVEY, A.: Central nervous control of prolactin secretion in the rabbit: effect of local oestrogen implants in the amygdaloid complex. J. Endocr. **37**, 279 (1967).

TOPKINS, P.: The histologic appearance of the endometrium during lactation amenorrhea and its relationship to ovarian function. Amer. J. Obstet. Gynec. **45**, 48 (1943).

TÖRÖK, B.: Structure of the vascular connections of the hypothalamo-hypophysial region. Acta anat. (Basel) **59**, 84 (1964).

TRENTIN, J. J., HURST, V., TURNER, C. W.: Thiouracil and mammary growth. Proc. Soc. exp. Biol. (N. Y.) **67**, 461 (1948).

TUCKER, H. A., MEITES, J.: Induction of lactation in pregnant heifers with 9-fluoro-prednisolone acetate. J. Dairy Sci. **48**, 403 (1965).

TURKINGTON, R. W., TOPPER, Y. J.: Stimulation of casein synthesis and histological development of mammary gland by human placental lactogen in vitro. Endocrinology **79**, 175 (1966).

TURNER, C. W.: The mammary glands. Sex and internal secretions; a survey of recent research, 2nd ed., p. 740. Ed.: E. ALLEN. Baltimore: Williams & Wilkins 1939.

— The mammary gland. In: The anatomy of the udder of cattle and domestic animals. Columbia (Missouri): Lucas 1952.

— Regulation of lactation. A conference on radioactive isotopes in agriculture, Jan. 1956, East Lansing, Michigan, p. 403. Washington D. C.: U.S. Atomic Energy Commission 1956.

— GRIFFITH, D. R.: Possible mechanism governing release of lactogenic hormone. Excerpta medica, international congress series no. 47, XXII, Congress of International physiological sciences, Leiden, September 1962, vol. I. Lectures and symposia, part II, Symposia XI—XXII, p. 740. Amsterdam: Excerpta medica foundation 1962.

TURNER, R. A., PIERCE, J. G., DU VIGNEAUD, V.: The purification and the amino acid content of vasopressin preparations. J. biol. Chem. **191**, 21 (1951).

VELARDO, J. T.: Induction of pseudopregnancy in adult rats with Trilafon, a highly potent tranquilizer of low toxicity. Fertil. and Steril. **9,** 60 (1958).
— DAWSON, A. B., OLSEN, A. G., HISAW, F. L.: Sequence of histological changes in the uterus and vagina of the rat during prolongation of pseudopregnancy associated with the presence of deciduomata. Amer. J. Anat. **93,** 273 (1953).
VOKAER, R.: Sur l'existence d'un stade fugace a vacuoles basales (lucid zone) dans le cycle endométrial de la rate adulte. Ann. Endocr. (Paris) **11,** 652 (1950).
VOOGT, J. L., CLEMENS, J. A., MEITES, J.: Stimulation of pituitary FSH release in immature female rats by prolactin implant in median eminence. Neuroendocrinology **4,** 157 (1969).
WALKER, S. M., MATTHEWS, J. I.: Observations on the effects of prepartal and postpartal estrogen and progesterone treatment on lactation in the rat. Endocrinology **44,** 8 (1949).
WELLER, C. P.: Isolation of hypothalamic prolactin releasing factor. Bull. Res. Coun. Israel (Exp. Med.) **10E,** 237 (1963).
— MISHKINSKY, J., SULMAN, F. G.: Non-specificity of prolactin assay based on luteotrophic effect in mature mice. J. Endocr. **42,** 485 (1968).
WELSH, A. L.: The newer tranquilizing drugs. Med. Clin. N. Amer. **48,** 459 (1964).
WHITE, C.: The use of ranks in a test of significance for comparing two treatments. Biometrics **8,** 33 (1952).
WILHELMI, A. E.: Fractionation of human pituitary glands. Canad. J. Biochem. **39,** 1959 (1961).
WILKINS, L.: A feminizing adrenal tumor causing gynecomastia in a boy of five years contrasted with a virilizing tumor in a five-year-old girl. J. clin. Endocr. **8,** 111 (1948).
WILKINS, R. W.: Clinical usage of *Rauwolfia* alkaloids, including reserpine (serpasil). Ann. N. Y. Acad. Sci. **59,** 36 (1954).
WINNIK, H. Z., SULMAN, F. G.: Hormonal depression due to treatment with Chlorpromazine. Nature (Lond.) **178,** 365 (1956).
— TENNENBAUM, L.: Apparition de galactorrhée au cours du traitement de largactil. Presse méd. **63,** 1092 (1955).
WIRTH, W., GÖSSWALD, R., HÖRLEIN, U., RISSE, K. H., KREISKOTT, H.: Zur Pharmakologie acylierter Phenothiazinderivate. Arch. int. Pharmacodyn. **115,** 1 (1958).
WOLTHUIS, O. L.: An assay of prolactin based on a direct effect of this hormone on cells of the corpus luteum. Acta Endocr. (Kbh.) **42,** 364 (1963).
WOOLMAN, A. M.: Reserpine in severe hypertension. Lancet **1956 I,** 690.
WRENN, T. R., BITMAN, J., DE LAUDER, W. R., MENCH, M. L.: Influence of the placenta in mammary gland growth. J. Dairy Sci. **49,** 183 (1966).

YAMAMOTO, R., FUJISAWA, S.: Photodegradation of phenothiazine derivatives in aqueous solutions. International Congress Pharmaceutical Sciences, Muenster, Germany, September 1963.

ZAKS, M. G.: The motor apparatus of the mammary gland. Springfield (Ill.): Charles C. Thomas 1962.

ZETLER, G.: Cataleptic state and hypothermia in mice, caused by central cholinergic stimulation and antagonized by anticholinergic and antidepressant drugs. Int. J. Neuropharmacol. **7,** 325 (1968).

Subject Index

Aceperone 128
Acepromazine 73
Acetophenazine 71
Acridon 107
Acriflavine 107
Ajmalicine 116
Ajmaline 116
Alimemazine 76
Alizarin 107
Aminopromazine 82
Amiperone 126
Amitriptyline 109
Anafranil 107
Andantol 105
Anisoperidone 127
Atabrine 107
Aventyl 109
AW-14,2333 109
Azaperone 127

Banthine 106
Benzperidol 124
Benzquinamide 116
Blutene 83
Body temperature drop (BTD) 69
BP 400 Sulfoxide 105
Brilliant Cresyl Blue 106
2-Bromopromazine 71
Brontine 109
Butyropipazone 126
Butyrylperazine 72

Carbamazepine 108
Cardiorythmine 116
Carphenazine 72
Chiari-Frommel Syndrome 189
Chloroesucos 74
1-Chlorophenothiazine 83
2-Chlorophenothiazine 75

3-Chlorophenothiazine 81
Chlorproethazine 75
Chlorpromazine 74
Chlorpromazine sulfoxide 86
Chlorprothixene 104
Ciatyl 104
Clofluperol 123
Clopenthisol 104
Clothera 82
Clothiapine 110
Clothiapine sulfoxide 110
Colostrogenesis 1
Combelen 81
Compazine 75
Corticolactin 47, 197
Curatin 109
Cyamepromazine 74
Cyproheptadine 109

Dartal 70
Decaserpyl 116
Deptropine 109
Deserpidine 114
Dibenzodiazepine 110
Diethazine 77
Dimethoxanate 82
Dimetiotazine 82
Diparcol 77
Dipyripromazine 77
Dixyrazine 81
Dominal 105
Dorevane 83
Doxepin 109
Doxergan 86
Droperidol 123

Ejection, milk 1
Elatrol 109
Elavil 109

Eosin Blue 106
Esucos 81
Ethylbutrazine 72
Ethymemazine 72

Floropipamide 125
Floropipeton 127
Fluanisone 125
Fluphenazine 71, 137
Fluphenazine, receptor competition 140
Forbes-Albright-Castillo Syndrome 188
FR-02 127
FR-33 128
Frenelon 73

G-22150 108
G-31406 108
G-32883 108
G-34586 107
Galactorrhea 187
Galactopoiesis 1
Gallocyanin 106
GP-33006 108
Gynecomastia 192

Halidol 196
Haloperidide 125
Haloperidol 124, 196
HF-2159 110
8-Hydroxychlorpromazine 77
3-Hydroxypromazine 77

Imipramine 107
Insidon 109
Isothipendyl 105

KS-33 79
KS-75 80

Lactation, adrenocorticotropin 44
—, butyrophenones 122
—, calcitonin 52
—, corticosteroids 44, 146
—, dissociation of mammotropic and sedative effects 61
Lactation, effect of suckling 25
—, estrogens 1, 23, 37, 43, 182 ff.
—, fluphenazine 97
—, glucagon 53
—, gonadotropins 42
—, growth hormone 37
—, human 182
—, hypophysectomized rats 64
—, inhibition 182
—, insulin 53
—, interrelation between pituitary tropins 24
—, mechanism 1, 36
—, oxytocin 49
—, parathormone 52
—, perphenazine 142
—, persistence 185
—, pharmacological regulation 59
—, phenothiazine derivatives 67
—, phenothiazinelike compounds 102
—, phenothiazine sulfodioxides 69
—, pineal gland 54
—, pituitary-adrenal axis 146
—, pituitary-ovary axis 141
—, pituitary-thyroid axis 150
—, post-partum amenorrhea 182
—, priming 4, 68, 69
—, problems 182
—, progestogens 1, 23, 37, 44, 182 ff.
—, promotion 191
—, psychopharmaca 59, 131
—, psychotropic drugs 59, 131
—, research problems 196
—, reserpine derivatives 113
—, sex steroids 1, 23, 37, 42, 182 ff.
—, sheep 196
—, sterility 186
—, steroids 23
—, thyroid hormones 48, 150
—, thyrotropin 48, 150
—, vasopressin 49
Lactogenesis 1
Largactil 74

Subject Index

Largon 83
Lauth Violet 83
"Let-Down" 1
Levomepromazine 81
Lispamol 82
Litracen 107

Majeptil 76
Mammogenesis 1
Mammotropic index (MTI) 68, 170
Melitracen 107
Mellaril 73
Mepazine 78
Merbromine 106
Mercurochrome 106
Mesoridazine 79
Methantheline 106
Methdilazine 80
Methdilazine sulfoxide 85
Methixene 104
Methophenazine 73
Methoridazine 79
Methoserpidine 116
2-Methoxyphenothiazine 77
Methylene blue 80
Methylperidide 126
MK-130 109
Moderil 115
Mopazine 81
Moperone 122
Mornidine 71
MSH-releasing factor 100

N 7001 107
N 7049 107
NC-123 79
Neuleptil 70, 196
Nitoman 115
Norimipramine 108
Nortriptyline 109
Noveril 110
Nozinan 81

6-OH-Chlorpromazine 83
7-OH-Chlorpromazine 77
Opipramol 109
Oxomemazine sulfodioxide 86
Oxyperpendyl 105

P-4657 B 104
Pacatal 78
Parsidol, R.P. 3326 80
Pasaden 72
Perazine 75
Periactin 109
Permitil 71
Perphenazine 70, 196
Perphenazine sulfoxide 84, 196
Pertofran 108
Pervetral 105
PIF 2, 11
— assay method 11
Phenergan 83
— sulfoxide 85
Phenothiazine 83
— sulfoxide 86
— derivatives, implantation 94
10-phenothiazinylpropionic acid 82
Phenothiazinylpropionitril 82
Pimozide 124
Pipamazine 71
Piperacetazine 70, 196
— sulfoxide 84
Plegicil 73
Priming, estradiol 4, 68, 69, 170
Prochlorperazine 75
Profenamine 80
Proflavine 107
Proketazine 72
Prolactin 29
—, amniotic fluid 55
— assay 162
— —, concentration methods 175
— —, in vitro methods 162
— —, in vivo methods 169
— —, pigeon tests 173
— —, rabbit tests 169
— —, rat tests 170
—, body fluids 32
—, chemistry 31
—, deciduoma formation 158
—, effect on mammary growth 33
—, estrus cycle 35
—, hypothalamic feedback 17
— -Inhibiting Factor (PIF) 7

Prolactin levels in lactation   34
—, luteotropic hormone   156
—, pituitary tumors   32
—, perphenazine   88
—, placenta   55
—, properties   30
—, radioimmunoassay   179
—, release   88, 153, 162, 164, 166, 167
— —, hypothalamus   162
— —, neurohormones   166
— —, oxytocin   167
— —, phenothiazines   166
— —, pituitary   164
— —, reserpine   167
— —, steroids   164
— —, thyroid   167
— -Releasing Factor (PRF)   2, 13
—, storage and release   30
—, synthesis   88
Prolixin   71
Promazine   79
— sulfoxide   86
Promethazine   83
— sulfoxide   85
Promoton   77
Propericiazine   70, 196
Propionylpromazine   81
Propiomazine   83
Prothipendyl   105
Protriptyline   109
Pyrrolazote   83

Quantril   116
Quebrachine   116
Quide   70, 196
Quinacrine   107

R-1625   124
R-1647   127
R-1658   122
R-1892   126
R-1929   127
R-2028   125
R-2498   122
R-2963   126
R-3201   125

R-3248   128
R-3264   128
R-3345   125
R-4006   127
R-4082   127
R-4457   125
R-4584   124
R-4749   123
R-5147   123
R-6238   124
R-7158   128
R-9298   123
R-12962   126
Radioimmunoassay   179
Randolectil   72
Raubasin   116
Receptor theory of mammotropic drugs   133
Rescinnamine   115
Reserpine   114, 196
RP 4909   75
RP 5293   86
RP 6140   75
RP 6484   72
RP 6549   76
RP 6710   77
RP 7204   74
RP 7843   76
RP 8307   108
RP 8580   84
RP 8599   82

Sandoz 10-798   105
— 11-18   104
— 11-223   110
— 11-445   105
— 11-665   105
— 12-039   105
— BC 105 (Sandomigran)   110
— BP 400   104
SCH 13361   85
Selvigon   105
Sergetyl   72
Serpasil   114, 196
Sheehan's Syndrome   191
SKF 2680-J   79
— 3072-A   78

Subject Index

SKF 3227 75
— 4260-A 86
— 4523 81
— 4611-A sulfoxide 85
— 5332-I sulfoxide 86
— 5419-A 85
— 5611-A 78
— 5612-A 80
— 5757-G 76
— 5910-I$_2$ 74
— 6045-A$_2$ 82
— 6051-A$_2$ 76
— 6058-I$_2$ 75
— 6063-A 72
— 6064-J 75
— 6108-J$_2$ 75
— 6262-A$_2$ 79
— 6693-A$_2$ 78
— 7009-I 76
— 7100-A 71
— 7136-A 72
— 7145-A sulfoxide 86
— 7221-I 71
— 7838 78
— 7887-A$_2$ 73
— 8448-I 73
— 8745 76
— 9062-A$_2$ 106
— 12,116-A 79
— 13,813-A 81
— 17,910-A 85
— 18,722 81
— 20,716 70
— 24,200 83
Sordinol 104
Sparine 79
Spiroperidol 123
SQ-15283 105
Stelazine 73
Structure-activity relationship of mammotropic drugs 133
Su-7064 115
Su-9064 115
Su-9300 116

Su-9673 115
Su-10,092 116
Su-10,704 115
Su-11,279 115
Sum-3170 110, 196
Surmontil 107

Tacaryl 80
Taractan 104
Taxilan 75
Tegretol 108
Temaril 76
Tetrabenazine 115
Thiethylperazine 73
Thionin 83
Thioperazine 76
Thiopropazate 70
Thioridazine 73
Thorazine 74
Tindal 71
Tofranil 107
Tolonium 83
Toluidine Blue 83
Torecan 73
TPN 12 74
TPO-33 71
Tremaril 104
Trifludol 196
Trifluperidol 122, 196
Triflupromazine sulfoxide 85
Trifluorperazine 73
Trilafon 70, 196
Trimepramine 107

UCB 2493 74

Vivactil 109
Vontil 76

Winthrop Compound 22 79
Wy 1193 86

Yohimbine 116

Typesetting, printing and binding: Konrad Triltsch, Graphischer Betrieb, 87 Würzburg, Germany

Monographs on Endocrinology

Already Published:

Vol. 1: OHNO, S., Sex Chromosomes and Sex-linked Genes. With 33 figures. X, 192 pages. 1967. DM 38,—; US $ 9.50.

Vol. 2: EIK-NES, K. B., and E. C. HORNING, Gas Phase Chromatography of Steroids. With 85 figures. XV, 382 pages. 1968. DM 38,—; US $ 9.50.

In Preparation:

BAULIEU, E. E., Bicêtre: Current Problems in the Metabolism of Steroid Hormones.

BERSON, S. A./YALOW, R. S., New York: Immunoassay of Peptide Hormones.

BORTH, R., Toronto: Clinical Hormone Assays.

BREUER, H./RAO, G. S., Bonn: Metabolism of Estrogens.

CALDEYRO-BARCIA, R., Montevideo: Oxytocin.

COPP, D. H., Vancouver: Calcitonin.

EDELMAN, I. S., San Francisco: Aldosterone: Gene Action and Sodium Transport.

FEDERLIN, K., Ulm: Immunopathology of Insulin. Clinical and Experimental Studies.

GREGORY, R. A., Liverpool/GROSSMAN, M. I., Los Angeles: The Gastrins.

GURPIDE, E., Minneapolis: Tracer Methods in Hormone Research.

HORTON, W. E., London: Prostaglandins.

JENSEN, E./DESOMBRE, E., Chicago: Receptors and Action of Estradiol.

LUNENFELD, B., Tel Hashomer: Gonadotropins.

MCKENZIE, J. M., Montreal: The Pathogenesis of Graves' Disease.

MÜLLER, J., Zürich: Regulation of Aldosterone Biosynthesis.

NEUMANN/STEINBECK/ELGER, Berlin: Hormones in Sexual Differentiation.

RENOLD, A. E./STAUFFACHER, W., Geneva: Pathophysiology of Diabetes.

ROBERTS, S., Los Angeles: Subcellular Mechanisms in the Regulation of Corticosteroidogenesis.

SHORT, R. V., Cambridge: The Ovary and its Hormones.

WESTPHAL, U., Louisville: Steroid-Protein Interactions.

WILLIAMS-ASHMAN, H. G., Chicago: Androgen Action.

WITHDRAWN

MISS FRANCES RIDLEY HAVERGAL.
Photo by Elliott & Fry.

Famous Hymns
and their Authors.
By Francis Arthur Jones.

WITH PORTRAITS
AND FACSIMILES

DETROIT
Singing Tree Press
1970

This is a facsimile reprint of the 1902 edition
published by Hodder and Stoughton, London.

Library of Congress Catalog Card Number: 72-99067

To Ceil
In ever affectionate
Remembrance

Preface

IN the compilation of this volume (the work of some ten years) an immense number of authorities have necessarily been consulted. Among these should be specially mentioned the works of Dr. John Mason Neale, Canon Ellerton, the Rev. L. C. Biggs, the Rev. John Brownlie, the Rev. John Chandler, Dr. Phillip Schaff, Mr. G. T. Stevenson, and Dr. John Julian.

The " Lives " of many of our foremost hymnists have also been consulted, notably Bishop Ken, Bishop How, Bishop Heber, Bishop Christopher Wordsworth, Dean Alford, Dean Milman, Father Faber, T. T. Lynch, and others.

To the following who have rendered me invaluable assistance as regards special information, portraits, photographs, or manuscripts, I gladly acknowledge my indebtedness :—His Grace the Archbishop of York, the Right Rev. Bishop E. H. Bickersteth, Prebendary Thring, Dr. John Julian, the Rev. F. M. Bird (for much interesting information regarding American hymn writers), the Rev. H. N. Bonar, the Rev. W. St. Hill Bourne, the Rev. John Brownlie, the Rev. R. H. Bullock, the Rev. H. E. T. Crusoe, the Rev. Frank Ellerton, the Rev. P. E. L. Holland, the Rev. Charles H. Kelly, Sir Herbert Oakeley, and the Rev. W. H. Whiting; Miss J. Baker, Miss S. Gurney, Miss K. Hankey, Mrs. G. S. Hodges, Mrs. J. Luke, Mrs. M. F. Maude, Mrs. E. H. Miller, Mrs. F. A. Shaw, Mrs. E. M. Synge, and Miss G. Wordsworth; W. Beck, Esq., J. Potter Briscoe, Esq., H. C. Camp, Esq. (New York), C. E. Conder, Esq., H. Daniell, Esq., C. W. Lock, Esq., Arthur Milman, Esq., G. E. Newman, Esq., T. Viccars, Esq., and S. Young, Esq.

From the following hymnists who have passed away while this book was in the making I also received much valued assistance:—The Right Rev. Bishop W. Walsham How, the Rev Canon Bright, the Rev. Canon Twells, the Rev. A. G. W. Blunt, the Rev. S. J. Stone, the Rev. Laurence Tuttiett, W. Chatterton Dix, Esq., and Mrs. C. F. Alexander.

For permission to use the many copyright photographs included in this volume my thanks are due to those whose names appear below the portraits.

F. A. J.

LONDON, *Nov.*, 1902.

Contents

	PAGE
PREFACE	v

I
MORNING AND EVENING HYMNS 1

II
ADVENT HYMNS 40

III
CHRISTMAS HYMNS 56

IV
HYMNS SUITABLE FOR THE NEW YEAR . . . 77

V
HYMNS ON THE PASSION 87

CONTENTS

VI
Easter Hymns 119

VII
Processional Hymns 141

VIII
Communion Hymns 162

IX
Hymns for Holy Matrimony, Missions, and "Those at Sea" 193

X
Funeral and Harvest Hymns, All Saints' Day 230

XI
Hymns for Children 254

XII
Some General Hymns 293

Illustrations

Plate I . . . . *Frontispiece*
    Miss Frances Ridley Havergal.

Plate II . . . . *To face page* 9
    1. The Rev. F. W. Faber, D.D.
    2. The Right Rev. Bishop Thomas Ken, D.D.
    3. Mr. John Byrom, M.A.
    4. Kersal Cell.
    5. The Rev. G. R. Prynne, M.A.

Plate III . . . . *To face page* 43
    Berryhead House.
    The House in which Harriet Auber wrote
      "Our blest Redeemer, ere He breathed."

Plate IV . . . . *To face page* 71
    1. The Rev. T. J. Potter.
    2. Mrs. M. F. Maude.
    3. The Very Rev. Henry Hart Milman, D.D.

Plate V . . . . *To face page* 121
    Mrs. Cecil Frances Alexander.
    The Rev. Horatius Bonar, D.D.

ILLUSTRATIONS

Plate VI . . . . *To face page* 127
    1. Miss Catherine Winkworth.
    2. Mr. James Montgomery.
    3. Mrs. Emily Huntington Miller.
    4. Mr. B. S. Ingemann.
    5. Mr. Henry Kirke White.
    6. The Rev. Archer T. Gurney.

Plate VII . . . . *To face page* 151
    1. The Rev. J. J. Daniell.
    2. The Rev. T. T. Lynch.
    3. The Rev. G. S. Hodges, B.A.
    4. The Rev. S. J. Stone, M.A.
    5. The Rev. Thomas Binney, D.D.
    6. The Rev. Canon Ellerton, M.A.

Plate VIII . . . . *To face page* 169
    The Rev. Canon Bright, D.D.
    Mr. Albert Midlane.

Plate IX . . . . *To face page* 180
    The Rev. John Wesley, M.A.
    Mr. William Cowper.

Plate X . . . . *To face page* 200
    The Rev. Isaac Watts, D.D.
    The Right Rev. Bishop Reginald Heber, D.D.

Plate XI . . . . *To face page* 248
    The Right Rev. Bishop Christopher Wordsworth, D.D.
    Mr. Charles Wesley, M.A.

ILLUSTRATIONS

Plate XII . . . . *To face page* 260
   1. The Rev. Canon Twells, M.A.
   2. Mrs. J. Luke.

Plate XIII . . . . *To face page* 275
   The Right Rev. Bishop W. W. How, D.D.
   The Rev. John Keble, M.A.

Plate XIV . . . . *To face page* 296
   1. Mr. Richard Baxter.
   2. Sir John Bowring, LL.D.
   3. Mr. William Williams.
   4. The Right Rev. Bishop Richard Mant, D.D.
   5. The Rev. George Matheson, D.D.

Plate XV . . . . *To face page* 310
   Facsimile of "I heard the Voice of Jesus say."

Plate XVI . . . . *To face page* 317
   1. The Rev. W. Bullock, D.D.
   2. The Rev. W. St. Hill Bourne.
   3. Mr. William Whiting.
   4. Mr. W. Chatterton Dix.
   5. The Rev. Henry Francis Lyte, M.A.
   6. The Rev. Henry Alford, D.D.

I

Morning and Evening Hymns

FROM the days of St. Ambrose to the present time hymnists have found in the beginning and close of the day a favourite subject for religious verse. As a result those hymns specially written for morning and evening service not only form an mportant section of our hymnals, but also constitute some of the finest compositions of the kind in the language.

To Bishop Ken we are indebted for two hymns which, for a hundred and sixty years, have found a place in every English hymnal. " Awake, my soul, and with the sun " is not, perhaps, as often sung to-day at morning service as it was, say fifty years ago, but the Doxology, which concludes this hymn, as well as the same author's " All praise to Thee, my God, this night," is probably more frequently sung than any other single verse in our hymnals.

Bishop Ken wrote these two hymns, together with one for midnight, for the scholars of Winchester College, and it is said that before being published several copies were written in printed letters on large sheets and hung on the walls of the dormitories, where the boys could see them the first thing in the morning and the last thing at night. When they came to be published, in 1674, the author appended to them the following note:

"Be sure to sing the morning and evening hymn in your chamber devoutly, remembering that the Psalmist, upon happy experience, assures you that it is a good thing to tell of the loving kindness of the Lord early in the morning and of His truth in the night season."

The hymn for midnight, which consists of thirteen verses, is not very frequently sung, possibly owing to the rather inconvenient hour for which it was intended. However, in various forms it is to be met with in many hymnals. Six verses, together with the Doxology, form a complete hymn by themselves, and may, without any very great outrage on the intentions of the author, be sung as an evening hymn.

Thomas Ken was born at Berkhampstead in 1637, and was brought up under the guardianship of his brother-in-law, Isaak Walton. He was educated at Winchester, became a Fellow of New

College, Oxford, in 1657, Rector of Little Easton in 1663, and Rector of Woodhay and Prebendary of Winchester in 1669. It was during his last-mentioned preferment that he wrote his famous hymns. He is said to have had a beautiful voice, and was often heard in his chamber singing his morning and evening hymns to tunes which he had himself composed, accompanying himself on the viol or spinet.

In 1679 he became Chaplain to the Princess Mary, at The Hague, but was summarily dismissed from this post through expressing rather too openly his disapproval at certain Court proceedings. The following year he returned to Winchester and narrowly escaped a quarrel with Charles II by refusing, on the occasion of a visit from the King, to receive Nell Gwynne into his house. Instead of losing his temper, however, that extraordinary monarch, on being told of Ken's inhospitality, laughed good humouredly and shortly afterwards made him Bishop of Bath and Wells. When William III ascended the throne, Bishop Ken, for reasons which have never been very satisfactorily explained, refused to take the oaths, and was, in consequence, dismissed from office. A few years later, however, he was again offered the Bishopric of Bath and Wells, which he declined, preferring to end his days in retirement at Longleat, where he died in 1710.

From Keble's *Christian Year* we get two morning hymns which are to be found in all modern hymnals. "New every morning is the love," and "O timely happy, timely wise," are both taken from the same poem beginning "Hues of the rich unfolding morn." This poem, which was written in 1822, and published five years later, consists of sixteen verses, and from it several centos have been taken by various hymnal editors.

"Sun of my soul, Thou Saviour dear," acknowledged by many hymnologists to be the most frequently sung of all our evening hymns, was written by Keble in 1820, and also first published in the *Christian Year*. The original MS. of this work is preserved at Keble College, Oxford, and it is to the Warden of that College I am indebted for permission to give a facsimile reproduction of the first verse of this famous hymn:—

*Sun of my Soul! Thou Saviour dear,
It is not night if Thou be near:
Oh may no earth-born cloud arises
To hide Thee from Thy servant's eyes.*

As my readers are probably well aware, "Sun of my soul," as it appears in the majority

MORNING AND EVENING HYMNS

of our hymnals, consists of verses taken from the poem beginning " 'Tis gone, that bright and orbèd blaze." In the College Library may be seen two MSS. of "Sun of my soul," but the one I have chosen for reproduction here bears an earlier date than the other, and contains Keble's alterations. Though the *Christian Year* has not now so great a sale as formerly, it is still widely read, while at one time it had a larger circulation than that of any other work of a similar character. From the profits arising out of the sale of this book Keble built Hursley Church.

John Keble was born at Fairford, in Gloucestershire, in 1792, being the son of the Vicar of Colne. After a brilliant career at Oxford he took Holy Orders, his first curacy being that of East Leach and Burthorpe. The *Christian Year* appeared in 1827 and was an instant success, edition after edition being called for. Hymnal editors of all denominations begged for permission to make selections for their own particular collections, requests which were readily and cheerfully granted. Keble was a remarkably modest man, and probably thought less of his own work than did the least of his admirers. He once accompanied the vicar of a parish in the South of England on his visit to the Sunday School. The superintendent requested him to address a few words to the children, who were already acquainted with his hymns, so that they might the more easily remember them. He timidly shrank from complying, but the superintendent persisting, he then said: "May they sing something?" When they ceased, his face was beaming upon them as he said: "My dear children, you sang most beautifully in tune; may your whole lives be equally in tune, and then you will sing with the angels in heaven."

In 1829 Mr. Keble was offered the living of Hursley by Sir William Heathcote, which he declined for family reasons. Two years later he was elected Professor of Poetry at Oxford, and in 1833 laid the foundation of the Oxford Movement by delivering his now famous Assize Sermon. Two years afterwards he was again offered the living of Hursley, and accepted it. Here he lived for thirty years, greatly beloved by his parishioners. He died in 1866, on March 29th, his wife following him some six weeks later.

The number of tunes to which "Sun of my soul" has been set are many. Perhaps the most popular, certainly the most beautiful, is "Abends" by Sir Herbert Oakeley. The melody was so exactly suited to the words that it found immediate favour with the editors of every kind of hymnal, and to-day it would be difficult to find the collection which does not contain it.

Sir Herbert Oakeley is a rapid worker, and the tune to "Sun of my soul" was written in less than half an hour. In a letter which I received from the composer some time ago Sir Herbert says:

"There is not much to record *re* 'Abends.' I was, many years ago, impelled to set Keble's words to music for Sir Henry Baker, in consequence of the inadequacy if not vulgarity of the tune which had got into general use. I refer to 'Hursley,' which, however, is now less often sung than formerly.

"'Hursley,' strange to say, had been in use in Germany—where, as a rule, chorales (*Anglicè* hymn tunes) are so dignified and admirable—since *circiter* 1792, and is attributed to Paul Ritter.

"One of my reasons for disliking it is the resemblance it bears to a drinking song, 'Se vuol ballare,' in *Nozze di Figaro*. As Mozart produced that opera in 1786, he is responsible for the opening strain, which suits his Bacchanalian words very well. But to hear 'Sun of my soul, Thou Saviour dear,' sung to a lively tune, unsuitable to sacred words, often had the effect of driving me out of church."

No one can seriously think for a moment that the fact of a congregation joining in a hymn redeems a tune, musically, or at least ecclesiastically, bad. The public, however, will join in any easy melody if the words are good, and if the origin of the music is unknown. "The practice," continues Sir Herbert, "adopted by some hymnal editors, of associating tunes with hymns other than those for which they were written, is to be much regretted." This opinion was shared by the late Dr. Dykes, who on several occasions refused the use of a hymn tune on learning that it was to be sung to words other than those for which he had written it.

"Christ, whose glory fills the skies" was written by Charles Wesley, and first published in 1740. James Montgomery, who was more keenly alive to the true value of a hymn than many editors, averred that it was one of the finest of all Charles Wesley's compositions.

It has not, however, escaped alteration at the hands of editors, to whom, let us hope, John Wesley's appeal in the preface to his hymnal was unknown. "I beg leave," runs this characteristic paragraph, "to mention a thought which has been long upon

1 THE REV. F. W. FABER, D.D.
From a Photo.
2 THE RIGHT REV. BISHOP THOMAS KEN, D.D.
From an Engraving.
3 MR. JOHN BYROM, M.A.
From a Sketch made from life.
4 KERSAL CELL.
Photo by Mitchell, Cheetham.
5 THE REV. G. R. PRYNNE, M A.
Photo by Steer, Plymouth

my mind, and which I should long ago have inserted in the public papers, had I not been unwilling to stir up a nest of hornets. Many gentlemen have done my brother and me (though without naming us) the honour to reprint many of our hymns. Now they are perfectly welcome so to do, provided they print them just as they are. But I desire they would not attempt to mend them, for they really are not able. None of them is able to mend either the sense or the verse. Therefore I must beg of them one of these two favours: either to let them stand just as they are, to take them for better or worse, or to add the true reading in the margin, or at the bottom of the page, that we may no longer be accountable either for the nonsense or for the doggerel of other men."

"Sweet Saviour, bless us ere we go," one of the tenderest dismissal hymns for evening service that we possess, was written in 1849 by Frederick William Faber. Like the majority of Dr. Faber's hymns, it was composed expressly for use in the Brompton Oratory.

In several collections this hymn begins "O Saviour, bless us ere we go," and the last verse, appealing as it does more particularly to Roman Catholics, is generally omitted. It is one of the most frequently sung of all Dr. Faber's hymns and there are few collections of modern date in which it does not find a place.

Dr. Faber was the son of a Church of England clergyman and was born at Calverley Vicarage, Yorkshire, in 1814. After graduating at Balliol College, Oxford, he was ordained and became Rector of Elton in 1843. Three years later he left the Church of England and established the Brotherhood of St. Philip Neri, in King William Street, Strand, which was afterwards removed to the Brompton Oratory.

Dr. Faber was the author of several prose works as well as three volumes of hymns. It is in his *Jesus and Mary*, that many of his best hymns were first published. This collection, which is still in use at the Oratory, contains in the preface an account of the circumstances under which Dr. Faber wrote his hymns:—

"It was natural," he says, "that an English son of St. Philip should feel the want of a collection of English Catholic hymns fitted for singing. The few in the *Garden of the Soul* were all that were at hand, and, of course, they were not numerous enough to furnish the requisite variety. As to translations they do not express Saxon thoughts and feelings, and consequently the poor do not seem to take to them. The domestic wants of the Oratory, too, kept alive the feeling that something of the sort was needed: though at the same time the Author's ignorance of music appeared in some measure to disqualify him for the work of supplying the defect. Eleven, however, of the hymns were written, most of them, for particular tunes and on particular occasions, and became very popular with a country congregation. They were afterwards printed for the schools at St. Wilfrid's, and the very numerous applications to the printer for them seemed to show that, in spite of very glaring literary defects, such as careless grammar and slipshod metre, people were anxious to have Catholic hymns of any sort. The MS. of the present volume was submitted to a musical friend, who replied that certain verses of all or nearly all the hymns would do for singing; and this encouragement has led to the publication of the volume."

Numerous are the instances of hymn-writers who are remembered by single compositions only. Henry Francis Lyte is a case in point, for though he wrote many hymns, it is with "Abide with me" that his name will always be associated. This hymn, which was written at the little fishing port of Brixham, on the shores of Torbay, was the author's last composition, and it has probably brought as much, if not more, comfort and hope to stricken humanity as all the sermons ever uttered or written.

Brixham is celebrated in history as the landing place of William III. in 1688, "to uphold the religion and liberties of England." The stone on which His Majesty first set foot is still preserved in an obelisk at the head of the quaint little pier. Shortly after his accession, William IV paid a visit to Brixham and was met at the landing-stage by Mr. Lyte and a surpliced choir ; and the stone, on which his namesake had trodden so many years before, was carried down the steps in order that the King might also place his foot thereon. Berry Head House, about half-a-mile distant from the town, was a gift to Mr. Lyte from William IV. It is a solidly-built mansion, and was originally, when Berry Head was garrisoned, the military hospital. The ruggedness of its exterior is now toned down by yellow roses, clematis, and Virginia creeper. The scene from the verandah is magnificent, for the sea laps the very foot of the terraced gardens. It was here that "Abide with me" was written.

Some six or seven years ago I happened to be staying in Brixham and was fortunate enough to meet an old member of Mr. Lyte's choir, a worthy

MORNING AND EVENING HYMNS

gentleman who was credited with knowing more about the celebrated hymnist than any other living man. As we sat on the old pier one morning in early June, and watched the trawlers setting sail for the fishing grounds, my companion chatted animatedly about the late hymnist, evidently well pleased to find some one who took an interest in a man of whom he was palpably never tired of talking.

"I was a member of Mr. Lyte's choir," he said, "in 1846—I and a dozen others, all dead now. We were deeply attached to him. He had the gentlest expression and most winning manner possible, and yet I suppose we caused him more grief than all his trials of ill health. We left his choir and gave up teaching in his Sunday School, and though I should probably do the same thing to-morrow under similar circumstances, it gives me a feeling of intense sadness even now when I think of it.

"This is how it came about. A short while before he left us to go to Nice, where it was hoped the climate would benefit his health, some influential members of the Plymouth Brethren visited Brixham and persuaded ten of us to join them. After due deliberation we went in a body to Mr. Lyte and told him that we intended to leave his church. He took it calmly enough, though we practically constituted his entire choir, and said that nothing would be farther from his thoughts than to stand between us and our consciences. He bade us think the matter over very seriously and come to him again in a few days. We did so, but our decision remained unaltered. We left him, and never entered his church again. When 'Abide with me' came to be written, each of us was given a copy, and then we realized, perhaps more keenly than any one else, the true meaning of the words—

'When other *helpers* fail, and comforts flee,
Help of the helpless, O abide with me.'"

The story of how this hymn came to be written is an oft-told tale, and yet this little volume would be far from complete were it omitted altogether. Briefly, the story is this. In 1847 Mr. Lyte had become so weak and ill by his devotion to his flock that the doctors ordered his removal to Nice, where it was hoped the more genial climate would restore some of his lost health. The evening of the Sunday prior to his departure was a beautiful one, and after service he left his house and strolled, as was his custom, down the garden path to the seashore, alone. Here he walked up and down for perhaps half-an-hour, meditating sadly on the farewell words he had so

MORNING AND EVENING HYMNS

lately addressed to his congregation, being fully convinced that he had spoken to them for the last time on earth. When the sun had set and the night had closed in Mr. Lyte returned to his house and retired to his study. An hour later (it was thought that he had been lying down) he presented his family with the hymn

> *Abide with me! Fast falls the Eventide;*
> *The darkness thickens. Lord, with me abide.*
> *When other helpers fail, and comforts flee,*
> *Help of the helpless, O abide with me!*

accompanied by music which he had also composed. The next morning he left Brixham to return no more, for he died a few months later at Nice, where he now lies buried.

Whether Mr. Lyte intended "Abide with me" to be used solely as an evening hymn it is impossible to say, but Canon Ellerton in his collection places it among those for "General Use," with the following note attached :—

"It is sometimes (nearly always) classed among evening hymns, apparently on the ground of the first two lines, and their similarity in sound to two lines in Keble's 'Sun of my soul.' This is a curious instance of the misapprehension of the true meaning of a hymn by those among whom it is popular, for a very little consideration will suffice to show that there is not throughout the hymn the slightest allusion to the close of the *natural* day : the words of St. Luke xxiv. 29 are obviously used in a sense wholly metaphorical. It is far better adapted to be sung at funerals, as it was beside the grave of Professor Maurice ; but it is almost too intense and personal for ordinary congregational use."

After all it matters very little whether a hymn is sung at morning, at evening or at midday so long as it fulfils its purpose. " Abide with me " was written in the evening, and when published was accepted as an evening hymn, and will always be looked upon as an evening hymn by those who sing it. The close of the natural day is the evening of life to thousands, and so the connexion between the two is not so slight as Canon Ellerton would seem to suppose. It was written, we may be very sure, for a purpose, and that purpose was to bring comfort to the living. It is therefore hardly possible that the author intended it to be used as a funeral hymn only. In no hymnal save Canon Ellerton's does " Abide with me " appear in any but its proper place, namely, among those intended for evening service.

The original music to " Abide with me " is now seldom sung, having been supplanted by Dr. Monk's beautiful composition, " Eventide." Dr. Monk's manner of setting Mr. Lyte's hymn will serve as an example of the rapidity with which he could compose. Starting out one morning with the late Sir Henry Baker, his co-worker in the editing of *Hymns Ancient and Modern*, he suddenly recollected that there was no tune to No. 27, "Abide with me." He returned to the house, and, undisturbed by a music lesson that was going on, the doctor sat down and wrote the beautiful and appropriate melody in ten minutes.

An evening hymn which ranks with "Abide with me" and "Sun of my soul" in point of excellence and appropriateness is Canon Ellerton's:—

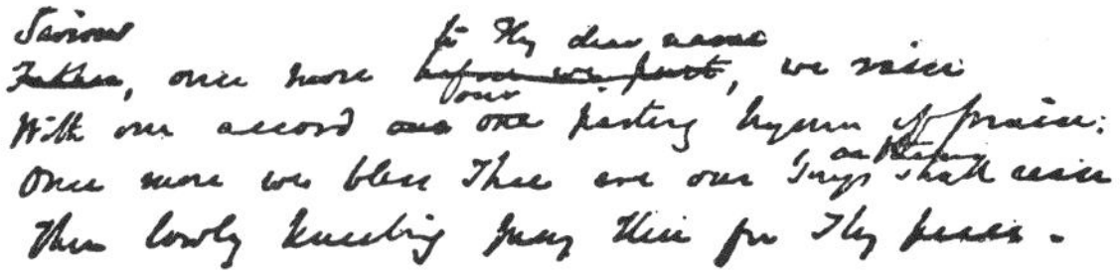

[handwritten verse]

It was written in 1886 for the Festival of the Malpas, Middlewich and Nantwich Choral Association of that year. Canon Ellerton had a fondness for writing hymns to tunes which took his fancy, and many of his best-known hymns owe their origin to melodies which captured his affections. In the case of "Saviour, again to Thy dear name" he had been much struck by a tune entitled "St. Agnes," in Thorne's collection, and when he was asked to write a hymn for the Malpas Festival it occurred to him that he would much like to write some words to that air. He therefore took a piece of sermon note, on one side of which, by the way, was a portion of the discourse he had preached the previous Sunday, and drafted out the first rough plan of the hymn. The MS., a portion of which we show here, is a most interesting one owing to the number of corrections made by the author. In comparing the hymn as first written by Canon Ellerton with the "fair" copy which he afterwards made, some difference is apparent. The opening verses originally read :—

> Father, once more before we part, we raise
> With one accord our parting hymn of praise :
> Once more we bless Thee, ere our songs shall cease,
> Then, lowly kneeling, pray Thee for Thy peace.
>
> Grant us Thy peace, Lord, through the coming night,
> Turn Thou for us its darkness into light :
> From harm and danger, fear and shame kept free,
> For dark and light are both alike to Thee.

Canon Ellerton originally wrote this hymn in six stanzas of four lines each. In most hymnals, however, the number of verses has been reduced to five and in a few to four. One verse in the MS., the fourth, has been omitted altogether, and, as it

MORNING AND EVENING HYMNS

is one of the most beautiful in the whole hymn, I cannot refrain from quoting it:—

> Grant us Thy peace—the peace Thou didst bestow
> On Thine Apostles in Thine hour of woe;
> The peace Thou broughtest, when at eventide
> They saw Thy piercèd hands, Thy wounded side.

For some years this hymn was sung to the tune to which it was originally written, but a short time before the publication of the first revised edition of *Hymns Ancient and Modern* the hymn was sent to Dr. Dykes, at Durham, with a request that he would set it to music. He did so and, as was his custom, took the manuscript down with him to St. Oswald's. After evening service he played it over to his children in order to obtain from them their opinion as to whether it was a good tune or not. The juvenile critics were unanimous in their approval, and the hymn was despatched to Sir Henry Baker. As may be well believed, Sir Henry was very much "taken" with the setting, and always referred to it as one of the most beautiful of Dr. Dykes' compositions. Shortly before her death Mrs. Ellerton, widow of the hymnist, remarked to me that her husband was also very much delighted with the tune and wrote Dr. Dykes a special letter of thanks

"Saviour, again to Thy dear name" has been translated into many languages and dialects, and is said to be in more extensive use than any other of Canon Ellerton's compositions.

"Saviour, breathe an evening blessing," by James Edmeston, made its first appearance in the author's *Sacred Lyrics*, published in 1820. For many years it remained in obscurity, unnoticed by hymnal compilers, until Dr. Bickersteth republished it in a little volume called *Christian Psalmody*. From thence it made its way into several collections, and soon began to take its place as one of the foremost hymns for evening service in the language. As originally written it consisted of two verses only, of eight lines each, but in most hymnals it is now given in four four-line stanzas.

"Saviour, breathe an evening blessing," may be said to owe its origin to Edmeston's love for books of travel. In 1819, happening to be reading Salte's *Travels in Abyssinia*, he came to the following passage: "At night, their short evening hymn, 'Jesus, forgive us,' stole through the camp." Laying aside his book, he took a sheet of paper and spontaneously penned those two simple verses which have since become so well known.

This hymn has undergone a good deal of

MORNING AND EVENING HYMNS

pruning and alteration at the hands of various editors, in some cases, perhaps, to its advantage. The following verse, dealing with sudden death, is in several hymnals omitted, probably for the reason that it is a somewhat unhappy conclusion to an otherwise beautiful hymn—

> Should swift death this night o'ertake us,
> And our couch become our tomb ;
> May the morn in heaven awake us,
> Clad in light and deathless bloom.

Prebendary Thring has rewritten this verse and cleverly overcome the rather unpleasant suggestion that our bed may also "become our tomb"—

> Be Thou nigh, should death o'ertake us,
> Jesus, then our Refuge be ;
> And in Paradise awake us,
> There to rest in peace with Thee.

James Edmeston, born in 1791, was by profession an architect and surveyor. He was a man of a peculiarly lovable disposition, and passionately fond of children, for whom he wrote many of his hymns. His collection, entitled *Infant Breathings*, contain compositions of a very tender and simple nature, admirably suited for the little pilgrims for whom he wrote. He took a great interest in all church work, and was for many years churchwarden of St. Barnabas', Homerton. His hymns were written at odd moments, and generally in the evening when he had laid aside his professional duties. He was a constant visitor to the London Orphan Asylum, and for the children there he wrote what is perhaps his second best-known composition—" Lead us, Heavenly Father, lead us." Though Mr. Edmeston has written between 1,500 and 2,000 hymns, only the two mentioned here can be said to have come into common use.

"The day is past and over" is one of Dr. John Mason Neale's many translations from the Greek. Dr. Neale attributes the authorship of this hymn to St. Anatolius, but there appears to be some doubt on the point according to many hymnologists. In the preface to his *Hymns of the Eastern Church*, where "The day is past and over" was first published, Dr. Neale says :—

"This little hymn, which, I believe, is not used in the public service of the Church, is a great favourite in the Greek Isles. Its peculiar style and evident antiquity may well lead to the belief that it is the work of St. Anatolius. It is, to the scattered hamlets of Chios and Mitylene, what Bishop Ken's evening hymn is to the village of our land, and its melody is singularly plaintive and soothing."

The original manuscript of Cardinal Newman's hymn:—

> Lead Kindly Light, amid the encircling gloom
> Lead Thou me on!
> The night is dark and I am far from home,
> Lead Thou me on!

owing to the circumstances under which it was composed, is probably not now in existence. The facsimile shown here is from a copy made by the late Cardinal on March 9, 1875, and sent with his prayers and best wishes to a friend. There are several such MSS. to be found among the autograph collections of private individuals, for the Cardinal, in reply to the very numerous requests for his autograph, thought so little of that which seemed to please his correspondents as to forward, instead of a simple signature, a verse of his celebrated hymn.

"Lead, kindly Light" was written during the summer months of 1833, at a time of much mental distress, and the words are a very echo of the author's own loneliness. In his *Apologia pro Vita*

Sua Cardinal Newman tells the story of how the hymn came to be written. While travelling on the Continent he was attacked by a sudden illness which necessitated a stay at Castle Giovanni. Here he lay weak and restless for nearly three weeks, the only friend at hand being his servant, who nursed him during his illness. This occurred early in May, and on the 27th of that month he was sufficiently recovered to attempt a journey to Palermo.

"Before starting from my inn," he wrote, "I sat down on my bed and began to sob bitterly. My servant, who had acted as my nurse, asked what ailed me. I could only answer, 'I have a work to do in England.' I was aching to get home, yet for want of a vessel I was kept at Palermo for three weeks. I began to visit the churches, and they calmed my impatience, though I did not attend any services. At last I got off in an orange boat, bound for Marseilles. We were becalmed for a whole week in the Straits of Bonifacio, and it was there that I wrote the lines, 'Lead, kindly Light,' which have since become so well known."

A great deal of controversy has taken place from time to time regarding the author's meaning in the lines—

MORNING AND EVENING HYMNS

And with the morn those Angel faces smile,
Which I have loved long since, and lost awhile.

Whole chapters have been written on the subject by people who endeavoured to elucidate the mystery. It is not improbable that these essays on the meaning of two lines caused the Cardinal, if he saw them, a good deal of amusement. The author himself, on being asked to solve the problem in 1879, replied that he was not bound to remember his own meaning, whatever it was, at the end of almost fifty years.

"Lead, kindly Light" was written specially as an evening hymn. It has been translated into more foreign languages than any other of Cardinal Newman's compositions. In the *Hymnal Companion* Dr. Bickersteth has added the following verse—

> Meantime, along the narrow rugged path,
>   Thyself hast trod,
> Lead, Saviour, lead me home in child-like faith,
>   Home to my God,
> To rest for ever after earthly strife
> In the calm light of everlasting life.

Cardinal Newman once paid a high compliment to the musical genius of the late Dr. J. B. Dykes, whose setting of his hymn added in no small degree to its beauty. He was being congratulated by a friend on having written so fine a hymn, when he silenced him with the remark, "It is not the hymn that has gained the popularity, but the tune. The tune is by Dykes, and Dr. Dykes was a great master."

The very beautiful evening hymn:—

> The sun is sinking fast,
> The daylight dies;
> Let love awake, and pay
> Her evening sacrifice.

is a translation by the Rev. Edward Caswall of the hymn from the Latin, "Sol praeceps rapitur proxima nox adest." As in the case of several other hymns obtained from a similar source the author is unknown. The Rev. L. C. Biggs, the editor of an annotated edition of *Hymns Ancient and Modern*, did his best to discover the original writer, but in vain. He wrote to Mr. Caswall, who, in reply, informed him that he also had made every effort to discover the original of this hymn, but without success. "It was," the translator believes, "in the possession of one of the former members of the Edgbaston Oratory, contained in a small book of devotions. It can scarcely be older than the eighteenth century."

The translations from the Latin by Edward Caswall are equal in point of merit to those by Dr. Neale. Though he wrote many original hymns, one only can be said to have become really familiar to the hymn-singing public. On the other hand his translations are widely sung and appreciated both in this country and in America.

Edward Caswall, son of the Rev. R. C. Caswall, was born at Yately in 1814. He was educated at Brasenose College, Oxford, and after graduating with honours he became, in 1840, incumbent of Stratford-sub-Castle, near Salisbury. In 1847 he resigned this living, and after seceding to the Church of Rome was received into the Oratory at Edgbaston, where he remained until his death. Mr. Caswall was a very devotional man, warm-hearted, wonderfully good to the poor, and passionately fond of children. Nearly all his hymns and other poems were written at the Oratory, Edgbaston. Though a considerable number of Mr. Caswall's original hymns are to be met with in Roman Catholic collections, few have found their way into Protestant hymn books.

Another instance of a hymnist who is known and remembered by a single composition is Canon Twells. His :—

> At even ere the sun was set,
> The sick, O Lord, around Thee lay;
> Oh, in what divers pains they met!
> Oh, with what joy they went away!

has gained for the author a foremost place in hymnody. "It was composed in 1868," wrote Canon Twells in a letter addressed to me, some few years since, "at the request of my friend, Sir Henry Baker, at that time Chairman of the Committee of *Hymns Ancient and Modern*, who said they wanted a new evening hymn. They were just about to bring out the first Appendix, and it was in this Appendix that the hymn was first published. I have been asked to insert it in 127 hymnals, and many more have taken it without asking me. No other of my hymns has attained a similar popularity.

"The hymn as I originally wrote it consisted of eight verses, but on the recommendation of Sir Henry Baker the fourth stanza—

> And some are pressed with worldly cares,
> And some are tried with sinful doubt:
> And some such grievous passions tear,
> That only Thou canst cast them out

was omitted."

MORNING AND EVENING HYMNS

In a second letter dealing with the apparent contradiction between the text on which the hymn was founded—"And at even when the sun did set, they brought unto Him all that were diseased, and them that were possessed with devils. And all the city was gathered together at the door"—and the opening line of the hymn —"At even ere the sun was set"—Canon Twells thus defends his reading of the Apostles' account of the healing of the sick.

"I should like to point out," he writes, "that there is no contradiction whatever, seeming or otherwise, between the first line and the text at the head. 'At even when the sun did set' (St. Mark) is surely not the same as 'At even when the sun had set.' There is no pluperfect either in the Greek or the English. The plain common sense meaning is that the incident took place at sunset, i.e., during sunset—not after it. If there were any doubt at all about the matter it would be settled by the corresponding passage in St. Luke, 'When the sun was setting.' The hymn merely states that they brought the sick before the sun was absolutely set—the simple fact, if we are to believe the scriptural narrative. There is no sort of discrepancy or shadow of discrepancy between :—

"When the sun did set" (St. Mark).
"When the sun was setting" (St. Luke).
"Ere the sun was set" (Hymn).

All are in perfect accord with the old painters, the glow of the setting sun resting upon the faces of the sick and infirm folk."

The strange idea that there is a contradiction was first started by Prebendary Thring. When compiling his *Church of England Hymn Book* Prebendary Thring wrote to Canon Twells asking for permission to insert " At even," at the same time pointing out to him that the opening line did not exactly coincide with the text on which the hymn was founded. Would the author kindly permit him to change the first line to " At even when the sun did set "? After some correspondence Canon Twells allowed him to make the alteration, though, as the author remarks in the letter quoted, he had never met with any one who agreed with Prebendary Thring as to the necessity.

" Now that the daylight fills the skies " is John Mason Neale's very beautiful translation of a hymn from the Latin, ascribed by many writers to St. Ambrose. It has never been conclusively proved, however, that the famous Milanese bishop was the author, and in the majority of hymnals the hymn is merely stated to be by an anonymous writer of the fifth century. Though there have been many translations of this morning hymn published—possibly thirty or more—that by Neale is by far the most popular and is to be found in nearly all hymnals published during the last fifty years. In most collections, however, it has been altered more or less by editors (in many cases to its disadvantage) and probably not one hymnal gives the text exactly as Neale wrote it. In Thring's collection the second verse—

> May He restrain our tongues from strife,
> And shield from anger's din our life;
> And guard with watchful care our eyes
> From earth's absorbing vanities,

has been altered by the Editor to—

> Would guard our tongue in every word,
> Lest sounds of angry strife be heard;
> From all ill sights would turn our eyes,
> And close our ears from vanities.

"The day is gently sinking to a close," one of the most beautiful of Christopher Wordsworth's evening hymns, is taken from his *Holy Year*. It was written about 1862 and first printed in leaflet form together with a companion hymn for morning service, "Son of God, Eternal Word."

As my readers are probably aware, these two

hymns, "Son of God, Eternal Word," and "The day is gently sinking to a close," now form the opening hymns to the *Holy Year*. Both compositions are well known in Great Britain and America, though neither has attained that popularity enjoyed by many of Bishop Wordsworth's other hymns.

It is somewhat unfortunate that apparently no hymn by this writer appears to have been preserved in manuscript. In a letter received from his son, dated June 6, 1895, the writer says:

"I am sorry to say that I have no certain recollection, even at the time of their composition (which I do recollect), of seeing the manuscript of my father's hymns. I was so far alive to such matters that I begged for a heap of 'copy' of his 'Commentary' and had it bound. I have some notion that the hymns were originally written on stray pieces of paper, very possibly half in pencil, with corrections, and then copied 'fair' by my mother or sisters for the press. My father was such a prolific writer that in house-movings there was nothing for it but large destruction of copy which had done its work."

In a further letter he says:

"I regret to say that my sister, Mrs. Steedman, tells me that my father destroyed the

MSS. when the hymns came back from the printers. My mother and I laid hands on some manuscript of another of his works, but that was rather later. I am afraid there can be no doubt that the originals of the *Holy Year* are no longer extant. As I have for many years, almost from childhood, been interested in the bibliographical side of my father's works I should almost certainly have found them if they existed."

From another source I learned that Bishop Wordsworth wrote his hymns at all sorts of odd moments and in all sorts of places—in the train, riding, or during a walk. If at night he was unable to sleep he would often get up and compose a few verses. The hymns were written on the backs of envelopes, small scraps of sermon paper, and even on the margin of the book he happened to be reading. He was an extremely rapid writer, but spared no pains in correcting until the composition satisfied him. Between forty and fifty hymns have been taken from the *Holy Year* by various hymnal editors and are now said to be in common use in this country. Bishop Wordsworth died in 1885 at the age of seventy-eight.

The hymnal compositions of the late Canon Bright have been growing in favour year by year. His beautiful hymn for the close of service, " And now the wants are told that brought," was written in 1865 and published the following year in the author's *Hymns and other Poems*. It soon attracted the attention of Sir Henry Baker, and when the Appendix to *Hymns Ancient and Modern* was under discussion he asked Canon Bright for permission to include it, a request which was readily granted. The hymn was originally written in six stanzas, Canon Bright subsequently adding the following Doxology :—

> All glory to the Father be,
> All glory to the Son,
> All glory, Holy Ghost, to Thee,
> While endless ages run.

Canon Bright had strong opinions with regard to the question of "copyright" in hymns. His compositions were always at the service of any one who cared to ask him for them. He averred that if a man wrote a good hymn, a hymn which had that in it which could bring comfort and consolation, the author had no more right to withhold it from the public than a publisher has to "copyright" the Psalms.

Canon Bright was extraordinarily painstaking in regard to his hymnal compositions. Though he frequently wrote spontaneously, he would spend hours of thought and care in altering and improving a single line. Fortunately, his hymns have escaped alteration and are published in the majority of hymnals pretty much as he wrote them.

Having called Canon Twells over the coals with regard to his "At even ere the sun was set," it was only right that Prebendary Thring should in his turn be brought to book in respect to one of his own compositions. The hymn in question is "The radiant morn hath pass'd away," the second verse of which originally ran :—

>   Our life is but a fading dawn,
>   Its glorious noon how quickly past ;
>   Lead us, O Christ, when all is gone,
>   Safe home at last.

A correspondent took the liberty of pointing out to the author the inconsistency of referring to the dawn as "fading," inasmuch as the dawn does not fade, but rather increases in brilliancy. Prebendary Thring, however, had already been struck with a similar idea, and he therefore altered the verse to the following :—

>   Our life is but an autumn day,
>   Its glorious noon how quickly past ;
>   Lead us, O Christ, Thou Living Way,
>   Safe home at last.

Subsequently Prebendary Thring again altered this verse to:—

> Our life is but an autumn sun,
> Its glorious noon how quickly past;
> Lead us, O Christ, our life-work done,
> Safe home at last.

This last is the author's revised version, and the one he would like to see copied by hymnal editors.

"The radiant morn hath pass'd away" is generally supposed to have been written for use at evening services, but this is not the case. In a letter received from the author a couple of years ago, Prebendary Thring gives the following particulars regarding the hymn and the object for which he wrote it.

"The hymn in question," he writes, "was composed as an 'afternoon' hymn, as in most of the country parishes in that part of Somersetshire in which I lived, the second service was nearly always held in the afternoon, and not in the evening, whilst all the hymns in the hymn books in common use were for the late evening or night. I wrote 'The radiant morn hath pass'd away' to supply this want. Several of my hymns were written in consequence of some want of this kind, felt either by myself or others; but most of

MORNING AND EVENING HYMNS

them, I think, though I have never made any calculations, arose almost spontaneously from thoughts that happened to be running in my mind at the time."

Prebendary Thring is the author of several volumes of poems, the most important being his *Hymns and Sacred Lyrics*. This work met with disaster soon after its publication, the whole edition being burnt in a great fire at the publisher's works. The author only discovered the fact of the fire some time afterwards, and then merely by accident — a stranger having written to him asking how he could get a copy, as he had been told by every bookseller to whom he had applied that it was "out of print!"

Prebendary Thring has written a great many hymns, fifty-nine of which appear in his *Church of England Hymn Book*. They have nearly all been written with an "object," and are, with very few exceptions, of great excellence.

"O Jesu, Lord of Heavenly grace" is John Chandler's translation of "Splendor paternae gloriae," by St. Ambrose. The exact date of its composition is uncertain, but the authorship is undoubted. In olden days it is said to have been invariably sung at Matins every Monday. Of the many translations which have been published

John Chandler's has long been the most popular, being found in a large number of English and American hymnals. John Chandler was for many years Vicar of Witley, and devoted much of his time to the translation of Latin hymns. In the preface to his *Hymns of the Primitive Church*, where " O Jesu, Lord of Heavenly Grace" appears next to Bishop Ken's " Awake, my soul, and with the sun," Mr. Chandler thus accounts for the publication of his translations :—

" My attention was a short time ago directed to some translations which appeared from time to time in the *British Magazine*, very beautifully executed, of some hymns extracted from the *Parisian Breviary* with originals annexed. Some, indeed, of the Sapphic and Alcaic and other Horatian metres seem to be of little value ; but the rest, of the peculiar hymn-metre, Dimeter Iambics, appear ancient, simple, striking and devotional—in a word, in every way likely to answer our purpose. So I got a copy of the *Parisian Breviary* and one or two other old books of Latin hymns, especially one compiled by Georgius Cassander, printed at Cologne in the year 1556, and regularly applied myself to the work of selection and translation. The result is the collection I now lay before the public. It will be observed that I have admitted no hymns but what appear to be expressly wanted for the purposes of our Church ; my aim in translating them has been to be as simple as possible, thinking it better to be, of the two, rather bald and prosaic than fine and obscure."

II

Advent Hymns

THOSE hymns written specially for the season of Advent are very numerous, and comprise some of the grandest examples of sacred verse in our hymnals. With very few exceptions, however, those which have taken firmest hold of the affections of the Church are translations from the Latin. That greatest of all Advent hymns, the *Dies Irae*, is generally supposed to have been written by Thomas of Celano in the thirteenth century. The number of translations which have been published, though very difficult to estimate, cannot fall far short of 150, and of these some twenty have become more or less familiar.

To Sir Walter Scott we owe what is, with one exception, probably the finest of all translations of the *Dies Irae*—" That day of wrath, that dreadful day." This hymn forms the concluding stanzas of the sixth canto of " The Lay of the Last Minstrel," and is headed "A Hymn for the Dead." The author leads up to these notable verses in the following lines:—

> Then mass was sung, and prayers were said,
> And solemn requiem for the dead;
> And bells toll'd out their mighty peal
> For the departed spirit's weal;
> And ever in the office close
> The hymn of intercession rose;
> And far the echoing aisles prolong
> The awful burthen of the song,—
> 	Dies irae, dies illa,
> 	Solvet saeclum in favilla;
> While the pealing organ rung.
> 	Were it meet with sacred strain
> 	To close my lay, so light and vain,
> Thus the holy Fathers sung:—
>
> That day of wrath, that dreadful day,
> When heaven and earth shall pass away!
> What power shall be the sinner's stay?
> How shall he meet that dreadful day?
>
> When, shrivelling like a parchèd scroll,
> The flaming heavens together roll;
> When louder yet, and yet more dread,
> Swells the high trump that wakes the dead!
>
> Oh! on that day, that wrathful day,
> When man to judgment wakes to clay,
> Be Thou the trembling sinner's stay,
> Though heaven and earth shall pass away!

Scott published his "Lay" in 1805, and within

a very short period this hymn was singled out by compilers and published in various collections. It soon became the most popular of all the translations, and few hymnals published during the last half-century omit it.

Scott himself was a fervent admirer of this great hymn in the original, and is said to have uttered a few lines of it a short while before his death.

Another translation of the *Dies Irae* which equals Scott's in excellence and popularity is "Day of wrath, O day of mourning," by W. J. Irons. It was written under somewhat remarkable circumstances.

It appears that Mr. Irons was in the French capital during the Revolution of 1848 when, among other atrocities committed, the Archbishop of Paris was murdered. Owing to the revolutionary spirit of the people it was many days before the funeral could take place with any degree of safety to the mourners. About a fortnight later a Memorial Service was held in Notre Dame, at which Mr. Irons was present. The Archbishop's heart, which had been severed from his body, was placed in a glass casket and reverently laid on a raised daïs in the choir so that all who desired to do so might gaze upon it. As the procession of mourners filed by, casting

BERRYHEAD HOUSE,
The birthplace of "Abide with me."
Photo by Upham, Brixham.

On a pane of glass in this house Harriet Auber wrote "Our blest Redeemer, ere He breathed."
From a Photo.

ADVENT HYMNS

looks of mingled terror and affection on the faithful heart which had so recently beat in their interests, the entire congregation sang in muffled tones the *Dies Irae*. As may well be believed the solemnity of the service made a deep and lasting impression on the mind of the English clergyman present, and when the congregation had dispersed he returned to his hotel and immediately made his now celebrated translation of the great Latin hymn:—

> Day of Wrath! O day of mourning!
> See once more the Cross returning,
> Heav'n and earth in ashes burning!"

Dr. William Josiah Irons was born at Hoddesdon—within a short distance of the house in which Harriet Auber wrote "Our Blest Redeemer, ere He breath'd"—in 1812, being the son of Joseph Irons, also a hymnist of some note. After taking his B.A. Degree at Queen's College, Oxford, he took Holy Orders in 1835, and two years later became Incumbent of St. Peter's, Walworth. After filling various livings he became Rector of St. Mary-Woolnoth and Prebendary of St. Paul's Cathedral. He died on June 18, 1883.

Dr. Irons wrote a great number of original hymns, but will be longest remembered by his translation of the *Dies Irae*.

"Hark the glad sound! the Saviour comes" is one of the few original Advent hymns qualified to be placed among those of the Mediaeval Church. It was written by Philip Doddridge in 1835, and first published ten years later in a Scotch hymnal. It was ten years more before it appeared in England, when it was published in a posthumous volume of hymns by Dr. Doddridge. Very few of Dr. Doddridge's hymns were published prior to his death in 1751. They were first circulated in manuscript, in the author's own clear handwriting, and, the number of copies being necessarily limited, were much prized by the fortunate possessors.

Another hymn by Philip Doddridge, which might very well be regarded as an Advent hymn, is "Ye servants of the Lord," written prior to "Hark the glad sound" and not published until after his death. It is written on the words of St. Luke:—"Blessed are those servants, whom the Lord when He cometh shall find watching." As a hymn of exhortation it could hardly be surpassed.

Perhaps the most unpopular hymn Doddridge ever wrote was one for early rising. The very object for which the good doctor wrote proved the death blow to the composition, for if there is one thing more than another to which members of the human race strongly object it is leaving their beds in the small hours. Dr. Doddridge was not unaware of this, and, in order that the hymn should not be entirely wasted, sang it himself. "At five o'clock," says a contemporary writer, "he prepared to leave his bed, repeating five stanzas before doing so; at the sixth he rose and dressed." It is not related how long the doctor lingered over those first five stanzas!

All his life Dr. Doddridge was more or less of an invalid. At his birth he was laid aside as dead, and would have been buried had it not been for the intervention of an affectionate nurse, who refused to believe that the child had been born but to die. He lived for forty-nine years and died at Lisbon from consumption brought on by overwork.

Few hymns have occasioned greater controversy as to their authorship than "Great God, what do I see and hear?" It appears to have been, like "Lo! He comes with clouds descending," the combined work of numerous hymnists and editors. It is closely associated with the great German hymnist, Bartholomäus Ringwaldt, and is said to be a translation of one of his hymns which was published about the year 1556, and frequently sung in his native country during the Thirty Years' War. This, however, is discredited by most hymnologists, as is also the assumption that it was written by Martin Luther.

No record exists as to the actual authorship of the first verse, which seems to have been published anonymously in 1802 in a small volume of hymns. It was there seen by Dr. Collyer, who included it in a hymnal he was compiling in 1812, adding three stanzas of his own composition. In 1820 it was republished by Thomas Cotterill, considerably altered, though, as in the case of Dr. Collyer, the first verse remained true to the original. From that time onward it has had very little rest, appearing in various forms. Few editors were found obliging enough to leave it alone, and in consequence scarcely two hymnals give it in exactly the same form. One of the last to alter it is Prebendary Thring in his *Church of England Hymn Book*, but he has been more merciful than many others. The tune to which this great Advent hymn is allied is called "Luther," and appeared in Joseph Klug's *Gesangbuch* in 1535.

"Thou art coming, O my Saviour," by Miss Frances Ridley Havergal, was written at Winterdyne in 1873. In connexion with this composition it is interesting to recall that it was the first junction with the verse being in the form of a question, perhaps an improvement.

The hymn was based on the sentence taken from the Lord's Prayer, "Thy kingdom come," which, it will be noted, also forms the opening words of the hymn. The very beautiful tune, "St. Cecilia," which was written for this hymn by the Rev. L. G. Hayne, is an especially appropriate one, and has materially helped to sustain the deservedly high reputation which the hymn enjoys.

Mr. Lewis Hensley, who is one of our few living hymnists, is Vicar of Hitchin, in the county of Hertfordshire.

"Hark! a thrilling voice is sounding" is Edward Caswall's translation of the Latin hymn "Vox clara ecce intonat," ascribed by some writers to St. Ambrose. There is, however, considerable doubt with regard to the authorship, and most hymnals content themselves by ascribing it to an anonymous writer.

This hymn has been a favourite one for translation with a great number of hymnists besides Caswall, a very fine rendering being that by Cardinal Newman, beginning, "Hark! a joyful voice is thrilling." Of the many translations, however, that by Caswall has long since been awarded the first place in public affection, and is to-day found in more hymnals than all the other translations put together.

The somewhat plaintive melody to which it is usually sung, called "Veni Emmanuel," is from an ancient Plain-song, the origin of which is lost in antiquity.

"Lo! He comes with clouds descending" may be said to combine the efforts of two hymnists and one editor—Cennick, Charles Wesley and Madan. This celebrated Advent hymn made its first appearance in printed form in Cennick's *Collection of Sacred Hymns*, published in 1752, the opening verse reading—

> Lo! He cometh, countless trumpets
> Blow before His bloody sign!
> 'Midst ten thousand saints and angels,
> See the Crucified shine.
> Allelujah!
> Welcome, welcome, bleeding Lamb!

In 1758 Charles Wesley published a small pamphlet containing forty hymns. In this little book Cennick's hymn was included, altered by Charles Wesley to the now familiar—

> Lo! He comes with clouds descending,
> Once for favour'd sinners slain;
> Thousand thousand Saints attending,

ADVENT HYMNS

Swell the triumph of His train:
Hallelujah!
God appears on earth to reign.

Two years later M. Madan again published it, giving portions by Cennick and portions by Wesley, with additions and alterations of his own. It is in this form that the hymn now appears in modern hymnals.

John Cennick was the author of a large number of hymns, of which a fair percentage are sung to-day. He was born at Reading in 1718, and was educated with the intention of his following the profession of a surveyor. At the early age of twenty-one, however, he came under the influence of the Wesleys, who persuaded him to leave the Church of England and assist them in their work. In 1840 he became a teacher at Kingswood, on the recommendation of John Wesley, his pupils for the most part consisting of the children of colliers. Shortly after his appointment, however, he left the Wesleys and joined George Whitefield. For a few years he worked under this leader, but again becoming restless, left him to join the Moravians. Much of his time he spent in Germany, where his preaching attracted many followers. He died in 1755 at the early age of thirty-seven.

Besides "Lo! He comes," and several other very fine hymns, Cennick was the author of the Graces " Be present at our table, Lord," and " We thank Thee, Lord, for this our food," which one occasionally hears even in these degenerate days, when " Grace " appears to be becoming more and more a custom of the past.

The tune with which " Lo! He comes with clouds descending" is inseparably associated, known as " Helmsley " or " Olivers," appears to have had as many vicissitudes as the hymn itself. In Grove's *Dictionary of Music* a writer says—

" The story runs that Thomas Olivers, the friend of John Wesley, was attracted by the tune which he heard whistled in the street, and that from it he formed the melody to which were adapted the words of Cennick and Wesley's Advent hymn. The tune heard by Olivers is commonly said to have been a hornpipe danced by Miss Catley in the 'Golden Pippin,' a burlesque by Kate O'Hara, but this seems inconsistent with chronology. The hymn tune appeared first, as a melody only, in the second edition of Wesley's *Select Hymns with Tunes Annexed*, 1765, under the name of 'Olivers.' In 1769 an improved version, in three parts, was published by the Rev. Martin Madan in *Hymns and Psalm Tunes*. It is there called

'Helmsley,' and under that name became widely popular. But at this time the 'Golden Pippin' was not even in existence. O'Keefe, who possessed the original manuscript, tells in his *Recollections* that it was dated 1771. The burlesque, in three acts, was produced at Covent Garden in 1773. It failed at first, but obtained some success when altered and abridged. The source from whence 'Olivers' was derived seems to have been a concert-room song commencing 'Guardian angels, now protect me,' the music of which probably originated in Dublin, where it was sung by a Mr. Mahone, and no doubt also by Miss Catley, who resided in the Irish capital in 1750. The melody of 'Guardian angels' was not in the 'Golden Pippin' as originally written, but, adapted to the words of the burlesque, was introduced into it in 1776 in the place of a song by Giordani, and was sung by Miss Catley in the character of Juno. The published score of the 'Golden Pippin' does not contain any horn-pipe, but such a dance may have been interpolated in the action of the piece."

The writer concludes by saying: "In 1765, when 'Olivers' was published, Miss Catley was in Ireland and did not return to London until five years afterwards, and, seeing that the hornpipe was not of earlier date than the 'Golden Pippin,' it seems to follow that instead of the hymn tune having been derived from the hornpipe, the latter was actually constructed from the hymn tune, which by that time had become a great favourite." This recollection may help to console us when next we sing, "Lo! He comes with clouds descending."

The late Rev. Canon Tuttiett has written many hymns for special occasions, his Advent hymn, "O quickly come, dread Judge of all" being remarkably fine. It is based on the words "Surely I come quickly. Amen. Even so, come, Lord Jesus Christ." In some respects it is not unlike a translation of the *Dies Irae*. It was written in 1854 while Mr. Tuttiett was Vicar of Lea Marston, and was one of the first hymns the author ever wrote. Indeed, it was always something of a surprise to the hymnist that its success was so much greater than many of the compositions he wrote in later life, when, it may be supposed, he had gained experience regarding the kind of hymns likely to prove most attractive.

"O quickly come, dread Judge of all" was published in the author's *Hymns for Churchmen*, after which it passed into the first Appendix to *Hymns Ancient and Modern*, and from thence

ADVENT HYMNS

into numerous hymnals both in Great Britain and America. It has been translated into several languages, including the Latin, and is probably the most popular of all Canon Tuttiett's compositions.

Soon after its publication it was seen by the Rev. J. B. Dykes, who wrote for it that very beautiful and solemn air which he christened " Veni cito."

Canon Tuttiett was born at the little village of Colyton, in Devon, in 1825, his father being a surgeon in the Royal Navy. It was intended that he should also follow the profession of a doctor, but in 1848 he abandoned this intention in favour of the Church. While acting as curate in the Isle of Wight his preaching attracted the attention of Lord Norton, who, in 1854, presented him with the living of Lea Marston, which he filled for sixteen years. Lord Norton used to affirm that Mr. Tuttiett's preaching equalled that of Bishop Wilberforce. In 1870 he became Incumbent of the Episcopal Church of St. Andrew's, and ten years later was preferred to a Canonry in St. Ninian's Cathedral, Perth. Canon Tuttiett remained at St. Andrew's until 1893, when owing to ill-health he was obliged to resign.

III

Christmas Hymns

IF a general consensus of opinion were taken as to which is the most popular of all Christmas hymns the result would probably be in favour of "Hark! the herald-angels sing." This famous hymn was first published by Charles Wesley in 1739, when it began "Hark! how all the welkin rings." From that date it has appeared in various hymnals with alterations by various editors. In one version, indeed, the opening stanza contained but a single word to be found in the original, that word being the exclamatory "Hark." As first published this hymn consisted of ten stanzas of four lines each, but was subsequently reduced to eight, then to six, and, finally, to three extended stanzas of eight lines each, with the refrain—

"Hark! the herald-angels sing
Glory to the new-born King."

Apparently no manuscript of this hymn exists, though Mr. Kelly, the Book Steward at the Wesleyan Conference Office, thinks otherwise. One day he hopes to make a systematic search among the many hundreds of Charles Wesley's manuscripts now under his care, when he feels confident the original of "Hark! the herald-angels sing" will come to light.

It is rather curious that these hymnal manuscripts of Charles Wesley have only been discovered within comparatively recent years. The story of how they were found is an interesting one. It appears that soon after Mr. Kelly was appointed Book Steward he was going on a tour of inspection through the cellars of No. 2, Castle Street, when he came to a small underground room which was boarded up. On having the barricade removed it was discovered that the compartment contained many things of interest connected with the Wesleys, not the least important being some fourteen volumes of manuscript hymns in the handwritings of Charles, John, and Samuel Wesley. Together with these volumes were numerous loose sheets of paper, on which the celebrated hymnists had written the first drafts of their compositions before copying them "fair" into the bound volumes. These loose sheets are extremely interesting, as they contain numerous corrections which convey some idea of what the first impressions of the hymnists really were.

Charles Wesley composed his hymns at all times of the day and night, never knowing when a fine line or verse would strike him. In order that these ideas should not be lost, he was in the habit of carrying about with him a set of tablets on which many of his hymns were written. He would then copy out these rough notes on a sheet of quarto paper, correct and finally copy "fair" into a manuscript book. Charles wrote a beautifully clear hand, bold, and, as was his character, fearless and straightforward.

"Hark! the herald-angels sing" is said to be found in more hymnals than any other of Charles Wesley's compositions, not even excepting "Jesu, Lover of my soul." It has been translated into numerous languages and dialects, and is as familiar to the converted natives of China and Africa as it is to the church-going population of our own country.

"Christians, awake, salute the happy morn" has been a favourite Christmas hymn for the last 150 years or more. It was written in 1745, and the story of its composition is a pretty one. John

CHRISTMAS HYMNS

Byrom, the author, had several children, but, like many another father, he had his "favourite." This child was a little girl named "Dolly," who afterwards became Mrs. Dorothy Byrom. A few days prior to Christmas, 1745, Mr. Byrom, after having had a romp with the favoured Dolly, promised to write her something for Christmas Day. It was to be written specially for herself, and no one else. The child, highly honoured and delighted, did not fail to remind her father of his promise each day as Christmas drew nearer. On the morning of the great day, when she ran down to breakfast, she found several presents waiting for her. Among these was an envelope addressed to her in her father's handwriting. It was the first thing she opened, and, to her great delight, proved to be a Christmas carol addressed to her, and to her alone :—

*Christians awake, salute the happy Morn
Whereon the Saviour of the World was born*

In the original the MS. is headed "Christmas Day for Dolly." This very document is now preserved in the archives of Cheetham's Hospital, Manchester, and, though a little creased and crumpled, probably from being carried about in

Miss Dolly's pocket, is in an excellent state of preservation. After remaining in the Byrom family for close upon a century it passed into the possession of James Crossby, at one time honorary Librarian to Cheetham's Hospital, and at his death was purchased by the authorities of that institution.

The carol—it is perhaps more a carol than a hymn—was first published in 1746 in Harrop's *Manchester Mercury*, where it soon attracted the attention of hymnists and others. Soon after the publication of the carol it was seen by the organist of the parish church (now the pro-cathedral), a clever musician called John Wainwright, and a composer of some note. He was struck by the words, and wrote for them the beautiful and popular setting known as "Stockport" or "Wainwright," to which they have ever since been sung. The composer sent a copy of the tune to the author of the hymn, who was delighted with its appropriateness, and this introduction began a friendship between the two men which was only broken by death.

John Byrom had a chequered career. In a letter received some time ago from Cheetham's Hospital, Mr. W. E. A. Axon, of Manchester, thus writes of him—

"He was born in Kelsal in 1691. His father, a linen draper of Manchester, sent him to Cambridge, where he graduated M.A. and became a Fellow of Trinity College; but, declining to take Orders, he resigned this provision, and soon after married his cousin, Mrs. Elizabeth Byrom, against the consent of both families. Being without a profession, and pressed by the *res angusta domi*, he repaired to the metropolis, and supported himself by teaching shorthand, of which he had invented the best system then before the public. In 1723 he was elected Fellow of the Royal Society, and in the following year succeeded, by the death of his elder brother, to the family estates, when he returned to Manchester, where he remained till his death in 1763."

Byrom's early boyhood was passed at the King's School, Chester, a fact of which Cestrians are not a little proud, for though the old school can boast a goodly list of famous men whose education has been carried on within its walls, the author of "Christians, awake" will not be the first forgotten.

To the library which now contains most of his MSS. John Byrom was a frequent visitor, and numerous are the hymns which he wrote for the Cheetham Hospital boys. In a letter to Thyer, the then librarian, he says he preferred that employment to being Laureate to Frederick II., who was then occupied in the Seven Years' War.

The portrait of this hymnist, ugly at first sight, but kindly and good-humoured on closer examination, is from a print contained in the College "Scrap-book." It is said to have been executed by a friend.

Only one other hymn by Byrom has come into general use, and is to be met with in our hymnals. It is a tender little composition and somewhat original. The manner in which the last line of one verse forms the first line of the following verse is quaint and effective. As it is short, consisting of four stanzas only, I quote it—

    My spirit longeth for Thee,
    Within my troubled breast,
    Though I unworthy be
    Of so Divine a guest :

    Of so Divine a guest
    Unworthy though I be,
    Yet has my heart no rest
    Unless it come from Thee :

    Unless it come from Thee,
    In vain I look around ;
    In all that I can see
    No rest is to be found :

> No rest is to be found
> But in Thy blessed love;
> O let my wish be crowned,
> And send it from above.

This little hymn deserves to be far better known than it is at present.

"As with gladness men of old," though, strictly speaking, an Epiphany hymn, is frequently sung at Christmas, and therefore no apology is needed for giving it a place under this chapter. It was written by the late Mr. William Chatterton Dix about the year 1860. In a letter received from the author shortly before his lamented death in 1900, Mr. Dix informed me that there was little of interest to record respecting its composition. He was unwell at the time, slowly recovering from a rather serious illness. One evening, when he felt somewhat stronger than he had for several days, the lines of the now well-known hymn gradually formed themselves in his brain, and, asking for writing materials, he committed them to paper. The following year it was published in a small hymnal, which had a very limited circulation. From thence it made its way into more popular collections, and to-day its reputation has become world-wide. The late Lord Selborne, who was a great admirer of Mr. Chatterton Dix's hymns, considered

"As with gladness men of old" one of the finest compositions of the kind in the language.

The wonderfully appropriate melody to which this hymn is invariably sung was composed by Conrad Kocher in 1838. This tune was first connected with Mr. Dix's hymn in *Ancient and Modern* and will now probably always be associated with it. It has been suggested that Mr. Chatterton Dix wrote the hymn to fit the music, but this is not the case.

"While shepherds watch'd their flocks by night" has for a hundred and fifty years or more been the standard carol of the "Waits" at Christmas time. It was written by Nahum Tate some time near the close of the seventeenth century, and first published in Tate and Brady's *Psalter*, 1702. There is little doubt that this hymn was the work of Tate alone, though in some hymnals it is ascribed to these writers jointly.

Nahum Tate was an Irishman, and born in Dublin in 1652. He was educated at Trinity College and at an early age became Poet Laureate. He collaborated with Brady in producing in 1696 *A New Version of the Psalms of David*. The different hymns in this volume were written to certain tunes already familiar in the churches, and though the book was subjected

CHRISTMAS HYMNS

to a good deal of adverse criticism it soon began to be used in a considerable number of churches. How the work was divided between the two, what portion one wrote and what the other, will never be known, and therefore those versions of the Psalms which have passed from this work into our hymnals—"Through all the changing scenes of life," "As pants the hart for cooling streams," etc., have come to be regarded as by Tate *and* Brady. One of the few hymns conclusively proved to be by Tate alone is "While shepherds watch'd." After leading anything but a temperate life, Tate died in London at the age of sixty-three.

"O come, all ye faithful, Joyful and triumphant" is Canon Oakeley's translation of the *Adeste Fideles*. It is by far the most popular version we have, and was first published in 1852. It soon became one of the foremost hymns for Christmas and is now to be found in nearly all hymnals published during the last forty years. Translations of Canon Oakeley's version have been published in the languages of nearly all countries where missions have penetrated, and the hymn is a general favourite among the Malays and Dyaks. In most cases the translator has adhered to the original metre so that the music to which the translation is sung is the same

F.H.

as that with which we are all familiar—*Adeste Fideles*.

This is not the only translation of this hymn made by Canon Oakeley. When he was Incumbent of Margaret Street Chapel in 1841 he made another and an earlier one beginning " Ye faithful, approach ye." This, however, was written exclusively for his own congregation, amongst whom it was distributed in manuscript. Many years afterwards, however, it appeared in one or two collections, though it has never gained much favour with hymnal editors. Indeed, Canon Oakeley may be said to have himself cast in the shade his first translation by the brilliancy of his second.

Canon Oakeley is but one of many celebrated hymnists who have left the Church of England in order to enter the Church of Rome. He was born in 1802 and educated at Christ Church, Oxford, taking Holy Orders in 1828. After becoming a Prebendary of Lichfield Cathedral he was in 1839 made Incumbent of Margaret Street Chapel, London. Here he remained for six years, but in 1845 the great change of his life came and he seceded to the Church of Rome. For many years he worked among the poor of Westminster, leading an uneventful but exemplary life. He died in 1880 at the age of seventy-eight.

CHRISTMAS HYMNS

Among the many other translations of the *Adeste Fideles*, which have been published during the last fifty years or so, the only one which comes into competition with Canon Oakeley's in the matter of public favour is by the Rev. William Mercer, beginning with the same line, " O come, all ye faithful." It was written later than Canon Oakeley's version, and the two translations are not dissimilar. In order that readers may compare the two, I give the first verse of Mr. Mercer's translation; that by Canon Oakeley will be found in any ordinary modern hymnal—

> O come, all ye faithful,
> Joyfully triumphant.
> To Bethlehem hasten now with glad accord;
> Lo! in a manger
> Lies the King of angels;
> O come, let us adore Him, Christ the Lord.

Apart from his translation of the *Adeste Fideles*, Mr. Mercer is little known as a hymn writer. He gained great reputation, however, as an editor, his *Church Psalter and Hymn-book* being at one time the most popular collection in the language. He was for many years Vicar of St. George's, Sheffield, where he died in 1873 at the age of sixty-two.

A Christmas hymn, which is immensely popular in America, and growing in favour in England, is

"It came upon the midnight clear." It is the work of an American writer, and was written in 1849. It was sent to the editor of the *Christian Register*, and first appeared in that magazine early in 1850. From thence it passed into various American hymnals, and soon made its way to England. The author, Edmund Hamilton Sears, has also written another hymn for Christmas, which is gaining in popularity in this country. Though not as perfect a composition as " It came upon the midnight clear," this hymn, beginning, "Calm on the listening ear of night," is a very beautiful Christmas hymn, and deserves to be more widely used. Mr. Sears was pastor of various churches in America, and died at Weston, Massachusetts, in 1876.

"From heaven above to earth I come" is Miss Catherine Winkworth's translation of Martin Luther's celebrated Christmas hymn. It was written specially for his little son, Hans, when the child was five years old, during the latter part of 1531, but not published until four years later. No man observed Christmas with more punctiliousness than did Martin Luther, and he educated his children to regard the season with similar veneration.

On every Christmas Eve a festival took place

in Martin Luther's house, and it was for the children's use at these festivals that " From heaven above to earth I come" was written. Luther himself was on these occasions the master of the ceremonies, and for many days before Christmas he conducted rehearsals, training his children to sing the carol perfectly when the festival night came. The first seven verses of the hymn were always sung by a man dressed as an angel. When he had finished, the children greeted him with the remaining verses, beginning—

>   Welcome to earth, thou noble guest,
>   Through whom e'en wicked men are blest!
>   Thou com'st to share our misery;
>   What can we render, Lord, to Thee?

In early editions of Luther's *Hymnal* this hymn is headed "Children's Song from the Second Chapter of St. Luke. Drawn up by Dr. M. L." It is still sung from the dome of the Kreuzkirche in Dresden before daybreak on Christmas morning.

The tune to which this hymn is usually sung is also ascribed to Martin Luther, but whether he really wrote it or not it is difficult to say.

"Of the Father's love begotten" is a translation of a portion of a poem by Prudentius, written in the fifth century. This poem, which is of considerable length, deals with the miracles of Christ, and has long been a favourite with translators. By far the most popular version, however, is that beginning " Of the Father's love begotten," which first appeared in *Hymns Ancient and Modern*. It is the combined work of John Mason Neale and Sir Henry Baker. The manner in which the two hymnists made their translation is interesting, Each translated the hymn independently, afterwards comparing the translations. The best verses were then chosen from each translation, and thus the hymn was built up. Each freely criticised the other's work, the result being a very fine hymn. I might mention that Neale had the honour of contributing the first verse. Of the entire hymn, Neale contributed five stanzas, and Baker four. It is usually sung to a very beautiful Plainsong, " Corde natus," so called from the first two words of the original Latin.

A hymn which should by rights be included in a section devoted to the season of Epiphany, is Bishop Heber's :—

Brightest and best of the Sons of the morning!
Dawn on our darkness and lend us thine aid,
Star of the East, the Horizon adorning,—
Guide where our Infant Redeemer is laid!

It is, however, becoming year by year more

1 THE REV. T. J. POTTER.
From a Photo.
2 MRS. M. F. MAUDE.
From a Photo.
3 THE VERY REV. HENRY HART MILMAN, D.D.
From a Photo.

often sung at Christmas, and I therefore give it a place in this chapter. This hymn was first published in a weekly paper called the *Christian Observer*, in 1811. It did not, however, make its appearance in any hymnal until after Heber's death, when it was published in a collection of his own hymns. It has gradually made its way in public favour until to-day it is side by side with the same author's "Holy, holy, holy, Lord God Almighty," in point of popularity. Curiously enough it has not yet been included in *Hymns Ancient and Modern*.

The manuscript of this hymn will be found in a small collection of hymns compiled by Bishop Heber, now deposited in the Manuscript Department of the British Museum. This collection, all the hymns of which are written in Bishop Heber's wonderfully clear handwriting, consists of two volumes made up of a couple of ordinary twopenny exercise books, which possibly belonged to one of his own children, for the backs are scribbled all over with problems in Euclid.

This MS. collection of hymns, made by Heber after seeing the Olney hymns, of which he was a great admirer, belonged to Dean Milman, having been presented to him by his friend the Bishop. There were no hymns Heber admired more than Dean Milman's, and the collection contains several by the latter hymnist. After remaining for many years in Dean Milman's family, it was presented to the British Museum.

The most successful Christmas hymn which James Montgomery ever wrote was "Angels from the realms of glory," which first appeared, like many others of his composition, in a Sheffield newspaper called the *Iris*, of which he was both owner and editor. It was printed in the Christmas Eve number for the year 1816, and attracted no more notice at the time than any other contribution in the paper. For many years it remained practically unknown, appearing in no hymnal of note until 1825, when the author republished it in his *Christian Psalmist*. After that date it began to find its way into numerous hymnals, both British and American, and to-day it ranks with the best of his compositions.

Montgomery is the author of between 400 and 500 hymns, many of which are worthy to rank with the very best of Watts, Wesley, Cowper, and Doddridge. His hymns were first circulated in manuscript, and a great number of these interesting documents have been preserved. He usually wrote on half sheets of writing paper, and each

CHRISTMAS HYMNS

composition bore at the foot the date when it was written, and his signature. Though some of his hymns came spontaneously, he corrected a good deal; and even after publication he could seldom resist making further alterations when fresh editions were needed. He was extremely self-critical, and was as hard on his own compositions as those of other writers. A man of sterling qualities, his character was as devotional, simple and refined as his hymns. He passed away in his sleep in 1854, his funeral taking place with full public honours at Sheffield.

"Waken, Christian children" is a Christmas hymn which I cannot omit from mention in this little volume. It was written when the author, Samuel Collingwood Hamerton, was barely twenty years of age. It appeared first of all, I believe, in a magazine and was afterwards republished in a small book of carols in 1872—the year the author died. It was based on the text: "They saw the young Child with Mary His mother, and worshipped Him: and they presented unto Him gifts." This hymn, or carol, as the author called it, is specially suited to children by reason of the following verses—

> Fear not then to enter,
> Though we cannot bring

> Gold, or myrrh, or incense
> Fitting for a King.
>
> Gifts He asketh richer,
> Offerings costlier still;
> Yet may Christian children
> Bring them if they will.
>
> Brighter than all jewels
> Shines the modest eye;
> Best of gifts, He loveth
> Infant purity.

Mr. Hamerton was for many years Vicar of St. Paul's, Warwick, where he died on January 6, 1872, in the fortieth year of his age. "Waken, Christian children" is the only hymn by Mr. Hamerton to be found in our hymnals. The music to which it is allied is also by the author—a simple melody admirably fitting the words.

Dean Farrar has not written a great number of hymns, in fact they might all be counted on the fingers of one hand, but he is the author of several carols, at least one of which should be in every hymnal. I refer to his:—

*In the fields with their flocks abiding
They lay on the dewy ground,*

of which the following is the first verse—

In the field with their flocks abiding,
  They lay on the dewy ground,
And glimm'ring under the starlight
  The sheep lay white around,
When the light of the Lord streamed o'er them,
  And lo! from the heaven above
An angel leaned from the glory,
  And sang his song of love;
He sang, that first sweet Christmas,
  The song that shall never cease—
Glory to God in the highest,
  On earth good-will and peace.

This carol, Dean Farrar tells me, was written while he was an assistant master at Harrow School. It was composed expressly for the boys there, and was frequently sung in the chapel. It has been very effectively set to music by Mr. John Farmer in his "Christ and His Soldiers."

Dean Farrar is also the author of "Father, before Thy throne of light" and "God the Father, great and holy," two hymns of great merit, which are not in as many hymnals as they might be.

I cannot bring this chapter to a conclusion without quoting a little story relating to a Christmas carol sung in the olden days, which I came across while making researches in the British Museum. It is told in "Pasquil's Book of Jests" and headed "A Merrie Carroll sung by Women."

"There was sometime an olde knight," relates

the chronicler, " who, being disposed to make himselfe merrie in a Christmas time, sent for many of his tenants and poore neighbours with their wives to dinner ; when, having made meat to be set on the table, would suffer no man to drinke until he that was master ouer his wife should sing a carroll to excuse all the company. Muche niceness there was who should be the musician, yet with muche adoe, looking one upon the other, with a dry hemme or two, a dreamy companion drew out as muche as he durst towards an old fashioned-ditty, when, having made an end, to the great comfort of the listeners, at last it came to the women's table, where likewise commandment was given that there should no drinke be touched until she that was master ouer her husband had sung a Christmas carroll, whereupon they all fell to such a singing that there was never heard such a catterwalling peece of musicke, whereat the knight laughed so heartilly that it did him as muche good as a corner of his Christmas pie."

IV
Hymns Suitable for the New Year

IT is somewhat remarkable that the number of hymns specially composed for the New Year should constitute the smallest section of our hymnals; yet so it is. I shall, therefore, make no apology for including in this chapter many hymns which, while not written specially for this season of the year, are regarded by the majority of hymnologists as more or less suited for congregational singing on New Year's Day.

"Our God, our help in ages past," or, as it is given in most hymnals, "O God, our help in ages past," is particularly suited for the New Year, forcing upon us, as it does, the remembrance that earthly life is, at best, but short. The hymn is a paraphrase of the 90th Psalm and was first published by Dr. Watts in his *Psalms of David* in 1719. Of all Isaac Watts' compositions this is, perhaps, the finest, outrivalling even his "When I survey the wondrous Cross."

Whatever hymns of Dr. Watts may cease to be sung, "O God, our help in ages past" will live so long as the Church endures. In the original it consisted of nine verses, but these have long since been reduced to six. It was Charles Wesley who changed the opening line to "O God, our help in ages past," besides making many other alterations.

Dr. Watts is rightly looked upon as the founder of English hymnody. He composed many of his hymns "to order," which probably accounts for a great number of them being very poor stuff. For two years he is said to have written a new hymn every week, the majority of which were sung in the church in Southampton where he was accustomed to worship. A story is told that one day Watts complained to one of the deacons respecting the poorness of the hymns which were then in use, and that the deacon, rather nettled at the remark, ironically suggested that he should write something better himself. Watts, not at all offended, decided to do so, and the following Sunday arrived with his first hymn, which was sung by the congregation, being repeated after the clergyman. This hymn was "Behold the glories of the Lamb," and was so favourably received that the author was requested to write another for the following

Sunday. It was in this way that Dr. Watts' reputation as a hymnist began.

The tune to which "O God, our help in ages past" is almost universally sung, known as "St. Anne," was in existence some thirty years before the hymn was written, and first appeared in *Barber's Psalm Tunes* 1687, the composition being ascribed to "Mr. Denby." It is not at all improbable, therefore, that the hymn was sung to this same tune in the days when Watts himself joined in the singing of his own compositions.

Probably the greatest favourite of all hymns specially written for New Year's Day, and the one which is most often sung on that occasion, is the late Canon Tuttiett's "Father, let me dedicate." It is a hymn full of promise and hope for the coming year. In America it is equally popular, being invariably sung at every New Year's service, both in chapel and church. The American versions, however, nearly all begin with "Father, here we dedicate," and the number of alterations are remarkable. The final verse in some hymnals has been altered from the original to—

> If we must in grief and loss
> Thy behest obey,
> If beneath the shadowy Cross
> Lies our homeward way ;

> We will think what Thy dear Son
> Once for us became,
> And repeat till life is done,
> Glorify Thy Name.

It is interesting to note that in this verse the only line which adheres to the original is the last—"Glorify Thy Name."

This hymn was written sometime during the year 1864, while Mr. Tuttiett was Vicar of Lea Marston, in Warwickshire. In a letter received from the author in 1895, Canon Tuttiett tells me that there is no particular incident connected with the writing of this hymn. The author had been struck for some time by the fact that hymns for New Year's Day were particularly scarce, and he therefore determined to try if he could supply what he felt was a very evident want. "Father, let me dedicate" was the result.

A very beautiful hymn, written specially for the Old and the New Year, is Horatius Bonar's "A few more years shall roll." This hymn was composed while Mr. Bonar was superintendent of the Sunday School attached to the Church of St. James', Leith. It was one of the first he ever wrote, and was composed expressly for his Sunday School scholars.

Dr. Bonar wrote his hymns at odd times and

under various circumstances. The Rev. R. H. Lundie, an intimate friend of the late hymnist, said in a memorial service preached at Grange, Edinburgh :—" His hymns were written in very varied circumstances, sometimes timed by the tinkling brook that babbled near him; sometimes attuned to the ordered tramp of the ocean, whose crested waves broke on the beach by which he wandered; sometimes set to the rude music of the railway train that hurried him to the scene of duty; sometimes measured by the silent rhythm of the midnight stars that shone above him."

" A few more years shall roll " was first printed in the form of a leaflet, and distributed among the congregation of St. James', Leith. Here it was sung for the first time on New Year's Day, 1843, only a few months before the author left the Established Church to become a minister of the Free Church of Scotland. The following year Dr. Bonar published it in his *Songs for the Wilderness*, after which it appeared in several hymnals, ultimately becoming one of the most widely sung of all Dr. Bonar's hymns. The appropriate music to which it is set was the work of the Rev. L. G. Hayne.

" Days and moments quickly flying " is one of

the few original hymns by the late Edward Caswall which have gained any great hold on the affections of our own Church. It was specially composed for use at "Watch-night" services on New Year's Eve or the day following. Like a great number of Mr. Caswall's other hymns, this composition was written after the author had seceded to the Church of Rome, and while he was at the Oratory, Edgbaston. It was first published in his *Masque of Mary and other Poems* in 1858, and rapidly made its way into public favour with Protestants as well as Roman Catholics.

"Days and moments quickly flying" appears in various hymnals in various forms. Caswall is the author of four stanzas only, the concluding stanza as given in *Hymns Acient and Modern*—

> Whence we came, and whither wending;
> Soon we must through darkness go,
> To inherit bliss unending,
> Or eternity of woe.

being by the compilers of that popular hymnal.

In the *Church Hymnary* the hymn is given in an extended form and divided into two portions, each part concluding with the following somewhat curious stanza—

HYMNS FOR THE NEW YEAR

> Life passeth soon;
> Death draweth near:
> Keep us, good Lord,
> Till Thou appear,—
> With Thee to live,
> With Thee to die,
> With Thee to reign through eternity.

This hymn is sometimes also sung as a funeral hymn. The exquisite melody to which it is allied is by Dr. J. B. Dykes—one of the most successful of his many beautiful compositions.

A hymn suitable for the New Year, though originally written for Advent, is Charles Wesley's "Thou Judge of quick and dead." It is by no means as often sung to-day as it was in Wesley's time, containing as it does verses of a rather terrible nature. A hundred and fifty years ago congregations gloried in such verses as—

> The solemn midnight cry,
> Ye dead, the Judge is come!
> Arise and meet Him in the sky,
> And meet your instant doom!

but since then opinions have changed and we have come to regard the Great Judge of all as merciful rather than severe.

This hymn appears in the MS. volumes left behind by Charles Wesley. The copy is a "fair" one and contains no corrections. In John Wesley's *Hymns for the Use of the People called Methodists* this hymn is given, as originally written, in four stanzas of eight lines each, but in many collections it is broken up into eight verses of four lines. The final verse is one of the finest Charles Wesley ever wrote :—

> O may we thus be found
> Obedient to His Word,
> Attentive to the trumpet's sound
> And looking for our Lord !
> O may we thus ensure
> A lot among the blest ;
> And watch a moment to secure
> An everlasting rest.

A hymn by the late Miss Frances Ridley Havergal, which was specially written for New Year's Day, and which is becoming year by year more universally sung, is "Jesus, blessèd Saviour," written at Leamington in November, 1872. This hymn was originally intended for children ; but after it was written Miss Havergal, thinking that it was perhaps quite as suitable for grown-up people as for little folk, scored through the words " For the little ones," and it now appears among those hymns specially devoted to the New Year, and may be sung with equal appropriateness by either children or adults.

The MS. of this hymn was found loose in a volume of the late Miss Havergal's hymns, together with others which are equally well known. It was written without difficulty, the lines being conceived almost as quickly as the authoress could write them down. There are very few corrections in the original manuscript, and the hymn as given in the majority of collections is exactly as the authoress wrote it. In this respect Miss Havergal was more fortunate than a great number of writers. On more than one occasion, however, she was asked by hymnal editors if she had any objection to a line being altered. She was very quick to acknowledge an improvement, and when an alteration helped to make a verse clearer or more direct in its meaning she cheerfully consented to its being made.

"Jesus, blessèd Saviour" is based on the words "Thou hast holden me by my right hand. Thou shalt guide me with Thy counsel," and was written to a melody of German origin which was a great favourite with the authoress.

"The year is gone beyond recall" is the Rev. Francis Pott's translation of a Latin hymn, the author of which is unknown. It is one of the most widely known of all New Year's hymns and is a great favourite in France and other European countries. The version as given in *Hymns Ancient and Modern* has been a good deal altered by the compilers from Mr. Pott's original translation, and the following Doxology added—

>All glory to the Father be,
>All glory to the Son,
>All glory, Holy Ghost, to Thee,
>While endless ages run.

This Doxology is also used at the conclusion of several other hymns.

Though Mr. Pott is the author of several original hymns it is as a translator that he is most widely known.

V

Hymns on the Passion

IT is only natural, perhaps, that those hymns which have been written commemorative of the death of our Lord should be among the saddest in our hymnals. Many of them have taken their place with the finest examples of religious verse in the language, and have gained for their authors names which will be forgotten only when hymns cease to be sung.

In this chapter I shall not only refer to those hymns which have been specially written on the Passion, but also to several which are more or less suited to that season. Among these latter may be mentioned Cardinal Newman's "Praise to the Holiest in the height," which was written in 1865 and formed part of a long poem entitled "The Dream of Gerontius."

This poem, as my readers are probably aware,

is descriptive of the journey of a soul to Paradise, and the many hymns introduced are supposed to be sung by different choirs of Angelicals. As the soul is conducted into the presence chamber of the Emmanuel, the " Fifth Choir of Angelicals " sing the magnificent lines which have now been incorporated into our hymnals :—

> Praise to the Holiest in the height,
> And in the depth be praise ;
> In all His words most wonderful,
> Most sure in all His ways.

Newman himself thought little of the poem at the time of writing it. Three years after its composition the editor of a religious magazine wrote to the cardinal asking him to contribute "something" to his paper. He was just about to write and decline when he bethought himself of the " Dream," and sent it. The composition was published and immediately attracted attention, though it never became really popular. The same year that saw its publication the compilers of one of our most popular hymnals requested permission to insert the hymn, as sung by the " Fifth Choir of Angelicals," to their appendix. This the cardinal readily granted, and Dr. Dykes wrote for it the very fine tune entitled " Gerontius," to which it is now invariably sung. As in the case of " Lead, kindly light," Cardinal Newman used to protest that the melody had a good deal to do with the popularity of the hymn.

There is no difference between "Praise to the Holiest in the height" as it appears in the "Dream" and in the hymnals, for Cardinal Newman never permitted compilers to tamper with or "improve" his compositions, preferring rather that they should be omitted altogether. The cutting words he is reported to have uttered regarding the editor who added a fourth verse to his "Lead, kindly light" may be remembered.

Some years ago, when Mr. Gladstone was asked to name his favourite hymns, he replied that he scarcely knew whether he had a "favourite" or not, and then, on the impulse of the moment, he mentioned "Lead, kindly light" and "Rock of Ages." On his deathbed, however, the great statesman found his greatest consolation and strength in "Praise to the Holiest in the height," and in a touching sermon preached in St. Paul's Cathedral during Mr. Gladstone's last illness, Canon Scott Holland referred to the late Premier as "spending his life in benediction to those whom he leaves behind in this world, and in thanksgiving to God, to whom he rehearses over and over again, day after day, Newman's hymn of austere and splendid adoration, 'Praise to the Holiest in the height.'"

The names of Gladstone and Gordon will long be linked together, and it is strange as it is touching to recall the fact that when all hope had been abandoned, the hero of Khartoum fortified himself for death with the very words which were the comfort and consolation of Mr. Gladstone during his last days on earth :—

> O Generous Love! that He who smote
> In man for man the foe!
> The double agony in man
> For man should undergo.

A hymn specially written for Passiontide, to which is attached a double interest from the fact that the words and music are by brother and sister, is :—

And now, belovèd Lord, Thy Soul resigning
  Into Thy Father's arms with conscious Will,
Calmly, with reverend grace, Thy Head inclining,
  The throbbing Brow and labouring Breast grow still.

This hymn was written in 1868 by Mrs. Eliza Sibbald Alderson, sister to the late Dr. Dykes. In a letter received from a relative of Mrs. Alderson's, I learn that the hymn was written at her brother's request, who used to affirm that his sister had unusual powers as a hymnist. After it was written

HYMNS ON THE PASSION

Mrs. Alderson handed a copy of the hymn to Dr. Dykes, who composed for it the well-known tune " Commendatio."

In the original this hymn consists of six verses, but many collections omit the following stanzas :—

O love ! o'er mortal agony victorious,
  Now is Thy triumph! now that Cross shall shine
To earth's remotest age revered and glorious,
  Of suffering's deepest mystery the sign.

The present, past, and future here are blending,
  Moment supreme in this world's history,
Mid darkness, opening graves, and mountains rending,
  New light is dawning on humanity.

This is the best known of all Mrs. Alderson's hymns, and appears in a great number of collections. There are, however, several others of her hymns for which Dr. Dykes wrote special tunes, the most popular being that for " Almsgiving," " Lord of Glory, who hast bought us."

Mrs. Alderson's husband was for many years chaplain to the West Riding House of Correction in Wakefield, and it was here that the greater number of Mrs. Alderson's hymns were written.

The most popular hymn in the English language, according to a general consensus of opinions, is " Rock of ages, cleft for me," which is essentially a hymn for Passiontide. It was written, as all the world knows, by the Rev. Augustus Montague Toplady, sometime curate-in-charge of the parish of Blagdon, on the Mendips, about eight miles from Wells. In an interesting letter published some time ago in the *Times*, Sir William Henry Wills says: "Toplady was one day overtaken by a heavy thunderstorm in Burrington Coombe, on the edge of my property (Blagdon), a rocky glen running up into the heart of the Mendip range, and there, taking shelter between two massive pillars of our native limestone, he penned the hymn :—

> Rock of ages, cleft for me,
> Let me hide myself in Thee.

Whether this story has a greater claim to credence than many others which naturally grow around world-known compositions, I cannot say, but I fear that it must be relegated to the same category as the story sometimes told regarding the sea-bird and "Jesu, Lover of my soul." So far as the history of its composition can be traced, it appears to have been written spontaneously, sent to the *Gospel Magazine*, where it first appeared in 1775, and afterwards republished in Toplady's collection of hymns. It was written only three years before the author's early death, which occurred on August 11, 1778.

"Rock of Ages" was the hymn which the late Prince Consort asked to be sung to him a few hours before his death, and in his very interesting little volume entitled *Hymns that have Helped*, Mr. W. T. Stead tells us that "when the *London* went down in the Bay of Biscay, January 11, 1866, the last thing which the last man who left the ship heard as the boat pushed off from the doomed vessel was the voices of the passengers singing 'Rock of Ages.'"

A missionary who returned from India the other day tells rather an amusing incident in connection with "Rock of Ages." He complained of the slow progress made in India in converting the natives on account of the difficulty of explaining the teachings of Christianity so that the ignorant people could understand them. Some of the most beautiful passages in the Bible, for instance, are destroyed by translation. He attempted once to have the hymn,

> Rock of ages, cleft for me,
> Let me hide myself in Thee,

translated into the native dialect, so that the natives might appreciate its beauty. The work was entrusted to a young Hindu Bible student who had the reputation of being something of a poet. The next day he brought his translation for approval, and his rendering, as translated back into English, read like this:—

> Very old stone, split for my benefit,
> Let me absent myself under one of your fragments.

It is not difficult to believe, after this, that many of the natives see little that is beautiful in our most cherished hymns and poems.

William Cowper's most famous and widely-known hymn is said to be the one he wrote for Passiontide:—

> There is a fountain filled with blood
> Drawn from Immanuel's veins;
> And sinners plunged beneath its flood
> Lose all their guilty stains.

This hymn was written about the year 1770, and was based on the text which usually heads the composition in the majority of our hymnals: "In that day there shall be a fountain opened to the house of David and to the inhabitants of Jerusalem for sin and for uncleanness." It was one of the first hymns he wrote after his first attack of temporary madness, and as this period of the poet's life always seems to have a fascination for his admirers, I quote a passage taken from an article published in the *North American Review* for January, 1834, which appears to me to give the best account of this distressing incident yet published. My readers are probably well aware that Cowper was of a painfully nervous and shy temperament, his extreme sensitiveness probably accounting in no small measure for his malady. He had been promised a post as Clerk of the Journal to the House of Lords, and was happy in the contemplation of his approaching appointment, when, to his utter dismay, he learned that it would be necessary to undergo a public examination before the House before he entered upon his duties.

"As the time drew nigh, his agony became more and more intense; he hoped and believed that madness would come to relieve him; he attempted also to make up his mind to commit suicide, though his conscience bore stern testimony against it; he could not by any argument persuade himself that it was right, but this desperation prevailed, and he procured from an apothecary the means of self-destruction. On the day before his public appearance was to be made, he happened to notice a letter in the newspaper, which to his disordered mind seemed like a malignant libel on himself. He immediately threw down the paper and rushed into the fields, determined to die in a ditch, but the thought struck him that he might escape from the country. With the same violence he proceeded to make hasty preparations for his flight; but while he was engaged in packing his portmanteau his mind changed, and he threw himself into a coach, ordering the man to drive to the Tower wharf, intending to throw himself into the river, and not reflecting that it would be impossible to accomplish his purpose in that public spot. On approaching the water, he found a porter seated upon some goods: he then returned to the coach and was conveyed to his lodgings at the Temple. On the way he attempted to drink the laudanum, but as often as he raised it, a convulsive agitation of his frame prevented its reaching his lips; and thus, regretting the loss of the opportunity, but unable to avail himself of it, he arrived, half dead with anguish, at his apartment. He then shut the doors and threw himself upon the bed with the laudanum near him, trying to lash himself up to the deed; but a voice within seemed constantly to forbid it, and as often as he extended his hand to the poison, his fingers were contracted and held back by spasms.

" At this time some one of the inmates of the place came in, but he concealed his agitation, and as soon as he was left alone, a change came over him, and so detestable did the deed appear, that he threw away the laudanum and dashed the vial to pieces. The rest of the day was spent in heavy insensibility, and at night he slept as usual; but on waking at three in the morning, he took his penknife and lay with his weight upon it, the point towards his heart. It was broken and would not penetrate. At day break he arose, and passing a strong garter round his neck, fastened it to the frame of his bed: this gave way with his weight, but on securing it to the door, he was more successful, and remained suspended till he had lost all consciousness of existence. After a time the garter broke and he fell to the floor, so that his life was saved; but the conflict had been greater than his reason could endure. He felt for himself a contempt not to be expressed or imagined; whenever he went into the street, it seemed as if every eye flashed upon him with indignation and scorn; he felt as if he had offended God so deeply that his guilt could never be forgiven, and his whole heart was filled with tumultuous pangs of despair. Madness was not far off, or rather madness was already come."

It was after this terrible experience, and when he had somewhat recovered, that he turned his attention to the *Olney Hymns*. For eight years he was well, mentally and physically, and during that period a great deal of his religious verse was written. Then his mind gave way once more and he made another attempt to commit suicide. He was, however, again frustrated, this time by the action of his coachman who had received orders to drive to the river. The man purposely lost his way and brought the poet back to his home again. His reason once more returned, and in one of those fits of contrition which appear to have swiftly followed on every period of madness, he wrote that hymn which has since been a comfort to countless thousands :—

> God moves in a mysterious way,
>   His wonders to perform ;
> He plants His footsteps in the sea,
>   And rides upon the storm.

To an American writer we are indebted for one of the most beautiful hymns in the language and one which is particularly suited to Passiontide. I refer to :—

> My faith looks up to Thee,
> Thou Lamb of Calvary,
>   Saviour Divine ;
> Now hear me while I pray ;
> Take all my guilt away ;
> O let me from this day
>   Be wholly Thine.

This hymn was written by the greatest of all American hymnists, the late Dr. Ray Palmer. Although Dr. Palmer has been dead but a few years this composition of his has been a general favourite among all denominations for close upon three-quarters of a century. From this it may be inferred that the author wrote it at a very early age. This is so ; in fact it was the first hymn Dr. Palmer ever composed, and it is as strange as it is true that this initial performance in hymnody should be by far his most successful. Dr. Palmer is the author of many other hymns, not one of which has attained, in this country at least, to the popularity enjoyed by the composition at present under consideration.

"My faith looks up to Thee" was written a few months after the author had graduated at Yale College, New Haven, and while he was acting as tutor in a school in New York. Though Dr. Palmer has left no record to the effect, I should be inclined to believe that he was in low spirits at the time, possibly feeling somewhat home sick owing to his being away from relations and friends. It is a well-known fact that he was greatly affected when composing the hymn, and he is said to have written the concluding verse with the deepest emotion and in tears.

Very soon after it was written the hymn was published in a little volume of *Spiritual Songs*, and was there set to a melody by the late Lowell Mason, called " Olivet," a tune to which it has ever since been sung both here and in America.

" My faith looks up to Thee " is unique in many respects. Though it was the first hymn the author ever wrote it became the most popular; it was written, practically, by a boy; it was set to music the year of its birth and has been sung to the same melody ever since; and, lastly, it was a popular hymn for nearly sixty years during the author's lifetime. Among American hymns, therefore, none surely has a more interesting record than " My faith looks up to Thee."

A hymn for Passiontide, which was written by Isaac Watts, but which is not now found in many collections, is :—

> Not all the blood of beasts
> On Jewish altars slain
> Could give the guilty conscience peace,
> Or wash away our stain.

There is a general impression to the effect that this hymn was written by Dr. Watts after a visit paid to Smithfield market. There he had stood for a few moments contemplating the newly-slain animals, and the incident soon suggested to the poet a subject for a hymn. Continuing his stroll round the market he thought out the first draft of "Not all the blood of beasts." Though the composition has always been a favourite with nonconformists it is found in comparatively few hymnals of the Church of England. It is somewhat strange that Watts, who had a remarkably fine ear for metre and rhythm should, as shown in the following verse, have allowed himself to pass so clumsy a rhyme :—

> My faith would lay her hand
> On that meek head of Thine,
> While as a penitent I stand,
> And here confess my sin.

This is almost as bad, though not quite, as Charles Wesley in his "Forth in Thy Name, O Lord, I go," the second verse of which, as all those who use the Methodist hymnal are probably aware, has a very awkward line of a different nature :—

> The task Thy wisdom hath assigned
> O let me cheerfully fulfil,
> In all my works Thy presence find,
> And prove Thy acceptable will.

In many hymnals the verse has been changed to :—

> The task Thy wisdom hath assigned
> O let me cheerfully fulfil;
> In all Thy works Thy presence find,
> And prove Thy good and perfect will.

This is a considerable improvement, but who is answerable for the alteration I am unable to say.

The concluding verse of "Not all the blood of beasts" is rather a curious one:—

> Believing, we rejoice
> To feel the curse remove;
> We bless the Lamb with cheerful voice,
> And trust His bleeding love.

Altogether the hymn is one which suits the eighteenth century better than the twentieth. Somehow it misses the delicacy and refinement which is so characteristic of the best of Watts' hymns. It cannot, for instance, be compared with his tender and beautiful:—

> There is a land of pure delight,
> Where saints immortal reign,
> Infinite day excludes the night,
> And pleasures banish pain.

In connection with this hymn, which is not of course one for Passiontide, it is related that it was written by Watts before he had reached his twenty-first birthday, and that the theme was suggested by the beautiful view of Southampton Water as seen from the Isle of Wight. There is strong reason to believe that this is correct, for if we study the hymn we shall find that there are many expressions contained in it which substantiate the theory—" Death like a narrow sea," "Sweet fields beyond the swelling flood Stand dressed in living green," etc.

Over 500 hymns by Isaac Watts are said to be in common use throughout the English-speaking world, but the estimate is probably exaggerated. Half that number would certainly be nearer the mark, and even this is remarkable when we consider that the average hymnal contains but 500 contributions. Like his contemporary hymnists Watts wrote much that was excellent but also much that was inferior, and it is only the winnowing of years which brings all the wheat to the surface. As generation follows generation the number of his hymns to be found in our collections may lessen until perhaps only thirty or forty remain, but these will stand the test of all times and be sung as long as Christianity endures— surely a splendid monument to the memory of any man.

A hymn which has been described as the most pathetic of the Middle Ages is the Stabat Mater

Dolorosa, rendered familiar to us by the translation :—

> At the Cross her station keeping
> Stood the mournful Mother weeping,
> Where He hung, the dying Lord ;
> For her soul of joy bereavèd,
> Bow'd with anguish, deeply grievèd,
> Felt the sharp and piercing sword.

This hymn has been ascribed to at least six authors, and, though no really satisfactory conclusion has ever been come to, it is generally considered to be the work of either Pope Innocent III or Jacobus de Benedictus. If the opinion of a dozen hymnologists were taken it would probably be found that the majority were in favour of the latter being the real author. But after so many centuries and on such slender testimony obtainable it is impossible to decide with absolute certainty, and like many another hymn the Stabat Mater will have to be ascribed, if safety be desired, to that mysterious individual—" an anonymous writer."

Of the translations of the hymn there have been many, the most popular being that by Edward Caswall which appeared in 1867. This version has appeared in more hymnals than all the other translations put together, though in some instances

HYMNS ON THE PASSION

in an altered form. In Protestant collections, of course, the verses having special reference to the Virgin Mary have been omitted.

A very interesting feature connected with the Stabat Mater is the numerous occasions on which it has been set to music by various famous composers. Palestrina, Haydn and Rossini are only a few of the celebrated musicians who have expended on the Stabat Mater their finest efforts of musical composition. Who does not know the melody to Cujus Animam taken from the latter's work? The majority of us learned it during the first year of our musical education! Last of all the great Bohemian composer Dvorak has given us in the Stabat Mater some of his loveliest music. Of all the poems which have come to us from the Latin none has been so frequently set to music as the Stabat Mater.

There are three melodies to which " At the Cross her station keeping " is usually sung, and they are all, curiously enough, entitled " Stabat Mater " which at times occasions some confusion. Perhaps the setting most often sung is the one by Dr. Dykes. The remaining tunes consist of an ancient plain song and a modern French melody, both of which suit the mournful and tender words very well.

Another great hymn which, though not perhaps originally written for Passiontide, has come to be generally used for the season commemorative of the death of our Lord, is "My God, I love Thee; not because," a translation of "O Deus ego amo Te," usually ascribed to St. Francis Xavier. That it was written by the great Roman Catholic missionary has never conclusively been proved, but the probability is that it was. It is a hymn which has been known and loved by many countries for many generations, and the translations are numerous. In Great Britain that made by Edward Caswall has gained the greatest amount of favour, and is to be found in the greatest number of English hymnals.

Exception has been taken by several editors to the first verse of Mr. Caswall's translation, and for this reason have omitted it from their collections. The verse, as my readers are probably aware, sets forth the terrible punishment meted out to those who do not love God :—

> My God, I love Thee; not because
> I hope for Heaven thereby,
> Nor yet because who love Thee not
> Must burn eternally.

In some collections this verse has been slightly altered, the last two lines reading :—

HYMNS ON THE PASSION

>Nor yet because who love Thee not
>Are lost eternally.

Perhaps the difficulty might be got over by changing the words "lost" or "burn" to "mourn":—

>My God I love Thee; not because
>I hope for Heaven thereby,
>Nor yet because who love Thee not
>Must mourn eternally.

This is taking a liberty with a hymn for which I have not the excuse of being a hymnist to offer, but it seems a pity that owing to a word unfortunately chosen a composition so beautiful should be omitted from any collection.

That very well-known Passiontide hymn, "Sweet the moments, rich in blessing," is the combined work of two hymnists, James Allen and the Rev. and Hon. Walter Shirley, grandson of Earl Ferrers. It was first published in the *Kendal Hymn Book*, but in a form very different from that now found in most modern collections. Had not Mr. Shirley taken it in hand and recast it, there is more than a possibility that to-day the composition would have been forgotten. From half a dozen somewhat commonplace verses he succeeded in producing what has become one of the most prized hymns in the language:—

> Sweet the moments, rich in blessing,
> Which before the Cross I spend,
> Life, and health, and peace possessing
> From the sinner's dying Friend.

James Allen must have been a somewhat remarkable man. After taking his degree at Oxford he appears to have become a lay reader of the Church of England, but owing to his peculiar temperament did not get on very happily with the clergymen with whom his work brought him in contact. He joined various meteoric sects, with none of which, however, he remained for any length of time, and eventually built a chapel of his own, where he conducted service according to his own notions. He was the author of several original hymns, and was the friend of the celebrated Lady Huntingdon. He assisted in the compilation of the *Kendal Hymn Book*, and after leading an uneventful life, died at Gayle, in 1804, at the age of seventy.

James Allen's fellow-hymnist, Walter Shirley, his senior by nine years, was an Irish rector, and a poet of some ability. He was a far better hymnist than Allen, being the author of some really fine religious verse, his Good Friday hymn, "Flow fast, my tears, the cause is great," being especially beautiful. It is doubtful, taking all things into consideration, whether Allen's name would be remembered to-day had not "Sweet the moments, rich in blessing" been so successfully recast. Shirley died April 7, 1786, eighteen years before his staunch friend James Allen.

If the tenderest hymn John Ellerton ever wrote be his one for evening service, "Saviour, again to Thy dear name we raise," certainly his grandest and most solemn is the one for Passiontide :—

> Throned upon the awful Tree,
> King of grief, I watch with Thee;
> Darkness veils Thine anguish'd face,
> None its lines of woe can trace,
> None can tell what pangs unknown
> Hold Thee silent and alone.

This hymn was written in 1875 at the request of Sir Henry Baker for the new edition of his hymnal, and was there set to music by the Rev. Sir F. A. G. Ouseley. This tune, appropriately named "Gethsemane" (not to be confounded with Monk's tune of the same name, and usually sung to Montgomery's "Go to dark Gethsemane"), was one of the most successful of all Gore Ouseley's hymn tunes, the peculiar manner in which the melody to each line rises half a tone rendering the hymn doubly solemn and impressive.

Canon Ellerton based his hymn on the text,

"My God, my God, why hast Thou forsaken Me?" The original manuscript of this very fine composition has been preserved, and is now in the possession of the late hymnist's son, the Rev. Frank Ellerton.

I understand that the hymn was originally written in three stanzas, the following verse being subsequently added before publication in Canon Ellerton's collection:—

> Lord, should fear and anguish roll
> Darkly o'er my sinful soul,
> Thou, who once wast thus bereft
> That Thine own might ne'er be left,
> Teach me by that bitter cry
> In the gloom to know Thee nigh.

In Sir Henry Baker's hymnal this composition is placed next to the Stabat Mater, "At the Cross her station keeping," and perhaps no higher praise could be given to a hymn than to say it is in every way worthy of its companion. Indeed it would be difficult to say which of the two hymns is the finer.

Of the many hymns which Dean Milman wrote, and which were published in 1827, all have stood the test of time, and are to-day in common use throughout the English-speaking world. Probably the only other writer of which the same can be said is his friend and fellow-hymnist, Reginald Heber. Dean Milman's hymns include two for Lent, two funeral hymns, two for Advent, and one each for Easter, "Those at sea," and Passiontide This last is a remarkably beautiful hymn, very original as to construction, and containing much fine thought finely expressed. It may not be very well known to some of my readers, and I therefore quote the first and last verses:—

> Bound upon the accursèd Tree,
> Faint and bleeding, who is He?
> By the eyes so pale and dim,
> Streaming blood and writhing limb;
> By the flesh with scourges torn;
> By the crown of twisted thorn;
> By the side so deeply pierced;
> By the baffled, burning thirst;
> By the drooping, death-dewed brow:
> Son of Man! 'tis Thou! 'tis Thou!
>
> Bound upon the accursèd Tree,
> Dread and awful, who is He?
> By the prayer for them that slew—
> "Lord! they know not what they do!"
> By the spoiled and empty grave;
> By the souls He died to save;
> By the conquest He hath won;
> By the saints before His throne;
> By the rainbow round His brow:
> Son of God! 'tis Thou! 'tis Thou!

This composition, which in the original consists of four stanzas of ten lines each, was written at Heber's request, and is to be found in the Bishop's book of manuscript hymns now located in the British Museum.

As is well known, Bishop Heber thought more highly of Milman's hymns than he did of those of any other writer who lived during his own times. From a letter written to me some years ago by Dean Milman's son I quote the following interesting paragraphs :—

"With regard to my father's manuscript hymns," writes Mr. Arthur Milman, "I have never even seen one, and I doubt very much whether they can have survived. I have in my possession several letters written to my father by Heber, and as they may possibly interest your readers I send you one or two extracts. Under date of May 11, 1821, which would be a couple of years before his appointment to the see of Calcutta, Heber writes to my father :—

"'I rejoice to hear so good an account of the progress which your saint (the Martyr of Antioch) is making towards her crown, and I feel really grateful for the kindness which enables you while so occupied to recollect my hymn-book. I have during the last month received some assistance

HYMNS ON THE PASSION

from——, which would once have pleased me much, but, alas! your Advent, Good Friday, and Palm Sunday hymns have spoilt me for all other attempts of the sort.'

"Again, December 28, 1821, Heber writes: 'You have indeed sent me a most powerful reinforcement to my projected hymn-book. A few more such and I shall neither need nor wait for the aid of Scott and Southey. Most sincerely, I have not seen any hymns of the kind which more completely correspond to my ideas of what such compositions ought to be, or to the plan, the outline of which it has been my wish to fill up.'"

It is somewhat sad to recollect that this collection of hymns in which Heber took such intense interest was not published until after his death, and then possibly not in the form in which he intended it to appear.

Between the hymns of Heber and Milman a good deal of similarity exists, the predominating features in the compositions of both writers being lyric fire and wealth of colouring.

To Thomas Kelly, a most voluminous writer, whose hymnal compositions are said to approach a thousand in number, we owe that very fine and original Passion hymn :—

HYMNS ON THE PASSION

> We sing the praise of Him who died,
> Of Him who died upon the Cross;
> The sinner's Hope let men deride,
> For this we count the world but loss.

This hymn was first published in 1815, and quickly found its way into the collections of every denomination. In Sir Henry Baker's hymnal the following verse has been added:—

> To Christ, who won for sinners grace
> By bitter grief and anguish sore,
> Be praise from all the ransom'd race
> For ever and for evermore.

Thomas Kelly, the son of an Irish judge, was born in Dublin in 1769. It had been his intention to follow in the footsteps of his father, but on leaving Trinity College he altered his mind and took Holy Orders. Owing, however, to constant friction with the Primate of Ireland he left the Established Church, and for many years preached in unconsecrated buildings. He was a man possessed of great magnetic powers, and his fine orations attracted considerable crowds. His hymns were first sung by his own congregation, but after publication were eagerly seized upon by compilers of every class of hymnal. The requests for the use of his compositions invariably met with a ready and cheerful response.

Singularly genial and kindly in disposition, Mr. Kelly was greatly beloved by the poor of Dublin, who could always count on his assistance when times were more than ordinarily bad. It is told in the Irish capital how on one occasion, when a worthy couple were passing through a period of exceptional hardship and privation, the husband endeavoured to cheer his disconsolate wife with the remark: "Hould up, Bridget, bedad, there's always Misther Kelly to pull us out of the bog afther we've sunk for the last time." This somewhat paradoxical remark is eminently characteristic of the faith which the poor had in Mr. Kelly; no deserving case ever appealed to him in vain, and his memory is still most affectionately cherished in the capital of his " most distresthful country."

Like Archbishop Maclagan, Mr. Kelly was also a composer of considerable ability, and many of his hymns were first set to music by the author himself. Soon after the publication of his collection of hymns in 1815 he issued a companion volume containing tunes suited to every kind of metre to be found in his hymnal. All these tunes he composed himself, and among them are many of great beauty and originality. Mr. Kelly died in Dublin in 1854 at the advanced age of eighty-five.

The late Rev. Frederick William Harris, for many years vicar of Medmenham, was the author of one hymn so excellent as to be worthy a place in every collection. It was written specially for Passiontide, and appeared first, I believe, in a magazine, after which it was inserted in Prebendary Thring's collection. Here is the opening verse :—

"It is finished—It is finished!" all the untold agony,
When with death and hell He wrestled all alone upon the Tree;
All alone—nor man nor angel near to comfort and sustain,
E'en the cry of mortal anguish, "Eli! Eli"! spent in vain.

In a note on this hymn Prebendary Thring says: "The first line of each verse as originally written began with the Greek word $T\epsilon\tau\epsilon\lambda\epsilon\sigma\tau\alpha\iota$, of which ' It is finished ' is the translation."

The final verse to this hymn differs somewhat from the rest, containing as it does an extra line, which renders necessary a change in the musical setting. It is very fine, however, and as the hymn is not so well known as it should be I quote this stanza :—

"It is finished—It is finished!" all by heaven decreed of old,
In the sacred volume written, or by ancient seer foretold:

> Wrath appeased, transgression finished, God and man again at one;
> Comes the night when no man worketh: let it come, the work is done;
> Hark, through heaven the word is echoed: "It is finished"—"It is done."

So far as I have been able to discover, Mr. Harris was the author of no other hymns, certainly none that were ever published. So evidently did he possess the true instinct necessary to the production of successful hymns that it is more than a pity that our hymnals have received no further contributions from his pen. Mr. Harris died in 1872 at the comparatively early age of fifty-eight.

I cannot conclude this chapter without referring to the many excellent metrical litanies written by the late Mr. Thomas Benson Pollock, for some years rector of Pluckley in Kent. His Litany on the Seven Last Words from the Cross, which is so frequently sung at the three hours' service on Good Fridays, is perhaps his best known. It was first published, together with several others, in 1870, in a small volume, and subsequently appeared in *Hymns Ancient and Modern.* To this collection Mr. Pollock also contributed a couple of hymns, "We are soldiers of Christ, Who is mighty to save" and "We have not known Thee as we ought," neither of which is very well known, even by those who use this hymnal. It is by his litanies that Mr. Pollock will be longest remembered. He died in 1896.

VI
Easter Hymns

THE Resurrection was a favourite subject with the majority of ancient hymnists, and as a consequence a large proportion of those hymns which we are accustomed to sing on Easter Day have come to us from the Latin. Foremost among these is "Jesus Christ is risen to-day," a hymn the authorship of which is shrouded in mystery. It has, however, been conclusively proved to be a composition of the fourteenth century. Curiously enough the name of the translator is also unknown.

This hymn, very much as we sing it to-day, first appeared in a book entitled *Lyra Davidica*, published in 1708, the first verse reading:—

> Jesus Christ is risen to-day,
>   Halle-Halle-lujah.
> Our triumphant Holy-day;
> Who so lately on the Cross,
> Suffer'd to redeem our loss.

No name of author was appended, and little inquiry, if any, appears to have been made as to who it was who had contributed to our national hymnary so fine a translation of the Latin hymn. Three stanzas only were given, the fourth, the Doxology, as published in some hymnals, being afterwards written by Charles Wesley. As this Doxology may not be in the hymnals used by some of my readers, I give it :—

> Sing we to our God above
>   Hallelujah !
> Praise eternal as His love ;
>   Hallelujah !
> Praise Him, all ye heavenly host ;
>   Hallelujah !
> Father, Son, and Holy Ghost ;
>   Hallelujah !

The really magnificent melody with which this hymn will always be associated, rendering it a veritable "triumphant song," appeared in conjunction with the hymn in *Lyra Davidica*. It seems quite in keeping with the mystery surrounding the authorship of hymn and translation that the music should also be by an unknown composer. And so it is. No name was attached to the setting, and though various composers have been credited with its authorship, it has never been conclusively proved who was the real com-

MRS. CECIL FRANCES ALEXANDER.
Photo by Elliott & Fry.

THE REV. HORATIUS BONAR, D.D.
Photo by Moffat, Edinburgh.

poser. The mystery surrounding the authors of hymn, translation, and music, will probably never be unravelled in this world.

Another hymn which has come to us from the Latin, and one of which nothing is known regarding authorship, is "The strife is o'er, the battle done." It is said to have been written during the twelfth century, but even this information is scarcely to be relied upon. Dr. Neale and Dr. Bonar have both made translations, but their versions have long since been cast into obscurity by Mr Francis Pott's spirited rendering :—

> The strife is o'er, the battle done !
> The victory of life is won !
> The song of triumph has begun !
>   Alleluia !

Very soon after Mr. Pott published his translation in 1861, it was included in a great number of hymnals, usually in an altered condition. In one popular collection the only verse given as the translator wrote it is the last. In many hymnals the following verse is omitted :—

> He closed the yawning gates of hell ;
> The bars from heaven's high portals fell,
> Let hymns of praise His triumph tell !
>   Alleluia !

The melody called "Victory," to which this

hymn has for many years been sung, and which commences, as my readers will recollect, with a triumphant trio of alleluias, is from one of Palestrina's oratorios.

Mr. Francis Pott, besides being a translator, is also known as a writer of original hymns. His "Angel voices ever singing," and "Lift up your heads, eternal gates," are to be found in many hymnals. In a characteristic little note received from Mr. Pott some time ago, I was informed that he could tell me nothing about his hymns, from the simple fact that there was nothing interesting to tell.

One of the finest and most beautiful original Easter hymns we possess is:—

> Christ is risen! Christ is risen!
> He hath burst His bonds in twain,

by the late Rev. Archer Gurney. This hymn was first published in a little volume of original and collected hymns, entitled *A Book of Praise*, compiled by Mr. Gurney in 1860. The compiler's own copy, which he used when chaplain of the Court Church, Paris, is beside me as I write, having been sent to me by a relative of the late hymnist. It is marked "Altar," and is a very small volume bound in cloth. The book contains

281 pieces, of which Mr. Gurney composed no fewer than 147. With the exception of "Christ is risen!" however, no hymns by Mr. Gurney have come into common use, though many he wrote call for the serious attention of hymnal editors.

"Christ is risen!" appears in the author's collection in a somewhat different form from that in many hymnals. Indeed, the frequency with which Mr. Gurney's fine hymn has been altered, sometimes with his leave, but more often without, was a source of considerable distress to the author. On more than one occasion he remarked that if an editor thought "Christ is risen!" a sufficiently good hymn to insert in his collection, he wished he would print it as written, or leave it alone altogether. The refrain in this hymn, which has been more often altered than any other part of the composition, appears in *A Book of Praise* in the following form:—

> Christ is risen! Christ is risen!
> He hath burst His bonds in twain:
> Christ is risen! Christ is risen!
> Cry of gladness, soar again.

This has been altered in the majority of hymnals to:—

> Christ is risen! Christ is risen!
> He hath burst His bonds in twain:

> Christ is risen! Christ is risen!
> Alleluia! swell the strain.

The melody, called "Resurrexit," which was specially written to the words of Mr. Gurney's hymn, is by the late Sir Arthur Sullivan, and is one of the finest tunes in our hymn books. It was written about the year 1874.

Mr. Archer Gurney, besides being a hymnist was also a very clever musician, and many of his hymns were first set to music by himself. Mrs. Dorothy Gurney, the authoress of "O perfect love, all human thought transcending," is the daughter-in-law of the late Mr. Archer Gurney.

"Jesus lives! no longer now Can thy terrors, death, appal us" is the late Miss Frances E. Cox's translation of the German hymn by C. F. Gellert. It is generally regarded by hymnologists to be the finest of the many fine hymns by this writer, and is to be found in nearly all German hymnals published during the last hundred years. In Germany, where it is as great a favourite at Easter services as in this country, it is also very often sung at funerals.

Since Miss Cox published her excellent translation of this hymn in 1840 it has become very popular in Great Britain and all English-speaking countries, few hymnals published during the last half century omitting it. In many collections, however, the opening lines have been altered to :—

> Jesus lives! thy terrors now
> Can, O death, no more appal us,

probably due to the fact that the first line is apt to convey a wrong impression unless due regard be paid to punctuation.

Though several translations of other hymns by Gellert have been made, it cannot be said that any one of them has gained a popularity in this country equal to that enjoyed by "Jesus lives!" Christian Fürchtegott Gellert died at Leipzig in 1769 at the age of forty-four.

From the Greek we get a very beautiful Easter hymn, which has been translated by John Mason Neale :—

> Come, ye faithful, raise the strain
> Of triumphant gladness.

This hymn is by St. John of Damascus, and was written some time during the latter half of the eighth century. Since Dr. Neale's translation appeared in 1862 this hymn has taken its place among those most frequently sung at Easter. In his *Hymns of the Eastern Church*, where the translation first appeared, it is given in four stanzas, but the following verse is usually omitted in most hymnals :—

> Neither might the gates of death,
>   Nor the tomb's dark portal,
> Nor the watchers, nor the seal,
>   Hold Thee as a mortal :
> But to-day amidst the Twelve
>   Thou didst stand, bestowing
> That Thy peace which evermore
>   Passeth human knowing.

In one hymnal the following verse is substituted :—

> Alleluia now we cry
>   To our King Immortal,
> Who triumphant burst the bars
>   Of the tomb's dark portal ;
> Alleluia, with the Son
>   God the Father praising ;
> Alleluia yet again
>   To the Spirit raising.

Dr. Neale was the first to open up to us the beauties of Oriental hymnody. Before his translations began to find their way into our hymnals the lyrics of the Eastern Church were practically unknown. In the Preface to his first volume of Greek translations the author says : " It is a most remarkable fact, and one which shows how very little interest has been hitherto felt in the Eastern

1 MISS CATHERINE WINKWORTH.
Photo by Fisher, Clifton.

MR. JAMES MONTGOMERY.   3 MRS. EMILY HUNTINGTON MILLER.
From an Engraving.   *From a Photo.*

4 MR. B. S. INGEMANN.   5 MR. HENRY KIRKE WHITE.
From a Photo.   *From the Painting by T. Barber, Esq., Nottingham.*

6 THE REV. ARCHER T. GURNEY.
Photo by Lock & Whitfield.

Church, that they are literally the only English versions of any part of the treasures of Oriental hymnology." This was a fact. Since Dr. Neale's work appeared, however, other writers have essayed to follow in his footsteps, and have given us further translations from the Greek, but the majority of them compare very poorly with Mr. Neale's excellent versions.

Another hymn which has come to us from Germany is "Christ, the Lord, is risen again," by Michael Weisse. Though translations of this hymn were published in England as early as 1750 it was not until Miss Catherine Winkworth issued her *Lyra Germanica* in 1858 that the hymn began to make its way into English collections. It has now become one of the most popular of all Easter hymns and is to be met with in most modern hymnals.

Michael Weisse was a Silesian, being born at Neisse about the year 1480. For many years he lived in a monastery in Breslau, where a considerable number of his hymns were written. He ultimately left Breslau and joined the Bohemian Brethren at Landskron. Here he spent the remainder of his life preaching and writing hymns. He died in 1540.

"Christ, the Lord, is risen again" is the only

hymn by Weisse which can be said to have gained any great popularity in this country.

"Alleluia, alleluia, Hearts to heaven and voices raise," an Easter hymn of great beauty, is by the late Christopher Wordsworth, Bishop of Lincoln. It first appeared in a collection of the bishop's hymns, after which it was included in several hymnals. It had the advantage of being set to music by the late Arthur Sullivan, who successfully interpreted the joyful nature of the words, the result being an inspiring hymn, both as regards music and words.

It is based on the words taken from the 15th chapter of the second book of Corinthians: "Now is Christ risen from the dead, and become the first-fruits of them that slept." In most hymnals it is given in an unaltered form with the exception of the following verse, which is usually omitted:—

Now the iron bars are broken, Christ from death to life is born ;
Glorious life, and life immortal, on the holy Easter morn :
Christ has triumphed, and we conquer by His mighty enterprise,
We with Him to life eternal by His Resurrection rise.

This hymn is full of praise and hope and is one of the most lyrical compositions Dr. Wordsworth ever penned.

A good translator seldom makes a very successful original hymnist. This was the case with John Mason Neale. Very few of his original hymns will live, though his translations will always be regarded as among the finest in our hymnals. One composition of an original nature by the late hymnist, however, has been accorded a good deal of favour, and is now to be found in a great number of collections. This was an Easter carol beginning, " The foe behind, the deep before." It differs considerably from any other hymn in the language, being written in a kind of irregular verse, if the expression be allowable. It was composed in 1853 and published the following year in his *Carols for Easter-tide.* In the original it consists of twelve stanzas, but these have been greatly reduced in the hymnal. It was set to music by Dr. Joseph Barnby, and was a great favourite among the Eton boys.

"Crown Him with many crowns," by the late Matthew Bridges, though not specially written for Easter, is eminently suited for that season of the year. It first appeared in the author's *Hymns of the Heart,* and was subsequently included in a great number of collections. Mr. Bridges based his composition on the words " And on His head were many crowns," and has succeeded in producing one of the finest sacred lyrics in the language. It has been altered by various editors and appears in few hymnals exactly as the author wrote it. In Thring's collection only one stanza is by Mr. Bridges, the remaining four being the work of the editor. The following verse is often omitted :—

> Crown Him the Virgin's Son,
>   The God Incarnate born,
> Whose arm those crimson trophies won
>   Which now His brow adorn :
> Fruit of the mystic Rose,
>   As of that Rose the Stem ;
> The Root whence mercy ever flows,
>   The Babe of Bethlehem.

Matthew Bridges was born at Maldon, in Essex, in 1800, and, though brought up a member of the Church of England, early in life seceded to the Church of Rome. He published two small volumes of hymns, one in 1848 and the other in 1852. From these the majority of his hymns have been taken, not one of which, however, can compare in point of popularity with his "Crown Him with many crowns." Mr. Bridges towards the close of his life lived in Canada. He died in Quebec in 1893.

"Rejoice, the Lord is King" is equally suited to either Easter or Ascension. It is by Charles

Wesley, and was first printed in 1746. In point of popularity and the number of hymnals in which it is to be found it compares favourably with the same author's "Hark, the herald angels sing," and "Jesu, Lover of my soul." In many collections the following verse is omitted :—

> He all His foes shall quell,
> Shall all our sins destroy,
> And every bosom swell
> With pure seraphic joy;
> Lift up your heart, lift up your voice,
> Rejoice, again I say, rejoice.

This hymn attracted the attention of Handel, who wrote for it the very fine melody called "Gopsal." The original of this setting is at present located in the Fitzwilliam Museum, together with many other of Handel's manuscripts. It is often thought that the small notes for the organ, which my readers will perhaps recollect are interpolated between the two lines of the refrain, were added afterwards by another composer, but this is not the case. The tune has been collated with the original and agrees with it in every particular. "Rejoice, the Lord is King," is almost the only hymn to be found in our hymnals which has been set to music by Handel.

"All hail the power of Jesus' Name," though

seldom placed among the hymns for Easter or Ascension, is distinctly suitable for either occasion. It was written by Edward Perronet, and published in the *Gospel Magazine* in 1780. It is somewhat curious that the tune by Shrubsole, which is as famous as the hymn itself, was published at the same time and in the same magazine. This tune, which might have been by the same hand as penned that to "Lo, He comes with clouds descending," received its name of "Miles Lane," it is said, from the chapel in Miles Lane, London, where Shrubsole was for many years organist.

The hymn as originally published contained eight stanzas. The following verse, however, is generally omitted:—

> Let highborn seraphs tune the lyre,
> And as they tune it fall
> Before His face who tunes their choir,
> And crown Him Lord of all.

This verse is rather a clumsy one, and does not add in any degree to the value or beauty of the hymn.

In Wesley's Hymn Book appears another verse which is not by Perronet at all. It is the last in the hymn, and reads as follows:—

EASTER HYMNS

> O that with yonder sacred throng
> We at His feet may fall,
> Join in the everlasting song,
> And crown Him Lord of all.

This verse, too, is a weak one, and compares unfavourably with the rest of the hymn. Though Edward Perronet wrote other hymns, some of them, perhaps, in merit equal to "All hail the power of Jesus' Name," it cannot be said that any one of them has become familiar to the church-going public. Like many another hymnist, his reputation rests on a single composition.

"Light's abode, celestial Salem," a hymn suitable for Easter or Ascension, though not specially written for either, is a translation from the Latin by John Mason Neale. It is by an unknown writer of the 15th century. The translation first appeared in Dr. Neale's *Hymns on the Joys and Glories of Paradise,* published a few months previous to his death in 1866. The little volume attracted attention by reason of the beauty of the translations, and many of the hymns contained in it soon began to make their appearance in a large number of collections. Perhaps the following extract, taken from the Preface to the first edition of Dr. Neale's little volume, helped in a measure to popularize his Latin translation :—

"I wish to add," he said,—"and this for the publisher as well as for myself—that any compiler of a future hymnal is perfectly welcome to make use of anything contained in this little book, only he will, perhaps, in that case, let us have a copy of his Hymnal when published. And I am very glad to have this opportunity of saying how strongly I feel that a hymn, whether original or translated, ought, the moment it is published, to become the common property of Christendom, the author retaining no private right in it whatever. I suppose that no one ever sent forth a hymn without some faint hope that he might be casting his two mites into that treasury of the Church, into which the 'many that were rich'— Ambrose and Hildebert, and Adam and Bernard of Cluny, and S. Bernard—yes, and Santeüil and Coffin—'cast in much.' But having so cast it in, is not the claiming a vested interest in it something like 'keeping back part of the price of the land'?"

The melody, "Regent Square," to which "Light's abode, celestial Salem" is usually sung, was composed by Henry Smart soon after the appearance of the translation.

An Easter hymn of great beauty and vigour is Robert Campbell's translation from the Latin,

"At the Lamb's high feast we sing." The author of the original is unknown, but it is generally supposed to have been written sometime during the 6th century. The translation was compiled, according to a manuscript which is in the possession of Mrs. E. Campbell, in 1849. It was printed the following year in a small volume of hymns which has long since been out of print.

Robert Campbell is not known as a writer of original hymns, though a few of these have been published. The popularity of his "At the Lamb's high feast we sing," however, is beyond that of any other translation, though there have been many. It is to be found in a very large number of hymnals and has been copied into a great many foreign collections. The stirring melody to which it is allied, known as "Salzburg," is from J. Sebastian Bach.

Robert Campbell was a Scotch Advocate, but devoted much of his time to the classics. He found relaxation from his professional duties in making translations of Latin hymns, many of which were published in a volume called the *St. Andrew's Hymnal.* In 1852, at the age of thirty-eight, he left the Episcopal Church of Scotland and became a Roman Catholic. He died in Edinburgh in 1868 at the early age of fifty-four.

"The Day of Resurrection" is a translation from the Greek made by John Mason Neale. The original of this beautiful Easter song belongs to the eighth century, and is generally supposed to be by St. John of Damascus. This hymn is sung every Easter Day at Athens, and in his book of translations Dr. Neale quotes the following account by a modern writer of one of those Easter ceremonies which he witnessed previous to making his translation:—

"As midnight approached, the Archbishop, with his priests, accompanied by the King and Queen, left the church, and stationed themselves on the platform, which was raised considerably from the ground, so that they were distinctly seen by the people. Every one now remained in breathless expectation, holding their unlighted tapers in readiness when the glad moment should arrive, while the priests still continued murmuring their melancholy chant in a low half-whisper. Suddenly a single report of a cannon announced that twelve o'clock had struck, and the Easter Day had begun. Then the old Archbishop, elevating the cross, exclaimed in a loud exulting tone, ' *Christos anesti*, Christ is risen!' and instantly every single individual of all the host took up the cry, and the vast multitude broke through and dispelled for ever

# EASTER HYMNS

the intense and mournful silence which they had maintained so long with one spontaneous shout of indescribable joy and triumph, 'Christ is risen! Christ is risen!' At the same moment the oppressive darkness was succeeded by a blaze of light from thousands of tapers, which, communicating one from another, seemed to send streams of fire in all directions, rendering the minutest objects distinctly visible, and casting the most vivid glow on the expressive faces, full of exultation, of the rejoicing crowds; bands of music struck up their gayest strains; the roll of the drum through the town, and further on the pealing of the cannon announced far and near these 'glad tidings of great joy'; while from hill and plain, from the seashore and the far olive grove, rocket after rocket ascending to the clear sky, answered back with their mute eloquence that Christ is risen indeed, and told of other tongues that were repeating those blessed words, and other hearts that leapt for joy; everywhere men clasped each other's hands, and congratulated one another, and embraced with countenances beaming with delight, as though to each one separately some wonderful happiness had been proclaimed—and so in truth it was—and all the while, rising above the mingling of many sounds, each one of which was a sound of gladness, the aged priests were distinctly heard chanting forth a glorious old hymn of victory in tones so loud and clear that they seemed to have regained their youth and strength to tell the world how 'Christ is risen from the dead, having trampled death beneath His feet, and henceforth the entomb'd have everlasting life.'"

' He is gone—beyond the skies" is one of the few hymns by the late Dean Stanley which may be said to have come into common use in this country. It first appeared in a popular magazine, signed with the Dean's initials, after which it was included in a large number of collections. A pretty story is told in connexion with the writing of this hymn. While in conversation with the Dean a friend happened to remark that his children had complained that there was no hymn really suitable for Ascension Day. They were also very much concerned as to what the disciples thought when "a cloud received Him out of their sight." The Dean seems to have been struck by the childish remarks, and replied that he would write such a hymn. "He is gone—beyond the skies," was the result. This story is related in a volume of poems entitled *Christ in Song*, by Dr. Phillip Schaff.

Of the immense number of hymns which Thomas Kelly wrote, the majority are of a joyful nature. He must have been a man possessed of a tremendous fund of good spirits and well able to look on the bright side of life, for though his career was anything but an untroubled one he seldom gave expression to feelings of melancholy, even in his writings. His hymns are characterized by unbounded faith, hope, joy and praise.

One of Mr. Kelly's most beautiful compositions is " The Head that once was crowned with thorns," evidently intended by the author for use at Ascension services. It was first published in a collection of his hymns in 1820, and subsequently in a great number of hymnals. This hymn has never had the advantage of having a very good tune written to it. That by Jeremiah Clark, to which it is usually sung, is a somewhat melancholy setting, not at all in keeping with the general character of the hymn. The words are certainly worthy the attention of our foremost composers.

"Alleluia! sing to Jesus" is not infrequently sung as a Communion hymn, but I give it a place under this chapter in consequence of the following verse, which stamps it, in my opinion, as an Ascension hymn :—

Alleluia ! not as orphans
 We are left in sorrow now ;
Alleluia ! He is near us,
 Faith believes, nor questions how :
Though the cloud from sight received Him
 When the forty days were o'er,
Shall our hearts forget His promise—
 "I am with you evermore"?

This hymn was written by the late William Chatterton Dix about the same time that he composed "Come unto Me, ye weary." The joyful nature of the composition indicates, however, that the author was in good health when he wrote it, and not, as in the case of the latter hymn, just recovering from a serious illness. This hymn has been fortunate in being given in the majority of hymnals as the author wrote it. The melody to which it is generally allied was composed by Dr. S. S. Wesley. It is a spirited tune, and has added in no small degree to the beauty of the hymn.

VII

Processional Hymns

AMONG those hymns specially suited for singing in processions of a religious nature none enjoys a greater degree of popularity than Dean Alford's " Forward ! be our watchword." This hymn was written for a choral festival held in connexion with the Canterbury Diocesan Union, and was first published in 1871. Mr. H. E. T. Crusoe, the late Dean's son-in-law, some few years since drew my attention to the following account of how "Forward ! be our watchword" came to be written :—

"It is related in the life of the Rev. J. G. Wood that Mr. Wood once asked the Dean to write a processional hymn for a choral festival, and to compose the music also. The Dean was at first a little overcome by the audacity of the proposal, but finally consented, and wrote a very admirable hymn. But, good as it was, it was not the kind of hymn wanted, and so Mr. Wood wrote off again to the Dean, pointing out that the hymn was not well adapted to sing on the march. Would he, therefore, go into his cathedral, walk slowly along the course the procession would take, and compose another hymn as he did so? The Dean, not in the least offended, did as he was bid, and the result was that grand hymn beginning :—

> Forward! be our watchword,
> Steps and voices joined;
> Seek the things before us,
> Not a look behind.

The MS. reached Mr. Wood with a humorous little note to the effect that the Dean had written the hymn and put it into its hat and boots, and that Mr. Wood might add the coat and trousers for himself! On looking at the music, Mr. Wood found accordingly that only the treble and bass had been supplied by the composer; the alto and tenor were added by Mrs. Worthington Bliss. The effect of the hymn when sung by the vast body of a thousand choristers was utterly beyond the power of words to describe."

The tune to which this hymn was originally sung, and which, as already stated, was composed by the author, is now seldom used, having been long since supplanted by Henry Gadsby's " St.

Boniface," or a very beautiful melody by Henry Smart. The hymn is very often broken up into two parts, being far too long for an ordinary service. When composing the hymn Dean Alford kept before him the words, "Speak unto the children of Israel, that they go forward." It was on these words that the hymn was based.

"The Royal Banners forward go" is Neale's translation of a processional hymn by Fortunatus, which dates back to the sixth century. This magnificent composition, known as the Vexilla Regis, was written under exceptionally interesting circumstances. The story goes that in the year 569, St. Radegund presented to the town of Poictiers a fragment of what was believed to be the true Cross. Fortunatus was the one chosen to receive the sacred relic on its arrival at Poictiers. When the bearers of the holy fragment were some two miles distant from the town, Fortunatus, with a great gathering of believers and enthusiasts, some carrying banners, crosses and other sacred emblems, went forth to meet them. As they marched they sang the hymn which Fortunatus had composed, the Vexilla Regis, now rendered familiar to us by the version beginning, "The Royal Banners forward go."

There have been many translations of this

hymn, but that by Dr. Neale has eclipsed them all. It was first published in his *Mediaeval Hymns*, from whence it was soon transferred to a great number of hymnals. Though it is distinctly a processional hymn, Fortunatus himself having that object in view when he wrote it, it is also very often sung on the fifth Sunday in Lent, otherwise known as Passion Sunday.

There are two well known melodies to this hymn, one called St. Cecilia, by the Rev. John Hampton, and the other, a plain song, appropriately named "Vexilla Regis," by an unknown composer. Both these tunes are somewhat melancholy and quite unworthy so great a hymn. This composition also deserves the serious attention of our best composers.

A hymn well known to even the smallest child attending Sunday school is:—

*Onward Christian soldiers
Marching as to war,
With the cross of Jesus
Going on before.*

In fact, it was written for children, though many compilers of works on hymnody affirm that the author had adults in his mind when he wrote it. The hymn was written by the Rev.

Sabine Baring-Gould in a great hurry for his mission at Horbury Bridge about the year 1865. Here the children had to march many a long mile to take part in what is dear to the heart of every true child—a school feast. Owing to the distance from the church to the scene of the festivities, an early start was necessary, and marching in procession with banners waving, colours flying, and a cross preceding them, the little ones sang lustily all the way. It was for these processions that "Onward, Christian soldiers" was written, and though it is many a year since Mr. Baring-Gould led those enthusiastic little pilgrims of Horbury Bridge, it is not improbable that the hymn is as great a favourite among the newer generation there as it was thirty-five years ago. It was then sung to Gauntlet's tune, for Sullivan had not then composed that stirring march which would have made his name popular had he never written another note. In connexion with Sullivan's setting, which he christened "St. Gertrude," it is interesting to learn that after writing it the composer remarked that he was afraid that it would be too "brassy" and martial for church singing. He was more than surprised at its popularity.

The fourth stanza of this hymn, which runs :—

> What the Saints established
> That I hold for true.
> What the Saints believèd,
> That believe I too.
> Long as earth endureth,
> Men the faith will hold,
> Kingdoms, nations, empires
> In destruction roll'd,

is now generally omitted. Some time ago I asked Mr. Baring-Gould if he had written the verse, and if so, why it was so often excluded from our hymnals, and he replied: "The verse to which you refer was written by me, but as the hymn has been used in many religious communities where such words would be absurd if sung, they have been omitted." Mr. Baring-Gould has not the slightest objection to this being done; indeed, he says he considers it very sensible.

Exception is sometimes taken to the following lines, which occur in the third stanza:—

> We are not divided,
> All one body we,
> One in hope and doctrine,
> One in charity.

Though no one would deny that all true Christians are one in hope and charity, or ought to be, it is more than probable that on occasions they are not one in doctrine. But then, again, it is also equally true that, whenever the hymn is sung, whether in church or chapel, that particular congregation is not only one in hope and charity, but also one in doctrine.

Rather a good story is told in connexion with this hymn, which may or may not be true. It is related that a certain rather low church vicar, though he liked processions, particularly when he headed them, stoutly objected to the cross being carried. The organist and the choirmaster both did their best to persuade him that there was nothing wrong in carrying a cross, but they might just as well have addressed their remarks to his pulpit. The vicar was adamant. At last, losing all patience, the choirmaster altered the first verse, and the procession began their march round the church to the words :—

> Onward, Christian soldiers,
> Marching as to war,
> With the Cross of Jesus
> Left behind the door.

Whether the vicar saw more clearly after that is not recorded.

Another hymn which has, through the translation made by Mr. Baring-Gould, become a general favourite in all English-speaking countries is, "Through the night of doubt and sorrow." This hymn, while not specially written as a processional, is, through the lesson it teaches of progress and unity, eminently suited for singing as a " March Militant." Mr. Baring-Gould's translation has become widely used and is to be found in most hymnals published during the last thirty-five years. It is not always given exactly as the translator wrote it, but the alterations have been made in most instances, I believe, with Mr. Baring-Gould's sanction. In some hymnals the following verse— one of the best—is often omitted :—

> Onward, therefore, pilgrim brothers,
> Onward with the Cross our aid ;
> Bear its shame, and fight its battle,
> Till we rest beneath its shade.

This hymn, I believe, was also written for the children of Horbury Bridge, and was first sung by them previous to being published in 1867. It has been set to music by many celebrated composers, but the tune with which it is most closely allied is that by Dr. J. B. Dykes, called "St. Oswald." Dr. Dykes had a fondness for naming tunes after places of which he had happy recollections, and his tune to "Through the night of doubt and sorrow" received its name in memory of his own church in Durham.

The original of Mr. Baring-Gould's translation is

by a Danish poet, Bernhardt Severin Ingemann, and is a general favourite in Denmark. It was written in 1825, and is the only hymn by this author which has found a place in English hymnals. Other translations of his hymns have been made, but it cannot be said that any one of them is known to the ordinary student of hymnology. Ingemann died in 1862 at the age of seventy-three.

One of the most lyric of all Dr. Neale's translations from the Latin is a hymn, by an unknown author of the fifteenth or sixteenth century, beginning :—

> To the Name of our Salvation
> Laud and honour let us pay,
> Which for many a generation
> Hid in God's foreknowledge lay,
> But with holy exultation
> We may sing aloud to-day.

This hymn, as given in very many collections, has been considerably altered from Dr. Neale's original version first published in his *Mediaeval Hymns*. In *Hymns Ancient and Modern* the translation is quite as much the work of the compilers as it is that of Dr. Neale. The translator, however, was consulted with regard to the alterations before they were made, and offered no objection. This version has become by far the most popular and has been copied into a great number of other collections.

To the following verse :—

> Jesus is the Name exalted
> Over every other name ;
> In this Name, whene'er assaulted,
> We can put our foes to shame ;
> Strength to them who else had halted,
> Eyes to blind and feet to lame,

exception has been taken. "The separation of the last two lines from their verb," writes some one, "makes it difficult to follow the sense, and 'Eyes to blind and feet to lame' is not English."

Dr. Neale is said to have been very rapid in making his translations, and in support of this the following anecdote is related by the late Gerald Moultrie :—

"Dr. Neale was invited by Mr. Keble and the Bishop of Salisbury to assist them with their new hymnal, and for this purpose he paid a visit to Hursley parsonage. On one occasion, Mr. Keble, having to go to another room to find some papers, was detained a short time. On his return Dr. Neale said, "Why, Keble, I thought you told me that the *Christian Year* was entirely original?" "Yes," he answered, "it certainly is." "Then how comes this ?" and Dr. Neale placed before him the

1 THE REV. J. J. DANIELL.
Photo by Lambert, Bath.

2 THE REV. T. T. LYNCH.
Photo by W. E. Debenham, Haverstock Hill.

3 THE REV. G. S. HODGES, B.A.
Photo by W. Plumbe, Maidenhead.

4 THE REV. S. J. STONE, M.A.
From a Photo.

5 THE REV. THOMAS BINNEY, D.D.
From a Photo.

6 THE REV. CANON ELLERTON, M.A.
Photo by Gillman, Oxford.

Latin of one of Keble's hymns. Keble professed himself utterly confounded. He protested that he had never seen this 'original'—no, not in all his life. After a few minutes of quiet enjoyment, Neale relieved him by owning that he had just turned it into Latin during his absence."

The late Archbishop of Canterbury, Dr. Temple, is said to have remarked on one occasion that whenever he was called upon to open a new church or preside at a dedication festival he could safely count upon two things — cold chicken and "The Church's one Foundation." His Grace's remarks, however, were not intended to detract in any way from the real and acknowledged merits of either, both being practically above criticism.

The late Mr. Stone's hymn, deservedly termed "great," has been for close upon forty years the "National Anthem" of the churches :—

> "The Church's one Foundation
> "Is JESUS Christ her Lord:
> "She is His new Creation
> "By water and the Word.

In an interesting correspondence which I had

with the author some years ago regarding his famous hymn, Mr. Stone told me that its origin might be traced to the interest he took in Bishop Grey's defence of the catholic faith against the teachings of Bishop Colenso. In the following verse Mr. Stone pointedly refers to this circumstance:—

> Though with a scornful wonder
> Men see her sore opprest,
> By schisms rent asunder,
> By heresies distrest ;
> Yet saints their watch are keeping,
> Their cry goes up, "How long?"
> And soon the night of weeping
> Shall be the morn of song.

"The Church's one Foundation" has a remarkable effect on some temperaments. I have been told by men whose natures could hardly be termed "gentle" that to listen to this hymn sung by a large congregation was almost more than they could stand ; it made them feel weak at the knees, their legs trembled and they felt as though they were going to collapse. It seems an absurd statement to make, and yet I think some of us can understand the sensation. Though the hymn is "triumphant," in a sense there is a certain sadness running through it which almost brings the tears to one's eyes. The melody with which it is inseparably associated, though not specially composed for it, also has a certain mournfulness which admirably suits the suppressed exultation of the words, and may perhaps in some measure account for the feeling of depression which the hymn sometimes produces.

Mr. Stone altered the hymn a good deal after its publication in 1866, changing certain verses, omitting others, and substituting stanzas which were written after the hymn had appeared in a considerable number of collections. The version as given in *Hymns Ancient and Modern* is the one most widely known. It has been translated into many languages and finds a place in nearly all missionary hymnals.

"The Church's one Foundation" has been sung on many memorable occasions, perhaps the most striking being in 1888 when special services were held, in connexion with the Lambeth Conference, at Canterbury Cathedral, Westminster Abbey, and St. Paul's Cathedral. At each of these services "The Church's one Foundation" was sung and produced a profound impression. By one who was present at St. Paul's on this occasion I was told that the effect was almost appalling. It made a more lasting impression on his mind than anything else connected with that historic service.

Only a very short time before his death in 1901 I received a letter from this great hymnist written in a trembling hand. He apologized for the poorness of the writing, excusing himself with the remark that he was dying and therefore unable to produce good penmanship. Two days later he breathed his last, and the Church was the poorer by the loss of a great man. He lies buried in the yard of the church over which he had presided for so many years—St. Paul's, Haggerston.

On a processional hymn by St. Joseph of the Studium Dr. Neale based his very beautiful :—

> O happy band of pilgrims,
> If onward ye will tread
> With Jesus as your Fellow
> To Jesus as your Head!

On Dr. Neale's own confession this hymn may almost be regarded as original, seeing that there is in it very little that can be traced to St. Joseph. In most collections it is headed " The fellowship of His sufferings," and given exactly as Dr. Neale wrote it, no alteration being made whatever. The melody with which this hymn has been associated almost from the year of its translation was composed by J. H. Knecht towards the close of the eighteenth century.

A hymn which was specially written with the object of its being sung as a children's processional is "Brightly gleams our banner." It is an immense favourite in the country where Sunday school processions take place almost every other day throughout the summer. It is probably quite as well known as "Onward, Christian soldiers" and as frequently sung. The following verse makes it particularly suited for children :—

> Pattern of our childhood,
> Once Thyself a child,
> Make our childhood holy,
> Pure and meek and mild.
> In the hour of danger,
> Whither can we flee,
> Save to Thee, our Saviour,
> Only unto Thee?
> Brightly gleams our banner,
> Pointing to the sky,
> Waving on Christ's soldiers
> To their home on high!

This hymn was written by the Rev. Thomas Joseph Potter about the year 1858. It has been considerably altered, the verses which had special application to Roman Catholics having been eliminated. The refrain to this hymn in many collections runs :—

> Brightly gleams our banner,
> Pointing to the sky,

Waving wanderers onward
To their home on high.

This is the only hymn of Mr. Potter's which is to be found in the ordinary hymnal. He wrote a few others, which have been published in collections of limited circulation, and which have long since been forgotten. The author was a man of much learning and the writer of many prose works. He dabbled in fiction as a recreation from his more serious work, and several of his short stories appeared in various journals. Though brought up a Protestant he at the early age of twenty joined the Roman Catholic faith, and subsequently took great interest in foreign missionary work. He died in Ireland in 1873 at the age of forty-six.

"Hail the day that sees Him rise," though primarily a hymn for Ascension, is so often sung as a processional that I think no apology is needed for placing it in this chapter. It is one of Charles Wesley's most successful compositions and was written before he had reached the age of thirty. It has probably undergone more alterations at the hands of editors than any other of Charles Wesley's hymns, very few collections giving precisely similar versions.

In four hymnals before me of wide circulation it

PROCESSIONAL HYMNS

is interesting to compare the different versions. In Charles Wesley's collection the first verse reads :—

> Hail the day that sees Him rise,
> Ravished from our wishful eyes !
> Christ, awhile to mortals given,
> Reascends his native heaven.

In another collection this verse has been changed to :—

> Hail the day that sees Him rise,
> Alleluia !
> To His Throne above the skies ;
> Alleluia !
> Christ, the Lamb for sinners given,
> Alleluia !
> Enters now the highest heav'n.
> Alleluia !

In the *Church Hymnary* it is given as Charles Wesley wrote it, but in stanzas of eight lines each instead of four. In Thring's collection, besides being considerably altered, each line ends with " Hallelujah ! "

A very successful processional hymn by a living writer is :—

> Saviour, blessèd Saviour,
> Listen while we sing ;
> Hearts and voices raising
> Praises to our King.

This hymn in the original consists of nine

stanzas of eight lines each, but in the majority of hymnals it has been reduced to six or seven. The following verse is very often omitted :—

> Farther, even farther
> From Thy wounded side
> Heedlessly we wandered,
> Wandered far and wide ;
> Till Thou cam'st in mercy,
> Seeking young and old,
> Lovingly to bear them,
> Saviour, to Thy fold.

There is no particular story connected with the writing of this hymn. In a letter received from the author some short time since, Prebendary Thring says :—

"I am sorry to say that I am quite unable to give you any account of the circumstances under which I wrote 'Saviour, blessèd Saviour,' as I made no note on the MS. It probably arose, like a great number of my compositions, from some thought which happened to be passing through my mind at the time. I wrote it a great number of years ago, as far back as 1862, and it was one of the first of the many hymns I have written, the very first having been composed in 1861."

Prebendary Thring has written a great number

PROCESSIONAL HYMNS

of hymns for special occasions, such as church consecration, synods, friendly societies, consecration of a lych gate, etc., with music separately. His *Church of England Hymn Book*, one of the finest collections in the language, was published without music.

The late Dean Edward Hayes Plumtre is the author of that very fine processional hymn :—

> Rejoice, ye pure in heart,
> Rejoice, give thanks and sing ;
> Your festal banner wave on high,
> The Cross of Christ your King.

This hymn was written in 1865 and first sung in Peterborough Cathedral. The history of its origin is interesting. It appears that when the Peterborough Choral Union were arranging for their annual festival in 1865 Mr. Plumtre, as he was then, was asked to write a special processional hymn to be used in Peterborough Cathedral on the occasion of the festival. He readily consented to do so, and forthwith penned the hymn which has since become so popular, " Rejoice, ye pure in heart." The hymn was received with considerable favour, so much so, indeed, that the author subsequently printed it in a volume of his poems entitled *Lazarus*. It was there seen by Sir Henry Baker, who was at that time contemplating the first appendix to *Hymns Ancient and Modern*. He wrote to the author requesting his permission to include it in his hymnal, a request which Mr. Plumtre cheerfully granted. Sir Henry immediately handed the hymn over to his musical editor, Dr. William Henry Monk, who wrote for it that very spirited and appropriate melody to which it has ever since been sung. In compliment to the place of its birth and the grand old cathedral where the lines were first publicly sung, the composer christened the tune " Peterborough." It may be mentioned that the composition was written with Dr. Monk's usual celerity, that is in some ten or fifteen minutes.

In some hymnals the following verse, one of the finest, is often omitted :—

> Bright youth and snow-crown'd age,
> Strong men and maidens meek,
> Raise high your free exulting song,
> God's wondrous praises speak.

"Rejoice, ye pure in heart" is not the only processional hymn Dean Plumtre wrote. His " March, march onward, soldiers true" is a very fine composition, especially when sung to the " March of the Israelites," for which it was written. It is not of course as well known as "Rejoice, ye pure in heart," though in many respects quite equal to it in merit. Another processional hymn which Dean Plumtre wrote is "O praise the Lord our God," which is specially suited for a thanksgiving hymn. It was written about the same time as "Rejoice, ye pure in heart."

VIII

Communion Hymns

INCLUDED in the very large number of hymns which have been specially written for the office of Holy Communion are compositions by many of our greatest hymnists. Dr. Doddridge's "My God, and is Thy table spread" was written for the administration of the Holy Sacrament. Curiously enough, though it is, perhaps, the best known of all Doddridge's compositions in England and all English-speaking countries, its use in Scotland is very limited. Why this should be so it is difficult to say, seeing that many of Dr. Doddridge's other hymns are widely sung and appreciated in northern Britain.

"My God, and is Thy table spread," like the majority of Doddridge's hymns, was not published until after his death. This is probably to be accounted for by the fact that during his lifetime when any of his compositions were sung they were usually given out line by line from the pulpit. There are, however, many manuscripts of this hymn in preservation, one being in the possession of Mrs. Allen, of Bristol. It is related that when his hymns became known Dr. Doddridge was asked by several members of his congregation for copies, and the good Doctor, with that simple-hearted kindliness which was his chiefest characteristic, would sit down and write out the special hymn required and afterwards present it to the applicant with a smile of peculiar sweetness. In this way many of Dr. Doddridge's manuscripts have come to be preserved and treasured for close upon one hundred and fifty years. Some of them are so fresh and clear that they might have been written but yesterday.

When Dr. Doddridge's Sacramental hymn was first published in 1755 it was given in six verses of four lines each, but in modern hymnals it now appears in a reduced form, one or other of the following verses being usually omitted :—

> Let crowds approach with hearts prepared ;
> With hearts inflamed let all attend,
> Nor when we leave our Father's board
> The pleasures or the profit end.

> Revive Thy dying Churches, Lord,
> And bid our drooping graces live;
> And, more, that energy afford
> A Saviour's love alone can give.

In the hymnal which is said to have the largest circulation in the world, the following Doxology is given. It is not, however, by Dr. Doddridge :—

> To Father, Son, and Holy Ghost,
> The God whom heaven and earth adore,
> From men and from the angel-host
> Be praise and glory evermore.

This hymn is usually sung to a very beautiful melody called "Rockingham," by Dr. E. Miller. It is the same tune which is associated with "When I survey the wondrous Cross."

Among living hymnists who have given us compositions which have already taken a firm hold on the affections of the Church is the Rev. Vincent Stuckey Stratton Coles, the present Librarian of Pusey House, Oxford. Mr. Coles is the author of several hymns which are included in many modern hymnals, his most celebrated being the one he wrote for Holy Communion, "We pray Thee, Heavenly Father." This hymn was written while the author was curate of Wantage. Here Mr. Coles used to have large classes of young people to prepare for confirmation, and it was for them that he specially wrote this hymn.

Since the inclusion of " We pray Thee, Heavenly Father" in many hymnals Mr. Coles has written a revised version, which has not, however, yet been published. He sends it to me with permission to reproduce, and as it differs in many respects from the original, I give it. Like the first version, it is in four stanzas of eight lines each. In some respects it is the finer hymn of the two :—

<pre>
    We pray Thee, Heavenly Father,
      To hear us in Thy love,
    And pour upon Thy children
      The unction from above;
    That so in love abiding,
      From all defilement free,
    We may in pureness offer,
      Our Eucharist to Thee.

    All that we have we offer,
      For it is all Thine own;
    All gifts by Thy appointment
      In bread and cup are shown;
    One thing alone we bring not,
      The wilfulness of sin;
    And all we bring is nothing,
      Save that which is within.

    Within the pure oblation,
      Beneath the outward sign,
    Through that His operation
      The Holy Ghost Divine,
</pre>

Lies hid the Sacred Body,
  Lies hid the Precious Blood
Once slain, now ever glorious,
  Of Christ our Lord and God.

Wherefore though all unworthy
  To offer sacrifice,
We pray that this our duty
  Be pleasing in Thine eyes;
For thanks, and praise, and worship
  For mercy and for aid,
The catholic memorial
  Of Jesus Christ is made.

It will be noticed in comparing this hymn with the original that the first verse remains unchanged. Mr. Coles based his composition on the words "I love them that love Me: and those that seek Me early shall find Me." The very beautiful and appropriate tune to which this hymn is allied, known as "Dies Dominica," is by the late Dr. Dykes.

The best known of all Josiah Conder's hymns was written for use at Holy Communion. This is his very beautiful "Bread of Heaven! on Thee we feed," or as originally written, "Bread of Heaven! on Thee *I* feed." Mr. Charles E. Conder, son of the celebrated hymnist, tells me that the hymn was first published in the author's *Star of the East*, which appeared in 1824. It is there headed

"For the Eucharist," and preceded with the words taken from St. John :—" I am the Living Bread which came down from heaven. Whoso eateth My flesh and drinketh My blood, hath eternal life. I am the true Vine."

The first verse of this hymn is usually given in an unaltered form with the exception of the change of " I " into " we." The second verse in the manuscript, however, reading as follows :—

> Vine of Heaven ! Thy Blood supplies
> This blest cup of sacrifice.
> 'Tis Thy wounds my healing give,
> From Thy veins I drink and live.
> Thou my life ! Oh, let me be
> Rooted, grafted, built in Thee,

has been changed in nearly all hymnals to :—

> Vine of Heav'n ! Thy Blood supplies
> This blest cup of sacrifice ;
> Lord, Thy wounds our healing give,
> To Thy Cross we look and live :
> Jesus, may we ever be
> Grafted, rooted, built in Thee.

This hymn is contained in a large MS. volume consisting of 360 pages of close writing. Inscribed on the opening page is the following title :—" The Star of the East, with other Poems chiefly Religious and Domestic, by Josiah Conder." It contains odes to various people, a large number of hymns, a very fine elegy on the death of Henry Kirke White, and a poem on Queen Caroline.

Among the hymns is one which has not been published. As it is quite as terse and beautiful as those which have already found their way into our hymnals, I give it with his son's permission :—

> O God, whose all-creating might
> Gave birth to Nature's laws;
> Whose Sovereign working day and night,
> Knows neither rest nor pause.
>
> Let Thy redemptive work of grace
> Like light its course pursue,
> Till earth's wide circle it embrace,
> Creating all things new.
>
> Throughout the universe of mind
> Let light and healing spread:
> Then, come, Deliverer of mankind,
> And wake the slumbering dead.

Mr. Conder tells me that his father was a very quick writer and very seldom altered his compositions when written. There is, unfortunately, no portrait of the late Mr. Conder in existence. His son possesses an old print depicting a group of people, in the midst of which the late hymnist appears, but it is so very minute that it is impossible to distinguish one from the other. An

THE REV. CANON BRIGHT, D.D.
Photo by Elliott & Fry.

MR. ALBERT MIDLANE.
Photo by F. Quinton, Newport.

attempt was made some time ago to make an enlarged portrait of the hymnist from this group, but it proved a complete failure.

The sad and solemn air to which " Bread of Heaven! on Thee we feed" is invariably sung is by the present Archbishop of York, Dr. W. D. Maclagan.

The late Canon Bright was the author of several Communion hymns, the best and most widely known being "And now, O Father, mindful of the love," and "Once, only once, and once for all." Both these hymns first appeared in his volume of poems published in 1866. They soon attracted the eagle eye of Sir Henry Baker and subsequently appeared in his hymnal. A short while before his death Canon Bright sent me a manuscript copy of "And now, O Father, mindful of the Love," which I had hoped to reproduce, but unfortunately it has been mislaid. In sending me the composition Canon Bright wrote that he could remember nothing of interest regarding the circumstances under which he wrote it. He composed many hymns in his early life and the Holy Eucharist had a peculiar charm for him. The hymn in question was written in 1864, and as given in his volume of poems it is of considerable length. He reduced it for congregational singing, and the hymn as it now appears really consists of verses taken from the longer poem.

"Once, only once, and once for all," was written about the same time as "And now, O Father," and is considered by some hymnists to be the finer of the two compositions, an opinion which I do not share. In some hymnals the following Doxology concludes the hymn; it is not, I believe, the work of Canon Bright:—

> All glory to the Father be,
> All glory to the Son,
> All glory, Holy Ghost, to Thee,
> While endless ages run.

A hymn which was originally intended for use at Confirmation services, but which is now very frequently sung at Holy Communion, is Mrs. Maude's "Thine for ever! God of Love." This hymn, the author of which is a resident of Ruabon, was written as long ago as 1847. In a correspondence which I had with Mrs. Maude some few years since I learned the story of its composition, which is an interesting one.

"The hymn in question," wrote the authoress, "was written for my own class of young women in my late husband's then parish, St. Thomas, Newport, Isle of Wight. In 1847 we had large Sunday

COMMUNION HYMNS

Schools, and I used to take a class of elder girls. They were being prepared for Confirmation by the Bishop of Winchester and I was helping them in their work when I was suddenly attacked by a serious illness. When I was somewhat recovered I went for a change to the sea-side, and while there I used to write many letters to my girls. In one of these letters I wrote, quite spontaneously, the hymn 'Thine for ever! God of Love.' It was no effort to me whatever, the words came unsought; and, without even correcting the lines, the hymn, together with the letter, was despatched. These letters were afterwards published in a magazine, and some months later on opening the new hymnal published by the Christian Knowledge Society I was very much astonished to see my own little composition. How it got there I have never found out to this day. It has been a matter of most genuine surprise to me that my simple lines should have met with such acceptance in many lands, and I can only feel humble and thankful that they have been of service to so many."

The MS. which Mrs. Maude sent me of this hymn consists of seven verses, but the following two stanzas, she explains, were never introduced into any hymnal :—

> Thine for ever in that day
> When the world shall pass away:
> When the trumpet note shall sound,
> And the nations underground
>
> Shall the awful summons hear,
> Which proclaims the judgment near.
> Thine for ever. 'Neath Thy wings
> Hide and save us, King of kings.

"It has been a real regret to me," continues Mrs. Maude's interesting letter, "and I have been for many years trying to draw attention to the matter, that in several collections the fourth verse has been altered without any reference to me. It often stands thus :—

> 'Thine for ever! Saviour keep
> Us Thy weak and trembling sheep.'

"Now the connection between Shepherd (as I wrote) and sheep is too obvious to need comment; but besides this, the title of Shepherd introduces a fresh and most endearing office of our Lord into the hymn, where Saviour had already occurred.

"Then 'us' is a most unmusical word to begin a line with, and moreover the thought of the verse is lost, for the first two lines are a prayer for the catechumens from the congregation :—

'Thine for ever! Shepherd keep
These Thy frail and trembling sheep;'

then the supplication reverts and embraces all present:—

'Safe alone beneath Thy care,
Let us all Thy goodness share.'"

Mrs. Maude's complaint is a very reasonable one, and if any editor contemplating the publication of a new collection of hymns should see these lines, it is to be hoped that he will respect Mrs. Maude's wish and give the verse as she wrote it.

"O food that weary pilgrims love" is a translation from a Latin hymn, the author of which is unknown. It is said to have been written about the seventeenth century, and has been ascribed to various writers. Little reliance, however, can be placed on statements which, at the best, are the merest guesses. There have been many translations of the hymn, by far the most popular being that made by the compilers of *Hymns Ancient and Modern*. The following verse is particularly fine and suggests Sir Henry Baker:—

Lord Jesus, Whom, by power Divine
Now hidden 'neath the outward sign,
We worship and adore,
Grant, when the veil away is roll'd,
With open face we may behold
Thyself for evermore.

Ray Palmer, the American hymnist, and author of "My Faith looks up to Thee," also made a translation of this hymn, the first line of which begins "O Bread to Pilgrims given." This version is not very well-known in England but in America it is a great favourite, and one of the most frequently sung of all Communion hymns.

James Montgomery never penned a more perfect or pathetic hymn than his one for Holy Communion. This composition :—

[handwritten manuscript: "The Lord's Supper. / According to thy gracious word, / In meek humility, / This will I do, my dying Lord, / I will remember Thee"]

was written about the year 1824, and published a few months later in a collection of his hymns. In the original manuscript, kindly sent to me by a correspondent, this hymn is headed "The Lord's Supper," and in one corner Montgomery has written "This do in remembrance of Me." The second verse :—

>   Thy Body, broken for my sake,
>   My bread from Heaven shall be ;
>   Thy testamental cup I take
>   And thus remember Thee

COMMUNION HYMNS

does not appear to have been adhered to in all hymnals. In Thring's collection it has been changed to :—

> Thy Body, broken for my sake,
> My bread from Heaven shall be;
> The cup, Thy precious Blood, I take,
> And thus remember Thee.

It is interesting to note in the original MS. that Montgomery evidently desired that the words "will" and "me" in the first and last verses should be emphasised as he underlines both, the verses reading :—

> According to Thy gracious word,
> In meek humility,
> This will I do, my dying Lord,
> I *will* remember thee.

> And when these failing lips grow dumb,
> And mind and memory flee,
> When Thou shalt in Thy Kingdom come,
> Jesus, remember *me*.

This interesting manuscript is signed J. M., and the word "dying" in the first verse has been substituted for "precious," which the author has scored through.

James Montgomery wrote about 400 hymns, of which some sixty may be said to have come into common use. He was also the author of many prose works and volumes of poems which have sunk into oblivion. Montgomery himself does not appear to have had any very great opinion of his poems if we are to judge from the reply he gave when a Whitby solicitor asked him which of his works would live. "None, sir," replied the outspoken poet, "nothing, except, perhaps, a few of my hymns." He was a true prophet, for to-day while his hymns are remembered his poems and prose works have long since been forgotten.

In the days of Pope Urban IV. the greatest of all Communion hymns was written—" Pange lingua gloriosi corporis mysterium," known to present time celebrants by the translation, " Now, my tongue, the mystery telling." This version is the work of the late Edward Caswall. It has, however, been considerably altered by various compilers. Other translations have been made by Neale, Isaac Williams, J. W. Irons and many others, but none of these has gained the popularity enjoyed by that based on Edward Caswall's version.

A hymn, not specially written for the service of Communion, but one which is admirably suited to that season, is " Jesu, my Lord, my God, my All," by the Rev. Henry Collins. This hymn, with its very beautiful refrain :—

> Jesu, my Lord, I Thee adore,
> Oh make me love Thee more and more,

has become a general favourite with all denominations, and is to be found in a considerable number of hymnals. Strangely enough, Faber also wrote a hymn beginning "Jesu, my Lord, my God, my All," in the same metre, and each verse concluding with the line "Oh make us love Thee more and more." Father Faber's hymn, however, does not appear to have found its way into any protestant hymnal.

Mr. Henry Collins is one of our living hymnists, though for a great number of years he has published nothing. So far as I have been able to ascertain only two hymns by this author have come into general use, the one already referred to and a very beautiful hymn for Passiontide, the first verse of which reads :—

> Jesu, meek and lowly,
> Saviour, pure and holy,
> On Thy love relying,
> Hear me humbly crying,

written on the text, "I, if I be lifted up from the earth, will draw all men unto Me."

Though brought up a member, and ordained a minister, of the Church of England, Mr. Collins, in 1857, seceded to the Church of Rome, and soon afterwards joined the Cistercian Order of Monks. In a short note received from him a year ago in answer to a request for a manuscript of his well-known hymn, I was informed that he was unable to send it as he now belonged to a cloistered order and therefore lived " apart from the world."

In placing Sir Henry Baker's exquisite version of the 23rd Psalm under this chapter, I do so by reason of the following verse, which appears to me to render it a suitable Communion hymn:—

> Thou spread'st a table in my sight;
> Thy Unction grace bestoweth;
> And oh, what transport of delight
> From Thy pure Chalice floweth.

"The King of Love my Shepherd is," was written by Sir Henry Baker in 1868 and published the same year in the appendix to his hymnal. It was one of the most popular of the new hymns, and Sir Henry was soon inundated with requests for permission to insert it in various collections, requests which he never refused. The author himself, I believe, wrote the first tune to this hymn, but, as it was not very successful, he sent the MS. to his friend, Dr. Dykes, who composed for it the very beautiful melody to which it has ever since been usually sung, and which he named "Dominus regit me." It will further be remembered that one of the late Charles Gounod's most successful sacred songs was a setting of "The King of Love my Shepherd is."

Canon Ellerton, who was for many years a personal friend of Sir Henry Baker, says in a note on this hymn: "It may interest many to know that the verse:—

> Perverse and foolish oft I strayed,
> But yet in love He sought me,
> And on His shoulder gently laid,
> And home, rejoicing, brought me,

was the last audible sentence upon the dying lips of the lamented author."

Sir Henry Baker's last work was an edition of the *Psalter and Canticles*, pointed and set to appropriate chants, ancient and modern, which he edited in conjunction with his friend, the late Dr. Monk. It is greatly appreciated wherever used, and deserves to be more widely circulated.

Besides being a hymnist, Sir Henry Baker was also something of a composer, several of the most popular tunes in his hymnal being from his own pen. Perhaps his most successful setting is that to "Art thou weary, art thou languid?" which he called "Stephanos," a beautiful melody, which is as widely known as the hymn itself. Though Sir

Henry possessed a wonderful ear for music and great creative gifts of melody, he was not a master of composition, and the harmonies of his hymn tunes were in every case made by Dr. Monk.

Hymns written on the Divine Love of Christ are specially suited for congregational singing during the administration of Holy Communion. Such a hymn is "Hark, my soul! it is the Lord," and a more tender and beautiful composition it would be impossible to find in the whole range of hymnody. Was ever a more pathetic verse penned than the following :—

> Can a woman's tender care
> Cease towards the child she bare?
> Yes, she may forgetful be,
> Yet will I remember Thee.

"Hark, my soul!" was written by William Cowper about the year 1765, and afterwards published in the *Olney Hymn Book*. It was composed during that period, or soon after it, when the author was passing through a phase of melancholy which threatened to end in madness. One of his delusions is said to have been a conviction that he did not love his Maker with sufficient fervour, and the last verse of "Hark, my soul!" rather supports the supposition :—

THE REV. JOHN WESLEY, M.A.
From an Engraving.

MR. WILLIAM COWPER.
From the painting by F. Bartolozzi, Esq., R.A.

> Lord, it is my chief complaint
> That my love is weak and faint;
> Yet I love Thee and adore;
> Oh for grace to love Thee more.

Of all Cowper's hymns this is the finest. It has been translated into many languages, and is to be found in every hymnal of any standing published during the last hundred years. It was one of the favourite hymns of the late W. E. Gladstone, who translated it into Italian a few years before his death.

As an instance of the curious ideas children sometimes associate with a hymn, I was told by a lady who was in the habit of singing hymns to her little girl, aged six, in order to coax her to go to sleep, that "Hark, my soul!" was the one which appeared to give the greatest amount of pleasure. One evening, on her mother singing a different hymn, the child complained and begged for "the other one" instead. Her mother, forgetting for the moment the hymn she meant, asked her how it began. The little one replied that she didn't know, but it was about the "she-bear!" This was how the childish mind had construed the meaning of the two lines:—

> Can a woman's tender care
> Cease towards the child she bare?

There is only one recognized tune to "Hark, my soul!" though it is not improbable that many have been composed. This is "St. Bees," by the late Dr. Dykes. In connexion with the writing of this exquisitely tender setting to Cowper's lines, Mr. Bennet Kaye, of Durham, who was at one time assistant organist at Dr. Dyke's church, tells me an incident which is not without interest. "Dr. Dykes," Mr. Kaye says, "used frequently to come to the boys' rehearsals before morning service and begin practising with them the music of the day. Presently he would drift into something fresh, and the boys would remain perfectly still and listen entranced. On one occasion he wandered into a particularly beautiful melody, playing it over several times. The air made a lasting impression upon me, and afterwards, when it came to be published, I recognized in the tune to 'Hark, my soul! it is the Lord,' the melody which had so greatly attracted me. Dr. Dykes named the tune 'St. Bees,' from a little place where he had passed many pleasant hours."

Henry Kirke White's "Oft in sorrow, oft in woe," or, as originally written, "Much in sorrow, oft in woe," I place among Communion hymns on the strength of the first verse, in which there appears to be an intention on the part of the author that the composition should be used as a Sacramental hymn:—

> Much in sorrow, oft in woe,
> Onward, Christians, onward go,
> Fight the fight, and worn with strife;
> Steep with tears the Bread of Life.

This hymn has a curious history. When it was written is unknown, for it was not seen until after the death of the author. Soon after his decease, in 1806, his unpublished compositions were gathered together by a relative. Among the numerous papers which he had left behind were several sheets of ordinary foolscap, on which the young poet had evidently been working out problems in algebra. On the back of one of these papers was the hymn beginning, " Much in sorrow, oft in woe."

The hymn, as given in present-day hymnals, has been very much altered and added to, and the history of this alteration and addition is as interesting as the story connected with the finding of the original composition. In 1827 Mrs. Fuller-Maitland published a volume of hymns and included in it a new version of "Much in sorrow, oft in woe," written by her daughter Frances, a little girl barely fourteen years of age. It is this version, partly Henry Kirke White and partly Frances Fuller-Maitland, which has become the most popular, being given in nearly all hymnals. The following are the verses which were composed by little Frances Fuller-Maitland :—

> Let your drooping hearts be glad ;
> March in heavenly armour clad ;
> Fight, nor think the battle long ;
> Victory soon shall tune your song.
>
> Let not sorrow dim your eye ;
> Soon shall every tear be dry ;
> Let not fears your course impede ;
> Great your strength if great your need.
>
> Onward, then, to battle move ;
> More than conquerors ye shall prove ;
> Though opposed by many a foe,
> Christian soldiers, onward go.

I had hoped to have been able to give a facsimile of this hymn, thinking that so interesting a MS. would have been certain of preservation. I therefore wrote to Mr. Potter Briscoe, the principal librarian of Nottingham Free Library, the town where Kirke White was born, and where he lived for many years, asking for information respecting the poet's MSS. In reply, however, Mr. Briscoe says : "After many years of watching, I have at last arrived at the reluctant conclusion that there are no Kirke White MSS. in existence. I have been collecting material for a new edition of

White's poems for many years, and my researches would certainly have brought them to light had they existed. Some years ago the housekeeper to the late Rev. Kirke Vivyan became possessed of MSS. of Kirke White and burnt them."

It is somewhat surprising to note how few are the hymns which have been specially written for the occasion—so impressive to the young—of first Communion. Nevertheless, though this is so, there are several hymns more particularly suited to those who are about to partake of their first Eucharist, and among these is one which, in hymnody, is accorded a very high place. I refer to :—

>O Jesus, I have promised
>To serve Thee to the end,

by the late Mr. J. E. Bode. In a little volume dealing with the " First Communion," which I saw in the hands of a young communicant the other day, the following verse, taken from this hymn, had been inscribed on the title-page :—

>Oh let me feel Thee near me :
>The world is ever near ;
>I see the sights that dazzle,
>The tempting sounds I hear ;
>My foes are ever near me,
>Around me and within ;
>But, Jesu, draw Thou nearer,
>And shield my soul from sin.

A more appropriate quotation could hardly have been chosen; it is one which should be inscribed in all Communion books presented to young celebrants.

Mr. Bode was the author of two volumes of verse, one of which contained hymns specially suited to the festivals of the Church. With the exception of "O Jesus, I have promised," very few, however, have gained any great hold on the public, and, in consequence, the author might be classed among the "one-hymn" writers. It is noticeable that, though Mr. Bode wrote in the original "O *Jesus*, I have promised," it has been changed in the majority of hymnals to " O *Jesu*, I have promised," but probably more for the sake of euphony than anything else. In all other respects the hymn appears exactly as the author wrote it —rather remarkable, considering the popularity to which it has attained.

"O Jesus, I have promised" was written about the year 1866, during the time when Mr. Bode was rector of Castle Camps, Cambridge. Like "Thine for ever, God of Love," and several other hymns, it was specially composed for the author's Confirmation classes It was not then sung to the melody which has become so familiar to us and which was specially written for it by Mr. J.

W. Elliott. This well-known tune, which the composer called "Day of Rest," has since come to be associated also with that very fine temperance hymn by the late Mr. S. J. Stone:—

> O Thou before whose Presence
> Nought evil may come in,
> Yet who dost look in mercy
> Down on this world of sin ;
> Oh give us noble purpose
> To set the sin-bound free,
> And Christ-like tender pity
> To seek the lost for Thee.

Mr. Bode, after fourteen years of unremitting labour at Castle Camps, died in that town in 1874 at the comparatively early age of fifty-eight.

Another hymn which is also very suitable for communicants who are attending their first celebration is:—

> O happy day that fixed my choice
> On Thee, my Saviour and my God !
> Well may this glowing heart rejoice,
> And tell its raptures all abroad.

This hymn, which is very frequently used at Confirmations, was written by Dr. Phillip Doddridge and published after his death. Though in the original it consists of five verses, these have in many collections been reduced to four, the verse generally omitted being the third :—

'Tis done, the great transaction's done,
I am my Lord's, and He is mine;
He drew me, and I followed on,
Charmed to confess the voice Divine.

Tennyson was once credited with having written this hymn, the mistake being due to a young reporter whose visits to church were probably few and far between. It appears that at the request of the late Queen Victoria "O happy day that fixed my choice" was sung at the Confirmation of one of the Royal children. The following day it was reported in one of the leading London journals that the hymn had been specially composed for the occasion by Tennyson, the Poet Laureate. Together with this startling announcement appeared some critical comments to the effect that if the Poet Laureate could write nothing better it was high time that objection was raised to his receiving national pay!

"O happy day that fixed my choice" is one of Doddridge's most successful hymns, and is to be found in a large number of hymnals though the one which is said to have the largest circulation omits it. It has been translated into numerous languages and is included in a great number of foreign missionary hymnals. It has never been very fortunate with regard to its

COMMUNION HYMNS

setting, the melody to which it is usually sung being somewhat commonplace.

"When I survey the wondrous Cross," though originally intended for Good Friday services, is equally suited to Holy Communion. Written early in the eighteenth century by Isaac Watts, it has long since come to be regarded as that author's greatest hymn. Strangely enough it is not found in so large a number of collections as many other hymns of lesser importance. In the original it consisted of five verses, but the following is now more frequently omitted than inserted :—

> His dying crimson, like a robe,
> Spreads o'er His Body like a tree ;
> Then am I dead to all the Globe,
> And all the Globe is dead to me.

A story is related in connexion with "When I survey the wondrous Cross" which may be taken as an example of the absolutely meaningless manner in which a congregation will sometimes sing a hymn. It is told, I believe, of the late Rev. C. H. Spurgeon. It was in his early manhood, before the days of those large offertories which were so ungrudgingly given during his ministry at the Tabernacle. His congregation had finished singing "When I survey the wondrous

Cross" and Mr. Spurgeon was in the pulpit preparatory to beginning his discourse when, gazing somewhat mournfully upon his flock, he said: "Brethren, we have just finished singing Isaac Watts' grand hymn: the last words you uttered were these,—

> Were the whole realm of Nature mine,
> That were a present far too small;
> Love so amazing, so Divine,
> Demands my soul, my life, my all."

Then softly repeating the lines—

> Were the whole realm of Nature mine,
> That were a present far too small,

he suddenly electrified his hearers by demanding whether they knew what the collection amounted to that morning. Without waiting for a reply, he quickly added, " I will tell you. Seventeen shillings and a penny. The whole realm of nature, of course, is not yours to give, but you can surely afford more than a paltry seventeen shillings and a penny. It is an insult to your Maker. Perhaps you did not realize what you were singing. I feel sure you did not, and in order that you may not go away unhappy there will be another collection at the close of the service."

Whether the story is true or not I cannot say

but it is certainly eminently characteristic of the great preacher.

In some collections the following Doxology has been added to "When I survey the wondrous Cross." It is not, however, by Watts—

> To Christ, who won for sinners grace
> By bitter grief and anguish sore,
> Be praise from all the ransomed race
> For ever and for evermore.

The greatest of all Charles Wesley's hymns, "Jesu, Lover of my soul" may, I think, be given a place in this chapter in consideration of the frequency with which it is sung during the celebration of the Holy Eucharist. Written for those who were passing through any period of special anxiety or temptation, it has been a comfort to countless thousands. The date of its composition indicates that it was one of the first of Charles Wesley's hymns, being written soon after the author had completed his thirtieth year. I had hoped to have been able to give a reproduction in facsimile of this hymn from the manuscripts in the keeping of the Rev. Charles H. Kelly, but though we both searched untiringly through fourteen volumes we failed to bring it to light. Mr. Kelly, who has had a further search since, has reluctantly come to the conclusion that it is not among those MSS. placed under his charge.

Few hymns have had prettier legends woven about their birth than "Jesu, Lover of my soul." Every one, of course, knows the story of the sea-bird which flew to Charles Wesley's breast for protection from the storm. Equally well known is that which tells how a dove hunted by a hawk sought safety in a similar refuge. These and many other incidents have been associated with the great hymn, and perhaps we should be glad to think that some of them are true. Whether they are or not it is impossible to say. Certainly they have never been conclusively proved absolutely devoid of foundation. There is an old saying that a man should be believed until he has proved himself unworthy of belief, and the same maxim, with a slight variation, may be applied to these pretty stories regarding the greatest of all great hymns.

IX

Hymns for Holy Matrimony, Missions, and "Those at Sea"

THE number of hymns specially written for use at marriage ceremonies is not great, and of these only four may be said to have come into general favour. The one which now appears to be universally sung at all classes of weddings is Mrs. Dorothy Gurney's :—

> O perfect Love, all human thought transcending,
>   Lowly we kneel in prayer before Thy Throne,
> That theirs may be the love that knows no ending,
>   Whom Thou for evermore dost join in one.

This hymn was written in 1883. "I wrote it," says the authoress in a letter before me, "for the wedding of my sister, now Mrs. Hugh Redmayne. It was written some few weeks before her marriage. The story of its composition is a very simple one and I gladly tell you the circumstances connected with it. We were all singing hymns one Sunday evening, and had just finished 'O Strength and Stay,' the tune to which was an especial favourite of my sister's, when some one remarked what a pity it was that the words should be unsuitable for a wedding. My sister, turning suddenly to me, said :—' What is the use of a sister who composes poetry if she cannot write me new words to this tune.' I picked up a hymn-book and said :— ' Well, if nobody will disturb me I will go into the library and see what I can do.' After about fifteen minutes I came back with the hymn 'O perfect Love' and there and then we all sang it to the tune of 'Strength and Stay.' It went perfectly and my sister was delighted, saying that it must be sung at her wedding. For two or three years it was sung privately at many London weddings and then it found its way into the hymnals. The writing of it was no effort whatever after the initial idea had come to me of the two-fold aspect of perfect union, love and life, and I have always felt that God helped me to write it."

In the great dearth of marriage hymns "O perfect Love" was so much welcomed that it is now to be found in almost every sort of hymnal of the Christian Church. The authoress has received letters regarding it from all parts of the world, and it has been translated into many languages. Sir

Joseph Barnby set it to music for the wedding of the Princess Louise of Wales, and few fashionable marriages are now celebrated without it being sung. Mrs. Gurney says she is thankful that what she intended for one family only has been so great a help and pleasure to others.

The melody for which this hymn was written is one by the late Dr. Dykes which he specially composed for John Ellerton's translation of the Latin hymn :—

> O Strength and Stay upholding all creation,
> Who ever dost Thyself unmoved abide,
> Yet day by day the light in due gradation
> From hour to hour through all its changes guide.

"O perfect Love" goes excellently to this tune, in fact it seems to suit it rather better than the one composed by Dr. Barnby.

"The Voice that breathed o'er Eden" has long since been a favourite wedding hymn. It is usually sung at the commencement of the ceremony and its plaintive melody gives to it a touch of what may be described as a happy solemnity. It was written by John Keble when he had reached the age of sixty-five, and he therefore never lived to see the day when it would be sung at nearly every wedding in the land. It has been slightly altered from time to time by various hymnal editors and the following verse is usually omitted :—

> For dower of blessèd children,
> For love and faith's sweet sake,
> For high mysterious union
> Which nought on earth may break.

In many cases the first line of the fourth verse " Be present, awful Father " has been changed to " Be present, Holy Father " which is an improvement.

The " Voice that breathed o'er Eden " is interesting not only by reason of the great popularity to which it has attained, but also from the fact that it was probably one of the last hymns the author ever penned. No other of Keble's hymns which has gained any degree of celebrity bears a later date than this one and it appeared in no work by him published during his life. It was written by special request and has proved but another instance of the success which occasionally attends the writing of a hymn " to order."

The melody with which this hymn is usually associated and which is known by the title of " St. Alphege," was not specially written for it though there is a general impression to the contrary. It appears that the late Dr. Gauntlett really wrote the tune to " Brief life is here our portion " but it was afterwards found to suit Keble's hymn so well that a new tune was thought to be unnecessary. It has therefore ever since been usually sung to "St. Alphege."

"How welcome was the call," another very popular marriage hymn, was written by the late Sir Henry Baker for the first edition of his hymnal. At that time there was actually only one other wedding hymn of any degree of popularity in existence, and that was Keble's "The Voice that breathed o'er Eden." Sir Henry saw, therefore, how essential it was that at least a couple of marriage hymns should appear in his hymnal and he wrote "How welcome was the call," which was set to music by his friend Dr. Gauntlett. Though the extreme scarcity of marriage hymns may be in some way accountable for its immediate success, it is undeniably one of the best of the author's many beautiful hymns and in every way deserving of its great popularity.

The hymn, of course, is based on the words taken from St. John: "Both Jesus was called, and His disciples, to the marriage," the author taking as his subject the wedding feast at Cana.

No MS. of "How welcome was the call" appears to be in existence. Indeed, without exception, all the hymnal manuscripts of Sir Henry Baker seem to have been either destroyed or lost. Probably the very fact of his being the editor of the hymnal for which his compositions were written would account for his carelessness regarding their preservation in manuscript, and it is not at all unlikely that when they were returned from the printers he threw them into the waste-paper basket. In a letter received some time ago from the Chairman to the Committee of the hymnal with which Sir Henry's name is inseparably associated I was informed that no manuscript of any of Sir Henry Baker's hymns was known to exist, and the same information has been given me since by many relatives of the late hymnist.

"How welcome was the call" is one of the few hymns which have escaped alteration, appearing in all collections exactly as the author wrote it.

Another marriage hymn which, like "The Voice that breathed o'er Eden," owes its origin to a request received from a friend, is the late John Ellerton's :—

> O Father, all-creating,
>   Whose wisdom, love, and power
> First bound two lives together
>   In Eden's primal hour,

> To-day, to these Thy children
> Thine earliest gift renew,—
> A home by Thee made happy,
> A love by Thee kept true.

A somewhat interesting incident is connected with this hymn. It was written when Canon Ellerton was Rector of Barnes, and at the special request of the late Duke of Westminster. It appears that a short while before the marriage of his daughter to the Marquis of Ormonde, the Duke either wrote to, or saw, Canon Ellerton and asked him if he would be willing to compose a hymn for the occasion. The hymnist cheerfully consented and a few days later sent the Duke the MS. of "O Father, all-creating" with which his Grace was very much pleased. This was in 1876 but it was not until four years later that it appeared in any hymnal, the first to publish it being Prebendary Thring in his well-known collection. In that hymnal it appears in a slightly altered form, the first lines having been changed from the original to :—

> O Father, all-creating,
> Whose wisdom and whose power
> First bound two lives together
> In Eden's primal hour.

The omission of the word "love" rather detracts from the beauty of the verse.

This is not the only hymn Canon Ellerton wrote "to order." Indeed, he was accustomed to receive requests for hymns on all sorts of subjects, sometimes from people he had never seen. In many cases when the subject suggested was one on which he considered a hymn would be useful he complied with the request, and in this way many of his hymns are said to have originated.

The most popular of all missionary hymns is undoubtedly "From Greenland's icy mountains" by Bishop Heber. It was written as far back as 1819 at Wrexham, where Heber's father-in-law, Dr. Shipley, Dean of St. Asaph, was vicar. The story of its composition is one which has been told many times but which will bear repetition. Briefly, the circumstances are these. On Whitsunday Dr. Shipley was to preach in Wrexham Church a sermon in aid of the Society for the Propagation of the Gospel in Foreign Parts, and Reginald Heber, then Vicar of Hodnet, happened to be staying at the vicarage at the time. On the Saturday preceding Whitsunday the Dean, Heber and a few friends were collected together in the library, when the Dean asked his son-in-law to write something for them to sing in the morning—something appropriate to the subject on which his discourse would be based. Heber, readily consenting,

THE REV. ISAAC WATTS, D.D.
From a Painting.

THE RIGHT REV. BISHOP REGINALD HEBER, D.D.
From the painting by T. Phillips, Esq., R.A.

retired to the further end of the room for the purpose.

After some fifteen minutes silence Dr. Shipley called out and asked what he had written, to which Heber replied by reading aloud :—

> From Greenland's Icy Mountains,
> From Indias Coral Strand.
> Where Africs Sunny fountains
> Roll down their Golden Sand,

and so on to the end of the third verse. His listeners were delighted and would have had the hymn remain without any addition, but Heber answered "No, the sense is not complete," and silence reigning once more he quickly wrote the fourth verse :—

> Waft, waft, ye winds, His story,
> And you, ye waters, roll,
> Till like a sea of glory
> It spreads from pole to pole;
> Till o'er our ransomed nature
> The Lamb for sinners slain,
> Redeemer, King, Creator,
> In bliss returns to reign.

He afterwards gave the hymn to the Dean, who turned a deaf ear to his subsequent requests to be allowed to write another verse. The next morning it was sung for the first time in Wrexham Church.

The original MS. of "From Greenland's icy mountains" was for many years in the possession of the late Dr. Raffles, of Liverpool, himself a hymn-writer of some note. Popular tradition round Wrexham has it that a compositor in the printing works sold the MS. for a pint of ale; but it is far more likely that Dr. Raffles obtained it direct from the printer who was a personal friend of his. A few years since Dr. Raffles' effects were sold and among other objects of interest put up for auction was this identical MS. After some spirited bidding it was "knocked down" to an unknown buyer for the sum of forty guineas. On the authority of the auctioneer the MS. is now in America.

"From Greenland's icy mountains" is one of the finest examples of spontaneous writing we possess. It was written in twenty minutes, which gives an average of five minutes to each verse of eight lines, and the only correction Heber ever made was in the second verse where he substitutes the word "heathen" for that of "savage."

Among the many fine hymns by the late Rev. S. J. Stone, none, to my thinking, has ever surpassed the one he wrote for Foreign Missions in 1871 :—

> Through midnight gloom from Macedon
> The cry of myriads as of one,
> The voiceful silence of despair
> Is eloquent in awful prayer,
> The soul's exceeding bitter cry,
> "Come o'er and help us, or we die."

It is equal in all respects to his "The Church's one Foundation," and some of the stanzas surpass it. Few finer verses were ever written than the fourth :—

> Yet with that cry from Macedon
> The very car of Christ rolls on ;
> "I come—who would abide My day
> In yonder wilds prepare My way ;
> My voice is crying in their cry ;
> Help ye the dying, lest ye die."

Mr. Stone based this hymn on the words taken from the Acts of the Apostles :—" And a vision appeared to Paul in the night ; There stood a man of Macedonia, and prayed him, saying, Come over into Macedonia, and help us. And after he had seen the vision, immediately we endeavoured to go into Macedonia, assuredly gathering that the Lord had called us for to preach the gospel unto them."

"Through midnight gloom from Macedon"

is not the only mission hymn Mr. Stone has written, though it is by far the best known. His "Far off our brethren's cry" and "Lord of the harvest, it is right and meet," both written about the same time as his more popular hymn, are very fine and deserve to be in every hymnal.

A hymn for Foreign Missions, written by the late Mr. Henry Downton takes a high place among such compositions. This is :—

> Lord! her watch Thy Church is keeping ;
> When shall earth Thy rule obey ?
> When shall end the night of weeping?
> When shall break the promised day?
> See the whitening harvest languish,
> Waiting still the labourers' toil :
> Was it vain—Thy Son's deep anguish?
> Shall the strong retain the spoil ?

This hymn was written at Geneva, where Mr. Downton was Resident English Chaplain, and it owes its origin to the intense interest the author took in all matters connected with Foreign Missions. He was a frequent visitor to the Society's meetings, and it was during one of these gatherings that the desire to write a hymn suited to such occasions came to him. In the year 1866, some time previous to the annual meeting of the Church Missionary Society, Mr. Downton wrote the hymn "Lord! her watch Thy Church is keeping." It was subsequently sung at the meeting for that year, though to what tune I am unable to say. When it came to be inserted in the hymnals it was usually associated with Smart's setting entitled "Everton," a tune which was originally written for Ellerton's "King of Saints, to whom the number."

Mr. Downton, who was for many years Rector of Hopton, was the author of several volumes of prose works as well as a book of hymns and short poems. From the latter several compositions have been taken for various collections, though not one of them has obtained the great popularity which his mission hymn enjoys. In America "Lord, her watch Thy Church is keeping" is an especial favourite and vies with "The Church's one Foundation" in point of popularity. It has been translated into various languages and dialects, and is to be found in many missionary hymnals.

To Isaac Watts we owe that very fine hymn, which has for so many years been almost indispensable at all meetings held in connexion with Foreign Missions:—

> Jesus shall reign where'er the sun
> Doth his successive journeys run;
> His kingdom stretch from shore to shore,
> Till suns shall rise and set no more.

This hymn forms the second part of Dr. Watts' metrical version of the 72nd Psalm, and must have been written very early in the eighteenth century. In most hymnals the following verses are omitted :—

> For Him shall endless prayer be made,
> And praises throng to crown His Head ;
> His name like sweet perfume shall rise
> With every morning sacrifice.
>
> Where He displays His healing power,
> Death and the curse are known no more :
> In Him the tribes of Adam boast
> More blessings than their father lost.

From Mr. G. J. Stevenson's *Notes on the Methodist Hymn Book*, I quote the following, which has special reference to " Jesus shall reign " :—

" Perhaps one of the most interesting occasions on which this hymn was used was that on which King George, the sable, of the South Sea Islands, but of blessed memory, gave a new constitution to his people, exchanging a heathen for a Christian form of government. Under the spreading branches of the banyan trees sat some thousand natives from Tonga, Fiji, and Samoa, on Whitsunday, 1862, assembled for Divine worship. Foremost amongst them all sat King George himself. Around him were seated old chiefs and warriors who had shared with him the dangers and fortunes of many a battle—men whose eyes were dim, and whose powerful frames were bowed down with the weight of years. But old and young alike rejoiced together in the joys of that day, their faces most of them radiant with Christian joy, love, and hope. It would be impossible to describe the deep feeling manifested when the solemn service began, by the entire audience singing Dr. Watts' hymn :—

> Jesus shall reign where'er the sun
> Doth his successive journeys run;
> His kingdom stretch from shore to shore,
> Till suns shall rise and set no more.

Who so much as they could realize the full meaning of the poet's words? for they had been rescued from the darkness of heathenism and cannibalism and they were that day met for the first time under a Christian constitution, under a Christian king, and with Christ Himself reigning in the hearts of most of those present. That was indeed Christ's kingdom set up in the earth."

Of all missionary hymns, " Jesus shall reign," is probably the best known. It has been written now for close upon two hundred years, and in all likelihood it was one of the first hymns introduced into heathen countries penetrated by English missionaries. It has been translated into more languages and dialects than any other of Dr. Watts' hymns, and it would probably be a difficult matter to find a present day missionary hymnal in which it is not included.

The favourite setting to which this hymn is usually sung, and which is known by the title of "Galilee," was composed by the late Dr. Philip Armes.

A mission hymn which has gained a good deal of popularity during the last half century is :—

> Thou, Whose Almighty Word
> Chaos and darkness heard,
> And took their flight ;
> Hear us, we humbly pray,
> And where the Gospel-day
> Sheds not its glorious ray,
> Let there be light.

The metre of this hymn naturally suggests the National Anthem, and on many occasions it has been sung to the same melody. Whether it was written with this idea it would be somewhat difficult to say. It is not at all unlikely, however, that the author considered it would be to the advantage of a hymn if it were written to a tune already known practically all over the world, and that by a happy inspiration he hit on the National Anthem. The hymn was written early in the nineteenth century by the Rev. J. Marriott, who was for many years Rector of Church Lawford. So far as I can discover Mr. Marriott was the author of very few hymns, and of these the only one which can be said to have come into general use is his "Thou, Whose Almighty Word." Mr. Marriott died in 1825 at the early age of forty-six.

"Lift up your heads, ye gates of brass" is James Montgomery's best known mission hymn. It is said to have been written while the author was in prison doing penance for an injudicious article published in his paper, the *Sheffield Iris*. Whether this is so or not it would be difficult to say; there is certainly no record to this effect in the poet's life. It is true that he underwent several terms of imprisonment, but that was during his early days, ere he had reached his twenty-fifth year, and long before any of his hymns were published. In a little volume which was issued soon after his terms of imprisonment had expired, and which he called *Prison Amusements*, no hymns appear. In all probability, therefore, "Lift up your heads, ye gates of brass" has no connexion whatever with his prison life.

Montgomery took the greatest interest in all

foreign missions, and by his lectures and influence did much to spread the good work of sending missionaries into foreign lands.

In some hymnals the following verses contained in "Lift up your heads" are omitted:—

> A holy war those servants wage,
> In that mysterious strife;
> The powers of Heaven and hell engage
> For more than death or life.
>
> Follow the Cross; the ark of peace
> Accompany your path,
> To slaves and rebels bring release
> From bondage and from wrath.

This hymn is to be found in most collections used by foreign missionaries, and is well-known in the mission fields of India, Western Africa and China. It is usually sung to Myles Foster's "Crucis Victoria" (originally written for Isaac Watts' hymn, "Give us the wings of faith to rise") or to a melody from Este's Psalter, and known by the title of "Winchester." This latter tune is more closely associated with "While shepherds watched their flocks by night," but suits the mission hymn very well. The melody dates back to the year 1592.

One of the finest missionary hymns in the language comes to us from an American source.

HYMNS FOR MISSIONS

This is "Saviour, sprinkle many nations," by the late Arthur Cleveland Coxe, Bishop of Western New York. This hymn was begun in America in 1850, one Good Friday. Bishop Coxe could not finish it at the time, and it was laid aside to await the day when the spirit should move him to complete it. The following year he visited England, and while walking in the grounds of Magdalen College, Oxford, the thought occurred to him that "Saviour, sprinkle many nations" was still uncompleted. He therefore took a scrap of paper from his pocket and with a pencil wrote the concluding verse to this very beautiful hymn.

I cannot refrain from quoting the second stanza of this hymn which appears to me to be one of the very finest eight line verses in the language :—

> Far and wide, though all unknowing,
>   Pants for Thee each mortal breast ;
> Human tears for Thee are flowing,
>   Human hearts in Thee would rest ;
> Thirsting, as for dews of even,
>   As the new-mown grass for rain,
> Thee they seek, as God of heaven,
>   Thee as Man for sinners slain.

Curiously enough this hymn, together with many others by the same author, though found in a great number of collections has been omitted from the American Episcopal hymnal. Of course there is a reason for this and it is a very simple one. Bishop Coxe unfortunately happened to be on the committee appointed to select the hymns for a new collection to be issued in connexion with his own diocese. The hymns were selected by ballot, and he begged the committee as a favour to abstain from voting for any of his own compositions. They respected his wish, and, in consequence, though his hymns are published in every other collection in America, they do not appear in that used by his own Church.

After fulfilling the arduous duties of Bishop of Western New York for thirty-one years, Dr. Coxe died at Buffalo in 1896, at the age of seventy-eight.

"God of mercy, God of grace," by the Rev. Henry Francis Lyte, is generally looked upon as a mission hymn, and is to be found in several collections used in the mission field. It was one of Mr. Lyte's earliest hymns and was written, not at Berry Head, the birthplace of "Abide with me," but at Burton House, about a mile distant from the more historic building. This house, which is situated in Lower Brixham (Berry Head is also in the lower town) was the residence of Mr. Lyte for many years. On each side of the entrance to the house are very fine weeping willows, and residents will tell you that they were planted by the late hymnist himself, and further, that they came originally from Napoleon's grave. Dr. Mayer, the present resident of Burton House, however, tells me that though it is quite true that Mr. Lyte planted there a couple of trees which he brought as saplings from St. Helena, he fears they are not the two which now flourish so healthily in the front garden of his house. I was shown the remains of two other trees which had long since shed their last leaves, and these, it is thought, were the ones planted by Mr. Lyte about the year 1825.

"God of mercy, God of grace," which is written in the same metre as the author's well-known "Pleasant are Thy courts above," with the exception that the verses consist of six lines instead of eight, is to be found in his volume of hymns called *The Spirit of the Psalms*. It has not been altered by hymnal editors and appears in the majority of collections exactly as it was written.

Mr. Percival H. W. Almy, the author of *Scintillae Carmenis*, a volume of poems which should rank high among the works of our minor poets, in an interesting and appreciative account of Mr. Lyte's life published some years ago, says that among the most attractive characteristics of the hymnist was his passionate love for animals.

"One of his pets," writes Mr. Almy, "was a tame eagle. This eagle 'chummed' with a certain jackdaw—an impudent, republican sort of jackdaw, who waxed insolent and publicly abused his feathered majesty. The royal bird upon the first opportunity seized the miscreant and killed and ate him.

"In a field that at one time belonged to Mr. Lyte, just outside the town of Brixham, reposes all that is mortal of honest 'Var.' The tomb is let into a high bank, and the tablet that marks the spot bears the following inscription:—

Here lies
"V A R,"
Lap-dog of the Rt. Honourable Lady Farnham.

Breathe, gentle spring, breathe on this grassy mound,
And sing ye birds, and bloom ye flowers around,
Ye suns and dews make green the resting-place
Of honest Var, the noblest of his race;
Gentle yet fearless, active, fond and true,
He reads, proud man, a lesson here to you,
And bids you (happy might you hear) to be
Guiltless in life and calm in death as he.
Go, and as faithful to your Master prove,
As firm in duty and as strong in love,
You will not find this moment here misspent
In musing o'er a spaniel's monument.

May, 1826.

"These lines are Mr. Lyte's own composition, and I had considerable difficulty in deciphering them. This genuine relic of a great man is in a sad state of decay."

A hymn which is a general favourite with home missions is Charlotte Elliott's "Just as I am, without one plea." There is a generally credited story that the writing of this famous hymn was due to the following incident :—

"One day the pastor of a small church met in the street a young member of his congregation on her way to be fitted for a new dress which she contemplated wearing at an approaching ball. After she had told him her errand he said to her, 'I wish you would give up your life of vanity and become a Christian, and lead a godly life. Will you not stay away from the ball because I wish you to do so?' She answered, 'I wish you would mind your own business,' and bidding him a cheerful good-bye, went on her way."

That girl is supposed to have been the authoress of "Just as I am"—Miss Charlotte Elliott. The story goes on to say that she went to the ball and danced through the night. When she returned, tired and weary, her conscience smote her, and she went the next day to her pastor and asked his forgiveness for the words she had spoken.

"'I am the most wretched girl in the world,' she said. 'What must I do to be saved?' The pastor directed her to come to God just as she was. 'What! just as I am?' she exclaimed in astonishment. 'Yes,' replied the pastor, 'just as you are.' On her return home she knelt beside her bed and prayed to God to accept her just as she was; then, rising from her knees, she got out her writing materials and wrote the well-known hymn:—

"Just as I am, without one plea,
But that Thy Blood was shed for me,
And that Thou bidd'st me come to Thee,
O Lamb of God, I come."

This little story, charming though it is, cannot be relied upon; for the hymn was written in 1834, when Miss Elliott was forty-five years old—an age at which she could hardly be referred to as a "young girl."

From Mrs. Synge, a niece of the authoress, I some time ago received an interesting and authentic account of the origin of this most popular hymn.

"In 1834," says Mrs. Synge, "Miss Elliott was residing at Brighton, in a house long since pulled down, called Westfield Lodge. Her brother, the Rev. H. V. Elliott, having conceived the plan of erecting a college at Brighton for the education of the daughters of the poorer clergy, a bazaar was held in order to assist in raising the necessary money. All the members of Westfield Lodge were busy—all except Charlotte, who was weak and ill. The night before the bazaar she lay tossing on her bed, consumed with the thoughts of her own apparent uselessness.

"The day of the bazaar came, and Charlotte still lay upon her sofa. She continued in deep thought long after every one had gone to the bazaar; then came a sudden feeling of peace and contentment. Taking a sheet of paper from the table beside her she wrote the verses by which her name is now held most dear, without any apparent effort.

"When her sister-in-law, Mrs. H. V. Elliott, returned, she found the hymn, neatly written out, lying on the table. She read it and then asked for a copy. In 1845 it was printed in the form of a leaflet, without the knowledge of the authoress. It was unsigned, and a short time afterwards a friend brought a copy and gave it to Miss Elliott with the words, ' I am sure this will please you.'

Like Dr. Doddridge, Charlotte Elliott was an invalid for the greater part of her life. She lived to the age of eighty-two, and her compositions are to be found in every hymnal published during the last fifty years. Referring one day to "Just as I am," the brother of the late hymnist stated:—" In the course of a long ministry I hope I have been permitted to see some fruit of my labour, but I feel that far more has been done by a single hymn of my sister's."

I much regret that I am unable to show a MS. of "Just as I am." I have corresponded on the subject with many of the late Miss Elliott's relatives, who all assure me that so far as they are aware no hymns in the handwriting of this author have been preserved.

Another hymn which is usually associated with home missions is :—

> There were ninety and nine that safely lay
> In the shelter of the fold,
> But one was out on the hills away
> Far off from the gates of gold ;
> Away on the mountains wild and bare,
> Away from the tender Shepherd's care.

This hymn was written by Miss Elizabeth Cecilia Clephane, the daughter of the Sheriff of Fife. It is said to have been composed while the authoress was still at school and was first scribbled in one of her exercise books. Whether this was so or not the hymn remained unpublished until after the authoress's death, which took place in 1869 before she had completed her thirty-eighth year.

"There were ninety and nine" has come to be associated in the minds of many with "Tell me the old, old story," and has, indeed, on more than one occasion been ascribed to Miss Hankey. Perhaps this is owing to its having been, like the latter, very successfully set to music by Mr. Ira D. Sankey. It is related that one day when Mr. Sankey was on his way to attend one of the great meetings with which his name is associated, and to which he had to travel by train, he bought a paper to idle away the time. While perusing this his eye lighted upon Miss Clephane's hymn. He was much struck by the simplicity and beauty of the poem, and determined to sing it at the meeting that night. Towards the close of the service he went to the piano, and doubling up the paper placed it before him on the music-rest. Then he struck a few chords and began to sing the hymn to a melody which he composed as he went along. When he was nearing the end of the first stanza the unpleasant thought struck him that possibly he might forget the original melody when he came to the second verse. His fears were groundless, however, and he sang the second verse with equal success. "After that," he is reported to have said, "the rest was easy, and the hymn made a deep impression on the vast assembly."

Soon after this it was inserted in Mr. Sankey's collection and was ever after a great favourite at all the meetings held by the celebrated American Missioners.

Miss Clephane was the author of several other hymns, none of which, however, has attained to the popularity enjoyed by "There were ninety and nine." She took the greatest interest in the welfare of children, and her early death was a great loss to the poor of Edinburgh, where her influence, strengthened by unfailing sympathy and charity, was immense.

Of the hymns specially suited to "Those at sea" one of the finest is that by St. Anatolius, rendered familiar to us by Dr. Neale's spirited translation:—

> Fierce was the wild billow,
>   Dark was the night;
> Oars laboured heavily,
>   Foam glimmered white;
> Trembled the mariners,
>   Peril was nigh!
> Then said the Lord of Lords,
>   "Peace! it is I!"

This hymn does not appear to be in as many collections as one would suppose, seeing that it is suited to occasions for which few really good hymns have been written. It is certainly one of the finest hymns we possess either for use at sea or on dry land when the wind is blowing and there is danger out on the deep. It first appeared in Dr. Neale's *Hymns of the Eastern Church*, together with many other compositions which are now among the prized possessions of our Church. That Dr. Neale experienced more difficulty in making his Greek translations than he did those from the Latin may be gathered from the preface to his little volume. He says:—

"Though the superior terseness and brevity of the Latin hymns renders a translation which shall represent those qualities a work of great labour, yet still the versifier has the help of the same metre; his version may be line for line; and there is a great analogy between the collects and the hymns, most helpful to the translator. Above all, we have examples enough of former translation by which we may take pattern. But in attempting a Greek canon, from the fact of its being in prose (metrical hymns, as the reader will learn, are unknown) one is all at sea. What measure shall we employ? Why this more than that? Might

222 HYMNS FOR "THOSE AT SEA"

we attempt the rhythmical prose of the original, and design it to be chanted? Again, the great length of the canons renders them unsuitable for our churches as *wholes*. Is it better simply to form centos of the more beautiful passages? or can separate odes, each necessarily imperfect, be employed as separate hymns? And above all we have no pattern or example of any kind to direct our labour." Further on Dr. Neale says: " My own belief is, that the best way to employ Greek hymnology for the uses of the English Church would be by centos."

In spite of these difficulties, and the fact that Dr. Neale was the first to make translations from the Greek, he still remains our most successful translator.

" Fierce was the wild billow " does not consist of centos, but is a translation of the complete hymn. Nearly all the hymnal compositions by St. Anatolius—and they number over a hundred—are short and terse, and therefore Neale's suggestion that it is better to employ centos was not always necessary in the case of this hymnist. The last verse of " Fierce was the wild billow " is exceptionally beautiful, and as it may not be known to all my readers I give it :—

HYMNS FOR "THOSE AT SEA" 223

  Jesu, Deliverer!
   Come Thou to me;
  Soothe Thou my voyaging
   Over life's sea:
  Thou, when the storm of death
   Roars, sweeping by,
  Whisper, O Truth of Truth,
   "Peace! it is I!"

Another hymnist, whose reputation rests on a single contribution, is the late Mr. William Whiting. He will be forgotten only when men cease to go down to the sea in ships, for he was the author of that very striking and popular hymn:—

Eternal Father! strong to save!
Whose Arm doth bind the restless wave,
Whose Will doth bid the Ocean deep
Within its Bounds appointed keep;

To the British sailor this hymn is as familiar as "Rule Britannia" or "Tom Bowling," and I would venture to say that it is quite as well known to the average Jack Tar as the Lord's Prayer or any one of the Ten Commandments. Neither is it a stranger in the French Navy, for a translation appears in the *Nouveau Livre Cantique* (the hymnal in use on the French men-of-war), with the tender and beautiful refrain :—

224   HYMNS FOR "THOSE AT SEA"

> Vois nos pleurs, entends nos sanglots,
> Pour ceux en péril sur les flots.

I remember some time ago showing my collection of hymnal manuscripts to a young sailor (on his expressing a wish to see them) who had just returned from a voyage round the world. He was a serious and excellent young fellow, but he tossed those manuscripts over in a manner which made me nervous; my most prized hymns he scarcely looked at. "Where's the sea-hymn?" he said at last. "What sea-hymn?" I answered; "There's 'Fierce raged the tempest,' if that's the one you mean." "No," he answered, "not that one, though it's a fine hymn and one we often sing on board; I mean the one for those in peril on the sea." I had not got it then, nor did I know that it was in existence until a correspondent wrote to me suggesting that I should write to the deceased hymnist's son. I did so, and by return of post came a volume of Mr. Whiting's original hymns, all written in the author's wonderfully clear, well-formed hand. This interesting volume of manuscripts contains three versions of "Eternal Father! strong to save." To one Mr. Whiting has added the date, February 9th, 1875, and the words, "This is my final version."

HYMNS FOR "THOSE AT SEA"

A friend of the late Mr. Whiting sends me the following particulars regarding his life :—

"On May 3rd, 1878, died at Winchester William Whiting, master of the Choristers at S. Mary's College in that city, and writer of the hymn, 'Eternal Father! strong to save.' He was born in London and came at an early age to Winchester to fill an appointment at the Training College.

"His principal prose works were *Rural Thoughts and Scenes* and *Edgar Thorpe*, besides which he wrote several odes, sonnets, etc. His death took place in the holidays, so that only eight of his sixteen pupils were present at the funeral. Two of these have since gained distinction in music, one being George J. Bennett, Mus. Doc., Cantab., and the other Clement Locknane, also a composer.

"The Warden of Winchester College and the Rev. A. J. Lowth read the burial service, and the hymns, 'Oh what the joy' and 'Jesus lives! no longer now' were sung by the College choir. The grave is on the north side and close to the walls of the burial chapel in the public cemetery at Winchester. Mr. Whiting never enjoyed very good health, but was invariably cheerful and possessed a fund of quiet humour. He was rather short in stature and wore spectacles."

Another very fine hymn for those at sea, which

was written about the same time as "Eternal Father! strong to save" is:—

Fierce raged the tempest o'er the deep,
Watch did Thine anxious servants keep,
But Thou wast wrapped in guileless sleep,
Calm and still.

This is probably the best-known of all Prebendary Thring's compositions. Like Mr. Whiting's hymn, it had the good fortune to be specially set to music by the late Dr. Dykes, who composed for it the very fine setting known as "St. Aelrëd." The author based his hymn on the words taken from St. Mark's Gospel: "And He arose, and rebuked the wind, and said unto the sea, Peace, be still."

The other day I was somewhat amused to read in an American paper the history of this hymn. "It owes its origin," wrote the imaginative scribe, " to the fact that on one occasion, the author being out by himself off the coast of Scotland in a small sailing-boat, was overtaken by a storm. His mast snapped, his tiny sail was torn to shreds, and he was himself in imminent danger of being drowned, when a vision of the scene on the Lake of Genesaret appeared before him and calmed his nerves. With renewed hope he began to bail out the water from his now rapidly-sinking little craft, and a few moments later was rescued by a steam yacht. When he had somewhat recovered he wrote, in the heartfelt gratitude which came upon him, the well-known lines beginning, 'Fierce raged the tempest o'er the deep.'" This account is a pretty one, but unfortunately it is not correct. The story of the vision appearing to him is true enough, but the rest is false. The fact is, Prebendary Thring was safe on dry land when the idea of the hymn was conceived. He was sitting alone at the time, doing nothing. With half-closed eyes he saw the raging sea, the terrified mariners, and our Saviour sleeping calmly and peacefully. Then, taking pen and paper, he wrote "Fierce raged the tempest" as rapidly and spontaneously as did Heber when he penned his immortal "From Greenland's icy mountains."

Dr. E. H. Bickersteth, the lately-resigned Bishop of Exeter, is the author of a great number of hymns, many of which were written with an especial object. Among these is one which he wrote for use at sea, and which is one of his best known :—

> Almighty Father, hear our cry,
> As o'er the trackless deep we roam ;
> Be Thou our haven always nigh,
> On homeless waters Thou our home.

Dr. Bickersteth tells me that there is no history attached to the writing of this hymn. He had long been struck by the fact that there were practically no hymns specially suited for use when in mid-ocean, and he therefore determined to write one. He is a great admirer of Whiting's "Eternal Father! strong to save," and gives it a place in his hymnal, but at the same time points out that it is not specially suited for singing *by* those at sea, being much more appropriate when sung *for* those at sea.

"Almighty Father, hear our cry," will be found in Dr. Bickersteth's volume of poems entitled *Two Brothers.* It there appears with the opening line in a somewhat different form to that usually met with in most collections. As some of my readers may be aware, the hymn, as given in the author's volume, commences, "Lord of the ocean, hear our cry."

Besides being a hymnist, Dr. Bickersteth has also done considerable work as an editor, having issued several collections. His most widely used is *The Hymnal Companion to the Book of Common Prayer*, which is used in a great number of churches. It was received with considerable favour the first year of publication, and now takes its place among the four most widely-circulated hymnals in England.

HYMNS FOR "THOSE AT SEA"

A hymn written by Bishop Heber, "To be sung by sailors in stormy weather," but one which is to be found in only a few hymnals, is "When through the torn sail the wild tempest is streaming." This hymn was written only a short while before the author's tragic end, and was not published until after his death. Though one of the best of Heber's hymns, it remains one of the least known. It is written with his usual lyric power, and as many of my readers may not know it, I give the three verses :—

When through the torn sail the wild tempest is streaming,
When o'er the dark wave the red lightning is gleaming,
Nor hope lends a ray the poor seaman to cherish,
We fly to our Maker, "Save, Lord, or we perish."

O Jesus, once rock'd on the breast of the billow,
Aroused by the shriek of despair from Thy pillow,
Now seated in glory, the mariner cherish,
Who cries in his anguish, "Save, Lord, or we perish."

And O! when the whirlwind of passion is raging,
When sin in our hearts his wild warfare is waging,
Then send down Thy grace Thy redeemèd to perish,
Rebuke the destroyer—"Save, Lord, or we perish."

This hymn is also suitable for the 4th Sunday after Epiphany, the Gospel for that day being the history of the stilling of the waves on the Lake of Genesaret.

X

Funeral and Harvest Hymns

ALL SAINTS' DAY

A HYMN which is said to have been a favourite with the late Queen Victoria is that for funeral services :—

> Now the labourer's task is o'er;
> Now the battle-day is past ;
> Now upon the farther shore
>   Lands the voyager at last.
>   Father, in Thy gracious keeping
>   Leave we now Thy servant sleeping.

This hymn was originally written in six stanzas of four lines each with the well-known refrain. It has, however, in almost all hymnals, been reduced, and the following verse omitted :—

> There the penitents, who turn
>   To the Cross their dying eyes,
>   All the love of Jesus learn
>   At His Feet in Paradise.

In his *Notes on Church Hymns*, for which

FUNERAL AND HARVEST HYMNS

collection this composition was specially written, Canon Ellerton tells us that the whole hymn, especially the third, fifth, and sixth verses, owes many thoughts, and some expressions, to a beautiful poem of the Rev. Gerald Moultrie's, beginning, "Brother, now thy toils are o'er."

Another well-known funeral hymn by Canon Ellerton is "God of the living, in whose eyes." This was one of the first hymns the author wrote, being composed before he had reached the age of thirty. It is equal in merit to many of his more popular hymns, and will probably increase in favour. Canon Ellerton wrote it, I believe, for the funeral of one of his Sunday school children at Brighton, where he was curate at the time. It soon found its way into several hymnals, one of the last editors to appropriate it being Prebendary Thring. In the original it does not appear in as many stanzas as we are accustomed to sing it, and the probability is that Canon Ellerton added to it afterwards. In many collections the following verse is omitted :—

> Released from earthly toil and strife,
> With Thee is hidden still their life ;
> Thine are their thoughts, their works, their powers,
> All Thine, and yet most truly ours ;
> For well we know, where'er they be,
> Our dead are living unto Thee.

This hymn was a favourite with the author, probably from its associations with his early days. It has been successfully set to music by various composers, one of the most attractive and appropriate melodies being "God of the living" by Mr. Everard Hulton.

Miss Sarah Doudney is, perhaps, better known as a novelist than a hymnist, probably from the fact that her hymns are very few in number. What she has written in this way, however, is very excellent, and her hymn for funeral services is now among the best for those solemn occasions :—

> Sleep on, belovèd, sleep, and take thy rest ;
> Lay down thy head upon thy Saviour's breast ;
> We love thee well, but Jesus loves thee best :
> Good-night !

This hymn, Miss Doudney tells me, was written after the death of a very dear friend of her girlhood. It was suggested by a custom of the early Christians who bade their friends "good-night" when they entered the arena for the final trial of their faith, so sure were they of their re-union. "Darkness and Dawn," and "Quo Vadis" have given us a sense of the reality and simplicity of their belief in the awakening to a better life. Christina Rosetti, in one of her poems, has the same idea, beautifully expressed :—

FUNERAL AND HARVEST HYMNS

We meet in joy, though we part in sorrow ;
We part to-night, but we meet to-morrow.

"The sixth verse of my hymn," writes Miss Doudney, "which reads :—

> Only 'good-night!' belovèd, not 'farewell!
> A little while and all His saints shall dwell
> In hallowed union, indivisible :
>   Good-night!

will hardly bear criticism: 'indivisible' does not rhyme well with 'farewell' and 'dwell'; but the verses have become so dear to many mourners that I did not like to make any change. They were set to music, when they first appeared in a religious monthly, by the late Mrs. Worthington Bliss ; but Mr. Sankey's setting is far better known."

In America, "Sleep on, belovèd" is the favourite funeral hymn, and has been sung at the obsequies of many public men and women. In England it is also very popular, and was sung at the Duchess of Teck's memorial service by command of the late Queen Victoria.

"Just before my mother died," continues Miss Doudney's interesting letter, "I was drawn to the open window by the sound of voices singing the hymn on a spring evening. The singers were hidden by the trees; darkness was closing in, and I never saw them come or depart. But the strain seemed even then to foretell the passing of the one I loved best; and one more sacred memory was added to those which cling to the lines, written so long ago."

The beautiful setting to which this hymn is invariably sung, known as "The Blessèd Rest" was specially composed for it by the late Sir Joseph Barnby.

From Dean Milman's " Martyr of Antioch " we get one of the loveliest of all funeral hymns :—

> Brother, thou art gone before us; and thy saintly soul is flown
> Where tears are wiped from every eye, and sorrow is unknown;
> From the burden of the flesh, and from care and fear released,
> "Where the wicked cease from troubling, and the weary are at rest."

To the many thousands who have listened to Sir Arthur Sullivan's musical version of "The Martyr of Antioch" it is needless to say how impressive this funeral hymn is when sung, as it invariably is, unaccompanied. The melody is exquisite in its solemnity, and it is surprising to find that " Brother, thou art gone before us " is in comparatively few hymnals. What could be finer than the subdued triumph of the closing lines :—

And when the Lord shall summon us, whom thou hast left behind,
May we, untainted by the world, as sure a welcome find !
May each, like thee, depart in peace, to be a glorious guest,
"Where the wicked cease from troubling, and the weary are at rest."

The argument of the " Martyr of Antioch " is given in the preface to Sullivan's adaptation. "Olybius is in love with Margarita, and she returned his love. This, however, was in her heathen days. She is now a Christian, and with her conversion, of which both her lover and her father are ignorant, she, though still not indifferent to him, rejects all ideas of union with a heathen. The piece opens with a chorus of Sun-worshippers, preliminary to a solemn sacrifice. The Prefect calls for Margarita to take her accustomed place and lead the worship. During her non-appearance, the Priest charges him with lukewarmness in the cause of Apollo, and he avows his intention to put all Christians to death.

" The scene changes to the Christian Cemetery, where one of the brethren is buried, and a hymn is sung over him ('Brother, thou art gone before us'). After the funeral, Margarita remains behind, and pours forth her feelings in adoration of the Saviour. Her father finds her thus employed, and learns for the first time of her conversion.

"The scene again changes to the Palace of the Prefect. The Maidens of Apollo sing their evening song. Olybius and Margarita are left together; he tells her of the happiness which will be hers when they are united. She then confesses she is a Christian; he curses her religion, and she leaves him for prison.

"The final scene take place outside the prison of the Christians on the road to the Temple of Apollo. The Maidens of Daphne chant the glories of the god, while from within the prison are heard the more solemn and determined strains of the Christians. Margarita is brought out and required to make her choice. She proclaims her faith in Christ. Her lover and her father urge her to retract, but in vain; and she dies with the words of rapture on her lips:—

"'The Christ, the Christ, commands me to His Home; Jesus, Redeemer, Lord, I come! I come! I come!'"

This fine musical drama contains some of Sullivan's finest inspirations, and it is rather remarkable that the public so seldom has an opportunity of hearing it. In setting the drama to music Sir Arthur Sullivan had to have several alterations made, in many cases blank verse being changed to rhyme. This latter work was carried out by Mr. W. S. Gilbert.

Another hymn for funeral services by Dean Milman and one which he wrote specially for Bishop Heber's collection is :—

> When our heads are bow'd with woe
> When our bitter tears o'er flow,
> When we mourn the lost, the dear,
> Jesu, Son of Mary, hear.

With the exception of the last line, which has been altered by innumerable editors, the hymn is usually given as written. It is one of the most popular of the late Dean's hymns, and is to be met with in a great number of collections. It is sung by all denominations and is sometimes used as a litany. The plaintive and soothing melody with which it is associated was written by the late Richard Redhead, the same musician who composed what is generally considered to be the most popular tune to "Rock of Ages, cleft for me." Curiously enough both these tunes are known by the same title, "Redhead," which, to say the least of it, is somewhat confusing. Mr. Redhead only died comparatively recently.

A hymn which is becoming year by year more often sung at funerals is the late Lord Tennyson's:

> Sunset and evening star,
> And one clear call for me!
> And may there be no moaning of the bar
> When I put out to sea.

This hymn, if hymn it can be called, is the only one which the author ever wrote. Like "Abide with me" the verses were among the last the poet ever penned and are likely to become his most widely known. The couple of stanzas were written very rapidly and under the following interesting circumstances. It appears that Dr. Butler, of Trinity College, Cambridge, when visiting Tennyson, asked how the poet came to write "Crossing the bar." Pointing to a nurse who had been with him some eighteen months, and had had great influence over him, he replied: "That nurse was the cause of my writing 'Crossing the Bar.' She asked me to write a hymn, and I replied, 'Hymns are often such dull things.' But at last he consented to write one, adding: 'They say that I compose very slowly, but I knocked that off in ten minutes.'" This little story was related by Canon Fleming in an address delivered at

York recently. As is well known, shortly before his death the poet called his son to him and told him that it was his desire that "Crossing the Bar" should appear at the end of all future editions of his works, an injunction which has faithfully been fulfilled.

Not written specially for funerals, but very often sung on these occasions, is Miss Charlotte Elliott's :

> My God and Father! while I stray
> Far from my home in life's rough way,
> O teach me from my heart to say,
> "Thy will be done."

This hymn is as popular as the same author's "Just as I am, without one plea." It was written some time before this latter hymn and published two years earlier. Various versions appear in different hymnals, and many editors have altered it to suit their own particular collections. The number of settings to which it is sung is almost as great as the number of versions. What makes it a particularly suitable hymn for singing at funerals is the lesson it teaches of resignation to those who are left behind.

"My God, my Father! while I stray" is not only a beautiful hymn but interesting from the fact that in it we can trace much of the author's own suffering and patience. As already mentioned, Miss Elliott was a confirmed invalid, and her constant prayer was that she should learn the lesson of patience. In the following verse how she strove for the mastery is beautifully expressed :

> Though dark my path, and sad my lot,
> Let me be still and murmur not,
> Or breathe the prayer divinely taught,
> "Thy Will be done."

Miss Elliott's hymns were never written with the intention of being sung in public, being, perhaps, more suitable for private use. After they had appeared in her *Invalid's Hymn Book*, however, they were taken, by various compilers, sometimes with her leave but more often without it, and inserted in numerous congregational hymnals. Of the 120 hymns which Miss Elliott wrote, about a sixth are in common use to-day and sung in all parts of the civilized globe.

Among our harvest hymns there are few more beautiful than Mr. St. Hill Bourne's "The sower went forth sowing." It was written in 1874 for a Harvest Festival at Christ Church, South Ashford, Kent, and first printed in a magazine. There is a touching little story connected with the writing of the music to this hymn which I have the composer's permission to record here. It appears that the Proprietors of *Hymns Ancient and Modern* having purchased from Mr. St. Hill Bourne the copyright of " The sower went forth sowing " and "Christ who once amongst us," sent them to Dr. Bridge of Westminster Abbey and Dr. Stainer of St. Paul's respectively, with the request that they would write special tunes to them. The MS. of "The sower went forth sowing" reached Dr. Bridge during a time of much trouble and anxiety, for his little girl, Beatrice, lay dying. Sitting by her bedside the composer read the words, and, as he watched the little life ebbing away, he composed the tune which so admirably suits the hymn. In memory of his little daughter the composer named the melody " St. Beatrice."

A hymn which is not only sung at almost every harvest festival in Great Britain but also at those gatherings in the country known as the" Harvest-home " is Dean Alford's " Come, ye thankful people, come." This hymn, which is published in Dean Alford's volume of poems, is very much longer in the original than it is in the ordinary hymnal collection. If Dean Alford had a fault, and there are few hymnists who had not, it was a tendency to make his compositions too long. Fifty-six lines for a single hymn is twice too many, but this is moderate compared with his " Forward!

be our watchward," which consisted of ninety-six lines!

"Come, ye thankful people, come" appeared in so many different forms and phases as on more than one occasion to seriously offend the author. In several instances the Dean wrote to complain and repudiate the various versions published, but it was of little use, and very few hymnals even to-day give it exactly as the author wrote it.

Compare this verse, as in the original :—

> Even so, Lord, quickly come
> To Thy final Harvest-home!
> Gather Thou Thy people in,
> Free from sorrow, free from sin;
> There, for ever, purified,
> In Thy Presence to abide;
> Come, with all Thine angels, come,
> Raise the glorious Harvest-home!

to that which appears in *Hymns Ancient and Modern* :—

> Come then, Lord of mercy, come,
> Bid us sing Thy Harvest-home:
> Let Thy saints be gather'd in,
> Free from sorrow, free from sin;
> All upon the golden floor
> Praising Thee for evermore;
> Come, with all Thine Angels come;
> Bid us sing Thy Harvest-home.

This hymnal was certainly the greatest sinner in the matter of alteration, and the Dean was much offended that his opinion regarding the various changes made was not even asked.

The very fine tune to this hymn was composed by Sir G. J. Elvey, who christened it "St. George's, Windsor," in memory of his connection with that historic chapel.

Our second most popular harvest hymn comes to us from a German source. This is :—

> We plough the fields and scatter
> The good seed on the land,

which was written by Matthias Claudius sometime about the year 1782. In the original it was a lengthy poem of many verses, only a selection from which is found in German hymnals. The author, who was born at Lubeck in 1740, was for many years connected with the Hamburg News Agency, and subsequently edited various German papers. He wrote poems and articles as well as a good deal of religious verse. Many of his hymns are to be found in German hymnals, and of these some two or three have been translated into English, the most popular being the one for harvest. He died during the latter part of the year which saw the Battle of Waterloo.

The translator of the most popular version of "We plough the fields and scatter" is Miss J. M. Campbell, who has given us many other excellent translations from the German. Miss Campbell, who was the daughter of a country vicar, used to teach in the schools attached to her father's parish, and it was for these children that she made her now famous translation. Miss Campbell scarcely lived sufficiently long to appreciate to the full the great popularity attained by her hymn, for she died in 1878. The very fine melody to which the hymn has always been sung is also from the German, having been composed by J. A. P. Schulz, and is contemporary with the original poem.

In Prebendary Thring's collection I notice a fourth verse is given. It is not a translation but is an original stanza by H. Downton. Like the fourth verse to "Lead, kindly light" there is no apparent reason for its being.

A very exultant harvest hymn is that by the late Chatterton Dix:—

> To Thee, O Lord, our hearts we raise
> In hymns of adoration,
> To Thee bring sacrifice of praise
> With shouts of exultation.

This hymn was written in 1863, some four years prior to the writing of his better-known " Come

FUNERAL AND HARVEST HYMNS

unto Me, ye weary." It was set to music by the late Sir Arthur Sullivan, who wrote for it the very excellent melody called "Golden sheaves." It is given in most hymnals without alteration and has attained great popularity. At harvest festivals it is very often used as a processional hymn, and in country villages I have known it to be sung by children bearing sheaves of corn. The concluding verse is one of the most successful Mr. Dix ever wrote :—

> Oh, blessèd is that land of God,
>   Where saints abide for ever ;
> Where golden fields spread far and broad,
>   Where flows the crystal river :
> The strains of all its holy throng
>   With ours to-day are blending ;
> Thrice blessèd is that harvest song
>   Which never hath an ending.

It is not altogether inappropriate, I hope, to place among the harvest hymns one which was written for a Children's Flower Service. The composition I refer to is that by the late Rev. Abel Gerald Blunt, " Here, Lord, we offer Thee all that is fairest." This hymn was written while Mr. Blunt was rector of St. Luke's, Chelsea, and when I last had the pleasure of meeting the author I was told that he wrote the hymn specially for the children of his parish, and that it was sung for two or three years at St. Luke's before the outside world had even heard of it. Mr. Blunt's flower services were events in the lives of the many hundreds of children who flocked to his Church. No child so poor but managed to bring a few blossoms as an offering, and some of the bouquets were really beautiful. There was keen competition among the little ones as to who should bring the prettiest posy and more than one instance is recorded of a child having saved her halfpence for many months in order that her " bunch " might be among the very best. Many of them knew the hymn " Here, Lord, we offer Thee " by heart, for they used invariably to sing it as they marched up to present their flowers. Mr. Blunt was for nearly forty years rector of St. Luke's, and to-day he is sadly missed, perhaps most of all by the children. Only a few weeks before his death he wrote out and sent me a MS. of his well-known hymn, which has been most successfully set to music by the Rev. P. Maurice.

A hymn which is usually associated with All Saints' Day, and which has come to us from the German, is :—

>Who are these like stars appearing,
>These before God's Throne who stand?
>Each a golden crown is wearing,

> Who are all this glorious band?
> Alleluia, hark! they sing,
> Praising loud their heavenly King.

This very fine translation was made by Miss Frances E. Cox, and has been included in a great number of hymnals. There are other versions, but they have all been cast into shadow by Miss Cox's brilliant translation. The author of the original hymn was Heinrich Theobald Schenk, the son of a German pastor, born in 1656. This is the only hymn associated with his name, but it has been sung in Germany for nearly two hundred years. It must have been written when he was an old man of sixty-one or sixty-two, and the melody, known as "All Saints" is contemporary with the hymn. Indeed, it appears very probable that the tune was in existence some years before Schenk's hymn was published, and it is not, therefore, altogether improbable that the hymn was written to the tune or, at all events, fitted to the tune after it was written. Miss Cox, the translator of this hymn, shares with Miss Winkworth the honour of being among the best translators from the German who have flourished during the last half century. Her translations are to be found in all modern collections.

"Hark! the sound of holy voices, chanting at

the crystal sea," by Bishop Christopher Wordsworth, is another favourite hymn for All Saints' Day. It will be found in his *Holy Year*, from whence it has been taken by many editors, and is now included in a great number of hymnals. Canon Ellerton makes an interesting reference to this hymn. He says :—" In the earlier editions of *Church Hymns* the fifth stanza of this hymn :—

Now they reign in heavenly glory, now they walk in golden light,
Now they drink as from a river, holy bliss and infinite ;
Love and peace they taste for ever, and all truth and knowledge see
In the Beatific Vision of the Blessèd Trinity,

was omitted in deference to the judgment of one of the Episcopal Referees of the Society for Promoting Christian Knowledge, who held that the verse was liable to be misunderstood as countenancing the popular error that the Blessed are already in the full fruition of their future and everlasting glory—the 'Beatific Vision.' It is scarcely needful to say that so accurate a theologian as the Bishop of Lincoln had no sympathy with this view. His Lordship, while pressing for the restoration of this verse, explained that the whole hymn, from beginning to end, was to be regarded as the

THE RIGHT REV. BISHOP CHRISTOPHER WORDSWORTH, D.D.
Photo by Elliott & Fry.

MR. CHARLES WESLEY, M.A.
From an Engraving.

utterance in triumphant song of a vision of the final gathering of the saints, not as an exposition of their present condition in the intermediate state. The Tract Committee of the society therefore desired that the verse should in subsequent editions be restored ; but should, in deference to those who might still think it liable to misconstruction, be bracketed for optional use."

This was accordingly done, and all those who use this hymnal are now let into the secret of the mysterious brackets.

Dean Alford's fine hymn for All Saints' Day, "Ten thousand times ten thousand," is very often used as a processional, and when sung to Dr. Dykes' spirited melody, "Alford," makes one of the finest religious songs to be sung on the march it is possible to conceive. It is one of the most lyric hymns in the language, and the glowing words as well as the alliteration render it a peculiarly impressive hymn when sung, as it used to be in Canterbury Cathedral, by an immense body of worshippers. A copy of the manuscript, which was sent to me many years ago by the late Dean's son-in-law, contains a note on the margin to the effect that in the original the hymn consisted of three stanzas only, but the Dean, thinking the final lines :—

250 FUNERAL AND HARVEST HYMNS

> Then eyes with joy shall sparkle
> That brimm'd with tears of late;
> Orphans no longer fatherless,
> Nor widows desolate

were somewhat sad and melancholy, added the following triumphant verse :—

> Bring near Thy great Salvation,
> Thou Lamb for sinners slain,
> Fill up the roll of Thine elect,
> Then take Thy power and reign:
> Appear, Desire of nations,
> Thine exiles long for home;
> Show in the heavens Thy promised sign;
> Thou Prince and Saviour, come.

In this form the hymn was sung at the author's funeral on January 17, 1871.

There is but little doubt that Archbishop Maclagan is best known by his hymn for All Saints' Day :—

> The saints of God! their conflict past,
> And life's long battle won at last,
> No more they need the shield or sword,
> They cast them down before their Lord:
> O happy Saints! for ever blest,
> At Jesus' feet how safe your rest!

This hymn, the author tells me, was first published in *Church Bells* in 1870. It was written in 1869. The year following its publication

FUNERAL AND HARVEST HYMNS 251

it appeared in the S.P.C.K. hymnal, and from thence it was taken for a considerable number of collections. The lovely melody with which it is associated was composed by the late Sir John Stainer, and is one of the most successful of all that composer's refined hymn tunes. It is generally supposed to have been written expressly for Archbishop Maclagan's hymn but this is incorrect. A few months before his death, having a desire to compare the published setting with the original, I wrote to Dr. Stainer asking him if he possessed the MS. He replied that he did not, but very kindly wrote me out a fresh copy, adding on the margin a humorous little note to the effect that I might judge from the manuscript that he could never earn his living as a music copyist! The melody was composed in 1873 for performance by the London Church Choir Association in St. Paul's Cathedral, and sung to the hymn:—

> Thou hidden love of God, whose height,
>   Whose depth unfathom'd, no man knows;
> I see from far Thy beauteous light,
>   Inly I sigh for Thy repose;
> My heart is pain'd, nor can it be
> At rest, till it finds rest in Thee.

By the express wish of Sir Henry Baker, Dr. Monk, and Dr. Dykes, however, it was, and ever will be, associated with "The Saints of God." It will be noticed that the name of the tune is "Rest."

Bishop Heber's fine hymn dedicated to St. Stephen is just as often sung on All Saints' Day, and therefore no apology is needed in giving it a place under this chapter. It is found in Heber's collection of manuscript hymns in the British Museum, and there begins "The Son of God is gone to war." Many fine settings have been written for this hymn, and it is sometimes given in stanzas of four lines and sometimes of eight. In the original it appears in four-line stanzas.

This hymn was brought prominently before the public some years ago by Juliana Horatia Ewing in her very beautiful *Story of a Short Life*. In that pathetic history of the troubles of a courageous little sufferer it will be remembered that "The Son of God goes forth to war" was the favourite hymn in the barracks, and was always referred to by the soldiers as the "tug of war" hymn. The hero of the story, one of the officer's sons who meets with an accident and is crippled for life, begs a few moments before his death that the soldiers may be allowed to sing their "tug of war" hymn once again before he dies. The soldiers are told of his desire, and they go beneath

FUNERAL AND HARVEST HYMNS

his window and sing the well-known lines. When they are in the midst of the verse :—

> A noble army, men and boys,
> The matron and the maid,
> Around the Saviour's Throne rejoice
> In robes of light array'd,

they glance towards the window and, seeing a hand stretched forth to pull the blind down, know that the last lines are falling on ears which will never hear them. It is a most beautiful story, and at the height of its popularity all the school children were asking for the "tug of war" hymn. It was Mrs. Ewing's husband who wrote the most popular tune to "Jerusalem the Golden," and the one which Dr. Neale declared most exactly suited the words.

XI

Hymns for Children

A STORY is told of an old man over eighty years of age, who, when he lay dying, endeavoured in vain to recall a single prayer or hymn which might help to comfort him in his journey into the unknown. He had led anything but a blameless life; since the age of twenty he had never once entered a place of worship or given a single thought to a future state; and now, as he stood on the threshold of a new life, his brain could frame no prayer to the God before Whom he was so soon to appear.

And then suddenly his vision cleared, and he saw himself a little lad again, kneeling at his mother's knee, repeating his evening hymn; and unconsciously from his lips issued those tender words which for nearly seventy years he had neither uttered nor heard—

HYMNS FOR CHILDREN

> Gentle Jesus, meek and mild,
> Look upon this little child;
> Pity my simplicity,
> Suffer me to come to Thee.

It is the same with a good many of us. We often remember most clearly the lessons we learned in childhood, and it is probable that there are few readers of this little volume who could not recall the days when they too knelt and repeated the same familiar lines. It was one of the earliest hymns Charles Wesley wrote, and he composed it expressly for children. It has, indeed, been stated that the author wrote it for his own children, but this of course is incorrect, inasmuch as Charles Wesley was not married until many years after its composition. There is little doubt, however, that in after years his own children loved this little hymn as much as any child who sings it to-day. It was written about the year 1740, and published two years later in the author's *Hymns and Sacred Poems*. This simple and beautiful composition is in two parts of seven verses each, the second part beginning—

> Lamb of God, I look to Thee,
> Thou shalt my Example be;
> Thou art gentle, meek, and mild,
> Thou wast once a little child.

Though Charles Wesley was fond of children and wrote many hymns for their benefit, it cannot be said that he was ever very successful as a writer for the young. The reason, as a contemporary has pointed out, is not very far to seek. "He started with the wrong idea, attempting to lift children up to the level of adults, merely adapting his compositions to them by simplicity of diction." With the exception of "Gentle Jesus, meek and mild," not one of the many hymns he wrote for children has lived to be sung to-day.

A children's hymn which has become almost a classic is :—

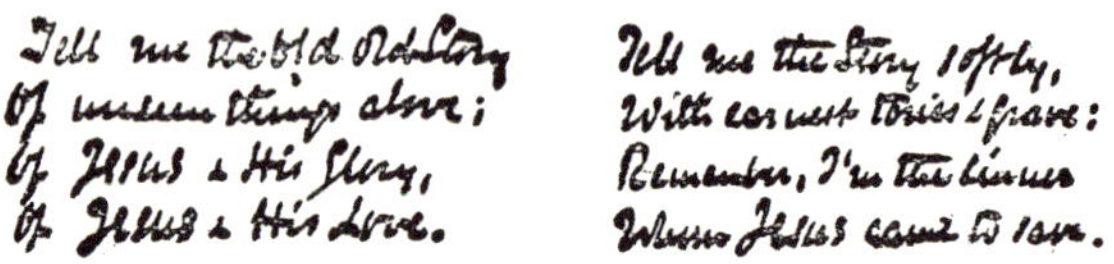

[handwritten musical verse:]

> Tell me the old old story
> Of unseen things above;
> Of Jesus & His glory,
> Of Jesus & His love.
>
> Tell me the story softly,
> With earnest tones & grave;
> Remember, I'm the sinner
> Whom Jesus came to save.

written by Miss Katherine Hankey some thirty-five years ago. It has probably been translated into more languages and dialects than any other child's hymn, and every year the author receives numerous requests from missionaries and workers in distant corners of the globe for permission to make fresh translations. The hymn has become so closely identified with Ira D. Sankey's *Sacred Songs and Solos*, as to give rise to an impression that the author wrote it specially for that hymnal. This, of course, is not the case. It is much more probable that soon after its publication it was reprinted in an American paper and there seen by Mr. Sankey, who, thinking it would make an attractive addition to his hymnal, sent it to his friend, Mr. W. H. Doane, of Preston, Connecticut, with a request that he would set it to music. This Mr. Doane did, but instead of conforming to the original he turned the four-lined verses into eight-lined stanzas and added the now well-known refrain—

> Tell me the old, old story,
> Tell me the old, old story,
> Tell me the old, old story,
> Of Jesus and His love.

The hymn "caught on" in America, and soon became the first favourite at Moody and Sankey's Meetings. This setting, however, Miss Hankey greatly deprecates, for, she argues, each verse is complete in itself, there being no connecting links between any two of the verses. However, Mr. Doane's setting is now the most popular, both in this country and America, and has done a good deal towards making the hymn known in all parts of the world.

The history of the origin of "Tell me the old, old story" I heard from the lips of the author herself some months ago, as she sat and wrote an autograph of the simple and beautiful hymn for reproduction here.

"The hymn as I first wrote it," said Miss Hankey, "consisted of fifty verses of four lines each. It was divided into two parts—'The Story Wanted' and 'The Story Told.' I wrote Part I. towards the end of January, 1866. I was unwell at the time—just recovering from a serious illness—and the second verse really indicates my state of health, for I was, literally, 'weak and weary.' When I had written the first part, which consisted of eight verses, I laid it aside; and it was not until the following November that I completed the whole hymn. It is, perhaps, strange that the plea for the story, and not the story itself, should become the favourite hymn; but of course the second part is far too long for congregational singing."

Miss Hankey also composed a musical setting for "Tell me the old, old story," which is very simple and beautiful. Though frequently sung, however, it has never attained the popularity enjoyed by that published in the American hymnal.

"What has always greatly surprised me," continued Miss Hankey, "is that so many people, including hymnal editors, should look upon it

HYMNS FOR CHILDREN

only as a children's hymn; I certainly had not children in my mind when I wrote it. However, if it answers its purpose, I suppose it matters very little whether it is sung by the young or the aged. I am sincerely grateful that my little hymn has proved a comfort and a blessing to so many."

Miss Hankey is the author of many other hymns, not one of which, however, has become very well known. Like many another hymnist, she will be remembered by a single composition.

A hymn which was perhaps more popular with children fifteen years ago than it is to-day is "There is a happy land, far, far away." It was written by Mr. Andrew Young in 1838. Mr. Young happened during that year to be spending his holiday in Rothesay, and one day called at the house of a friend, where he passed the afternoon. In the drawing-room a little girl began to play on the piano. The tune was a pretty little Indian melody, very simple, and Mr. Young, who was passionately fond of music, begged her to play it again. He remarked that it would make a capital tune for a children's hymn, and again asked to have it repeated. That night, as he slept, the tune still haunted him, and early in the morning he rose, and, while walking in the garden, wrote the hymn which has now become so well known. It has been translated into many languages and dialects, and is a general favourite among the converted natives of China.

Mr. Andrew Young was born in Edinburgh, where he was educated, in 1807. At the early age of twenty-three he was appointed Head Master of Niddry Street School, Edinburgh, where, in less than ten years, he raised the number of pupils from 80 to 600. In 1840 he became Principal English Master at Madras College, St. Andrew's, where his success as a teacher was no less remarkable. This appointment, however, he resigned in 1853, and became Superintendent of the Greenside Parish Sabbath Schools. He died on November 30, 1889.

Though Mr. Young was the author of numerous hymns, many of which he wrote for his pupils, only one has stood the test of time—" There is a happy land."

One of the most widely known and best loved of all children's hymns is :—

I think when I read that sweet story of old,
When Jesus was here among men,
How He called little children as lambs to His fold,
I should like to have been with them then.

written in 1841. The authoress, Mrs. Jemima Luke, is still happily with us, and is now (1902)

THE REV. CANON TWELLS, M.A.
Photo by Mayall.

MRS. J. LUKE.
From a Photo.

living a retired life in the Isle of Wight. This lady has recently published a delightful book of reminiscences, written in so fresh and interesting a manner as to suggest the work of a young and ingenuous girl rather than that of a lady between eighty and ninety years of age.

"The Child's Desire," as Mrs. Luke entitled her hymn, was written, as many people are aware, in a stage coach, between Taunton and Wellington. The story of its composition is well known, but will bear repetition. At the Normal Infant School, Gray's Inn Road, where Miss Thompson (as she was then) had gone to learn the system, the teachers had to march up and down the schoolroom singing the marching pieces provided for their future use, and amongst them was the air to which Mrs. Luke's hymn was subsequently adapted. The words set to it in the book of marching pieces were simple and pretty, but Miss Thompson thought the air would better adapt itself to a hymn, and tried in vain to find one to suit the measure. Just about this time she became seriously ill with erysipelas, and for days lay only half conscious. When she recovered she was sent to Taunton for change of air, and it was during this visit that "The Child's Desire" was written.

"In the small town of Wellington," to quote Mrs. Luke's own words, "five miles away, there was a little association in aid of the Society for Female Education in the East. One fine spring morning I went in a two-horse coach to see how the Society was prospering. It was an hour's ride. There was no other inside passenger. I took a letter from my pocket, and on the back of the envelope wrote two verses of the little hymn now so well known. The composition originally consisted of two verses only, but in response to a request from my father to make it a missionary hymn, the third verse was added.

"My father superintended the Sunday School at the little chapel belonging to the estate. He used to let the children choose the first hymn themselves. One Sunday afternoon they struck up their new hymn. My father turned to my younger sisters, who stood near him, and said, 'Where did that come from? I never heard it before.' 'Oh, Jemima made it,' was the reply On the Monday he asked me for a copy of the words and tune. This he sent, with name and address in full, to the *Sunday School Teachers' Magazine*, where it appeared the following month. But for my father's intervention the hymn would in all probability never have been preserved."

HYMNS FOR CHILDREN

Mrs. Luke, who sat up in bed to write me a manuscript of "The Child's Desire," added that she always considers the little hymn an inspiration from above, and not "in her," for she has never since been able to write another hymn of equal merit.

Another hymn for children which has become a great favourite in England, though by an American writer, is "Jesus loves me, this I know." It was written about the year 1858 by Miss Anna Warner, sister of the author of *Queechy* and other popular novels. Miss Warner, who has all her long life taken the greatest interest in the religious education of children, had, until quite recently, a very large Sunday School at West Point, and it was her invariable custom to write for her pupils a fresh hymn once a month. She used to take a tune which the children knew and liked, and then write words to fit the melody. One of these hymns was "Jesus loves me," and it was written for the very tiny members of her class. It soon became a favourite in America, ultimately finding its way into nearly all American hymnals. Very soon it was taken by English editors, and it would be difficult to say in which country it is now more often sung.

Miss Warner is the daughter of the late Henry

W. Warner, and was born in New York sometime during the latter part of the year 1821. She has written several volumes of poems and hymns, besides numerous novels which have had a large circulation in the States.

Miss Warner was especially successful in writing hymns for children, and many others of her compositions besides "Jesus loves me" are slowly making their way into English hymnals. The following, which the authoress calls "A Mother's Evening Hymn," might have been written by the author of "Sweet and low"—

> O little child! lie still and sleep;
> Jesus is near, thou need'st not fear;
> No one need fear whom God doth keep,
>   By day and night;
> Then lay thee down in slumber deep
>   Till morning light.
>
> O little child! be still and rest,
> He sweetly sleeps, whom Jesus keeps,
> And in the morning wakes so blest,
>   His child to be;
> Love every one, but love Him best,
>   He first loved thee.

"Hosanna we sing, like the children dear," by G. S. Hodges is found in most collections published during the last quarter of a century, and is an especial favourite on Palm Sunday. Mrs. Hodges tells me that, though her husband wrote a great number of hymns, the one referred to here is the best known and most appreciated. It was written in 1874, and published the following year, with music specially written for it by the late Dr. Dykes.

Mr. G. S. Hodges, who was Vicar of Stubbings, near Maidenhead, from 1882 to 1899, was passionately fond of children, and never happier than when he had a child perched on his knee. This love for children undoubtedly helped him to understand exactly the kind of hymn that would please them best, and accounts in a measure for the favour with which "Hosanna we sing" was received. Indeed, this hymn was particularly fortunate both as regards author and composer, seeing that Dr. Dykes was also devoted to children.

In sending me a MS. of her husband's hymn Mrs. Hodges writes: "You will see that it is not quite the same as in the hymnals. It must have been "touched up" afterwards, I think, but whether by author or editor I cannot say. My husband was not accustomed to speak much about the method he adopted in writing his hymns. He generally altered a good deal after making the first draft of the hymn, and would write it out several times before being perfectly satisfied."

Mr. Hodges is the author of a volume of sacred and secular poems, which he published in 1876. In this work appear many translations from the Latin as well as several hymns worthy the attention of compilers.

Two simple hymns by the late Dr. T. O. Summers find places in most collections for children. "The morning bright," based on the text "Whether we wake or sleep, we should live together with Him" was written about the year 1846, and intended to be used as a morning hymn, while "The daylight fades," founded on words taken from the Psalms: "The Lord is the strength of my life, of whom then shall I be afraid," was written some three years later. Curiously enough these two hymns appear to be the only compositions of the kind Dr. Summers ever wrote. Concerning them, the author says—

"My first child was born in January, 1845. When she was about a year old, as I was descending the Tombigbee River, in a little steamer, I wrote a morning hymn for her on the back of a letter, transcribed it when I reached Mobile, and sent it to her at Tuscaloosa. That was the origin of 'The morning bright.' When editing the *Southern Christian Advocate* I put it without name in the 'Child's Department.' It was copied into the religious papers generally, and into books. My second child was born in 1847, and for her I wrote 'The daylight fades,' as far as I can recollect, about the year 1849."

Sad to relate, neither of Dr. Summers' children grew up to womanhood, both dying when quite young, and within a short time of each other. Dr. Summers, who was born in Dorset in 1812, was for many years a missionary in North America. He died in 1882.

The following is a hymn which most of us can recall singing in our early days :—

There's a Friend for little children,
Above the bright blue sky,
A Friend who never changes,
Whose love can never die;

It was written by Mr. Albert Midlane between forty and fifty years ago at Newport, Isle of Wight, within a short distance of the house in which Thomas Binney penned those beautiful and well-known lines beginning "Eternal Light, Eternal Light."

Mr. Midlane began writing hymns as soon as he could read, and before his ninth birthday had composed a set of religious verses which so much impressed his parents and friends by their depth and fervour as to give rise to the cheerful belief that he was destined for an early grave. He has, however, conclusively proved that "prodigies" do not always die young.

"The very first hymn I wrote which was *used*," says Mr. Midlane, "was written on May 24, 1844, when I was nineteen years of age. It was published under the title of 'God bless our Sunday Schools,' and sung to the tune of the National Anthem. Fifty-seven years ago last summer it was first sung as our Anniversary Hymn, and still it finds expression from the lips, and, I trust, from the hearts of many little pilgrims."

But it is his hymn for children that will keep Mr. Midlane's memory green when the hand that penned it is still. "There's a Friend for little children" was written on February 27, 1859, and was first scribbled in his note book. The original manuscript is still preserved and the author sometimes looks at it and smiles to think how those few verses, coming, as they did, straight from the heart, sang themselves round the world in less than a decade. This hymn formed a contribution to a little serial called *Good News for the Little Ones*, edited by C. H. Mackintosh, and published by Broom, and was first printed in their publication as the final article for the year 1859, under the heading of "Above the bright blue sky." In the original MS., and as first printed, the opening verse began "There's a rest for little children," "Friend" being subsequently substituted for "rest."

Mr. Midlane has written an immense number of hymns, and has lately published a collection containing 400 of his own compositions. Not one of his hymns, however, has attained anything like the popularity enjoyed by "There's a Friend for little children." Mr. Midlane, who has been for more than half a century engaged in business at Newport, says that he first received poetic encouragement when he went, as a little fellow, to Sunday School. He used to write his verses, and after the lessons were over would show them to his teacher, an enthusiastic reader of poetry, who gave him valuable advice. His first printed hymn was written at the age of seventeen while on a visit to Carisbrook Castle. This hymn, beginning, "Hark! in the presence of our God," appeared in the *Youth's Magazine* for November, 1842.

Another hymn for children which has come to us from an American source is "Shall we gather at the river?" by Robert Lowry. The following rather quaintly-expressed history of the origin of this hymn is given by Mr. E. W. Long in his *Illustrated History of Hymns and their Authors*—

"On a very hot summer day in 1864 a pastor was seated in his parlour in Brooklyn, N.Y. It was a time when an epidemic was sweeping through the city, and draping many persons and dwellings in mourning. All around friends and acquaintances were passing away to the spirit land in large numbers. The question began to arise in the heart, with unusual emphasis, ' Shall we meet again? We are parting at the river of death: shall we meet at the river of life?' Seating myself at the organ," says he (Mr. Lowry), "simply to give vent to the pent-up emotions of the heart, the words and music of the hymn began to flow out as if by inspiration—

> Shall we gather at the river,
> Where bright angel feet have trod?"

Dr. Lowry wrote a great number of hymns, for several of which he also composed effective music which helped considerably to increase their popularity. A hymn by Dr. Lowry which has been taken by many authors as the foundation of pathetic stories is "Where is my wandering boy to-night?"

Mr. George Rundle Prynne, the present Vicar of St. Peter's, Plymouth, is the author of:—

HYMNS FOR CHILDREN

Jesu, meek & gentle,
Son of God most high,
Pitying loving Saviour,
Hear Thy children's cry.

one of the simplest and yet most perfect hymns for children ever written. It was composed in 1856, and first appeared in a collection of hymns edited by Mr. Prynne. In 1861 it was given in *Hymns Ancient and Modern* and subsequently in nearly all hymnals published in Great Britain and America. In 1881 Mr. Prynne published it in a volume of hymns entitled *The Dying Soldier's Vision*, but by a mistake in passing the proof sheets the opening line read "*Jesus*, meek and gentle" instead of "*Jesu*, meek and gentle" as originally conceived. In subsequent hymnals, however, the old form was adhered to.

With regard to this composition the author says—

"This little hymn has found its way into most hymn books. It is commonly thought to have been written for children, and on this supposition I have been asked to simplify the fourth verse. The hymn was not, however, written specially for children. When it is used in collections of hymns for children it might be well to alter the fourth verse, which in the original runs—

> Lead us on our journey,
> Be Thyself the Way
> Through terrestrial darkness
> To celestial day.

to the more simple—

> Lead us on our journey,
> Be Thyself the Way
> Through earth's passing darkness
> To heaven's endless day."

During a visit paid to St. Peter's Vicarage some time ago, I had the pleasure of hearing from Mr. Prynne the circumstances under which he wrote his now famous hymn.

"I composed it," said Mr. Prynne, "one summer's evening just forty-six years ago, and I don't suppose the entire composition took me more than half an hour. My wife, who was a very good musician, was playing to me from my favourite composers at the time, and as she played so the words of the hymn came into my mind. I did not at first think of reducing them to paper, and it was only after the entire hymn was conceived that I at last took an old envelope from my pocket and scribbled the verses on the back. Then I read them over to Mrs. Prynne, and as she seemed to like them they were preserved, and subsequently appeared in one of my own books. I have been rather surprised that most hymnal editors should take it for granted that I wrote the hymn for children. Of course I did nothing of the sort, but I daresay it makes just as good a hymn for little ones as for adults, and after all I suppose most of us are children only 'larger grown.'

"A short time after its publication I went for a holiday to Rome, and while there was asked to conduct the service at the English Church. When the time for giving out the hymns came I was rather startled to find myself delivering the first line of my own composition : Hymn No. — 'Jesu, meek and gentle.' The pleasure I experienced on finding that it had so soon made its way to Rome was in no way lessened on hearing from the Vicar that the name of the author had not occurred to him, in fact he had never heard it!"

Mr. Prynne has been Vicar of St. Peter's, Plymouth, for fifty-four years, having been appointed to the living in 1848.

A hymn for children, by the late Miss Frances

Ridley Havergal, which has become very popular both in this country and America is :—

> *Golden harps are sounding,*
> *Angel voices ring,*
> *Pearly gates are opened—*
> *Opened for the King.*

Her sister gives the following account of the writing of this hymn—

"When visiting at Perry Barr Frances walked to the boys' schoolroom, and, being very tired, she leaned against the playground wall while Mr. Snepp, a gentleman who was with her, went in. Returning in ten minutes he found her scribbling on an old envelope. At his request she handed him the hymn just pencilled, "Golden harps are sounding."

A few days later Miss Havergal composed a special tune for this hymn, and it was this same tune, "Hermas," that the gifted poetess sang a few moments before she died.

Miss Havergal was once asked by a correspondent how she composed her hymns, to whom she replied—

"I can never set myself to write verse. I believe my King suggests a thought and whispers me a

THE RIGHT REV. BISHOP W. W. HOW, D.D.
Photo by Elliott & Fry.

THE REV. JOHN KEBLE, M.A.
From the painting by George Richmond, Esq., R.A.

musical line or two, and then I look up and thank Him delightedly, and go on with it. That is how the hymns and poems come. The Master has not put a chest of poetic gold into my possession and said, ' Now use it as you like!' But he keeps the gold and gives it me piece by piece just when He will, and as much as He will and no more. Some day perhaps He will send me a bright line of verses on 'Satisfied' ringing through my mind, and then I shall look up and thank Him and say, ' Now, dear Master, give me another to rhyme with it, and then another'; and then perhaps He will send it all in one flow of musical thoughts, but more likely one at a time, that I may be kept asking Him for every line. There, that is the process, and you see there is no 'I can do it' at all. That isn't His way with me. I often smile to myself when people talk about 'gifted pen' or 'clever verses,' etc., because they don't know that it is neither, but something really much nicer than being 'talented' or 'clever.'"

Bishop Walsham How wrote a great number of hymns for children, one of the most popular being "Come, praise your Lord and Saviour," written in 1871, and based on the words "O come, let us sing unto the Lord," taken from the Psalms. The hymn consists of four verses of eight lines each, the first verse being sung by boys and girls together, the second verse by boys only, the third verse by girls only, and the concluding verse by boys and girls together again. The effect when sung by well-trained children's voices is very pretty, and the hymn is, naturally, a great favourite in Sunday Schools.

Another hymn for children, which was written about the same time as " Come, praise your Lord and Saviour," is—

> It is a thing most wonderful,
> Almost too wonderful to be,
> That God's own Son should come from Heaven,
> And die to save a child like me.

It was written with the intention of being sung at Sunday School anniversaries, and based on the text " Herein is Love."

Bishop Walsham How had more than an ordinary love for children, and it is therefore not surprising that his hymns for little ones are so successful. It is said of him that even when nearing that age when he might be said to be " getting old " he would sit down and write a set of nonsense verses for a favourite grandchild with the greatest enthusiasm and earnestness. His character was an extraordinarily lovable and unselfish one. Though Bishop How is the author of many works of great learning and value, it is not improbable that his hymns will outlive them all.

A hymn of American origin, which has found considerable favour with the editors of English hymnals, is—

> I love to hear the story
> That angel voices tell,
> How once the King of glory
> Came down on earth to dwell.

written by Mrs. Miller, the daughter of the Rev. Dr. Huntington, of Brooklyn, Connecticut. In a letter from Boonton, New Jersey, Mrs. Miller writes—

"I do not know that there are any circumstances of special interest connected with the writing of this children's hymn. I was at the time (1867) joint editor of a magazine for young people entitled *The Little Corporal*, for which I usually furnished a poem to be set to music each month. I had had a very serious illness and was slowly recovering, and, though too weak to do much literary work, the fact that *The Little Corporal* would be published without my usual contribution was something of a worry to me. I determined, if possible, that this should not happen, so one afternoon, when I felt a little stronger, I took pen and paper and began to write 'I love to hear the story.' Though it is now close upon thirty-five years ago, I remember that the words were suggested rapidly and continuously as if I were writing from dictation. In less than fifteen minutes the hymn was written and sent away without any corrections. Its popularity has always surprised me, as among the hundreds of hymns and songs which I have written, many seem to me to be of greater merit."

A very beautiful hymn by this author, but one which is not generally known, is the following, which should be included in all children's hymnals—

>Father, while the shadows fall,
>With the twilight over all,
>Deign to hear my evening prayer,
>Make a little child Thy care.
>>Take me in Thy holy keeping
>>Till the morning break;
>>Guard me through the darkness sleeping,
>>Bless me when I wake.
>
>'Twas Thy Hand that all the day
>Scattered joys along my way,
>Crowned my life with blessings sweet,
>Kept from snares my careless feet.
>>Take me in Thy holy keeping
>>Till the morning break;
>>Guard me through the darkness sleeping,
>>Bless me when I wake.

HYMNS FOR CHILDREN

> Like Thy patient love to me,
> May my love to others be;
> All the wrong my hands have done,
> Pardon, Lord, through Christ, Thy Son.
>   Take me in Thy holy keeping
>   Till the morning break;
>   Guard me through the darkness sleeping,
>   Bless me when I wake.

"Above the clear blue sky, In heaven's bright abode," by John Chandler, was first published in the author's *Hymns for Public Worship*, 1856, though when it was actually written is not very certain. It is one of the few original hymns by this author which has become what one might term "popular." It is an immense favourite with children, having had the good fortune to be allied with an inspiriting melody composed for it by the late Dr. E. J. Hopkins.

Among the MSS. in possession of the Church House, Westminster, there is a letter from John Chandler, in which special reference is made to this hymn. It was written to a correspondent, probably in reply to a question regarding Mr. Chandler's method of composition. The letter, which is dated March 20, 1875, reads—

"With the exception of 'Above the clear blue sky' I have composed no hymns since those published in 1837, which are translations. 'Above the clear blue sky' appeared first in some Irish collection of hymns some years ago; but that is all I can remember about it."

The late Mrs. Cecil Frances Alexander had every right to be considered par excellence the children's hymnist. Her hymns are known wherever Christianity is preached, and the translations that have been made are innumerable. The Rev. F. A. Wallis, of the Universities' Mission to Central Africa, says he has heard Mrs. Alexander's hymns sung by half-clad Africans in a language she had never known. Perhaps the best loved of all the hymns by this writer is :—

*There is a green hill far away
Outside a city wall,
Where the dear Lord was crucified,
Who died to save us all.*

It was written in 1847 and first published in the author's *Hymns for Little Children*, a tiny volume of some thirty leaves, illustrated with full-page coloured illustrations. Mrs. Alexander once told me that her hymns were usually written for her Sunday School class, and were nearly all read over to her small scholars before being published. It is related that the beautiful and pathetic "There is a green hill" was written while Mrs. Alexander was sitting by the bedside of a sick child. The little girl was dangerously ill, but recovered, and ever after referred to this particular hymn as her own property. Some of the greatest composers of modern times have set this hymn to music. Gounod, whose setting is perhaps the most widely known and appreciated, affirmed that it was the most perfect hymn in the English language, its greatest beauty being its simplicity. Between the author and composer an interesting correspondence took place, and one of Mrs. Alexander's most prized possessions was an autograph copy of the famous composition sent to her by Gounod himself. In speaking about Mrs. Alexander's hymns a short while before his death, Gounod remarked that many of them set themselves to music.

In the manuscript copy of this hymn it will be noted that the first verse reads—

> There is a green hill far away,
> *Outside* a city wall,
> Where the dear Lord was crucified,
> Who died to save us all.

This substitution of the word "outside" for "without" was made on the authoress being asked by a very small child what was meant by a green hill not having a city wall. This reminds one of the story of the little boy whose mother was teaching him a certain hymn, and when she came to the well-known lines—

> Satan trembles when he sees
> The weakest saint upon his knees,

stopped his parent with the paralyzing question, "Why does the weakest saint sit on Satan's knees?" The great beauty, however, of Mrs. Alexander's hymns is their simplicity, and probably fewer questions have been asked regarding their meaning (by children) than of those of any other writer.

Among other hymns for children by Mrs. Alexander is "Once in royal David's city," which ranks second in point of popularity to "There is a green hill." It might very well be sung as a Christmas hymn, being founded on words taken from the Creed—"Who was conceived by the Holy Ghost, born of the Virgin Mary." Mention must also be made of "We are but little children weak," a hymn founded on the proverb, "Even a little child is known by his doings"; "All things bright and beautiful," an exquisitely descriptive hymn based on the verse "God saw everything that He had made, and behold it was very good," and "Do no sinful action."

Mrs. Alexander was the daughter of Major John Humphreys, who fought at the battle of Copenhagen. In 1847 she married the Rev. William Alexander, who subsequently became Bishop of Derry and Raphoe, and ultimately Primate of All Ireland. She took the greatest interest in all religious and charitable works and was greatly beloved by the poor. Her best-known poem is "The burial of Moses," which had a wide circulation. Of this work Tennyson said that it was one of the poems by a living writer of which he would have been proud to be the author. Mrs. Alexander died at the Palace, Londonderry, in 1895.

A hymn by Mrs. Alexander which is not so well known as many of her others, though characterized by the same simplicity and beauty, is one she wrote on the text, "Do all things without murmurings, that ye may be . . . children of God"—

> Day by day the little daisy
> Looks up with its yellow eye,
> Never murmurs, never wishes
> It were hanging up on high.

The most successful hymn which John Hampden

Gurney wrote was one for children: "Fair waved the golden corn." It was written on the text, "The first-fruits of thy land thou shalt bring into the house of the Lord thy God," and first published in a collection of hymns compiled by the author in 1851. Why this hymn should be considered as specially suitable for children I do not know; it would be far more appropriate amongst the harvest hymns. However, it has now for close upon half a century been looked upon as a hymn of praise for children, and I have therefore included it in this chapter. It has been translated into several languages and its use is very extensive. John Hampden Gurney, who was no relation of Archer Thompson Gurney, the author of "Christ is risen, Christ is risen," was for many years rector of St. Mary's, Marylebone, and Prebendary of St. Paul's Cathedral. He died in 1862 at the age of fifty-nine.

Miss Janette Threlfall has not written a great number of hymns, but among them is one for children which takes a very high place. "Hosanna! loud Hosanna! The little children sang," was written about the year 1870 and first published in the author's volumes of hymns and poems, *Sunshine and Shadow*. This little book is prefaced by a few remarks from the pen of Bishop

Wordsworth, a great admirer of her compositions. "It is an occasion for great thankfulness," writes her fellow hymnist, "to be able to point to poems, such as many of those in the present volume, in which considerable mental powers and graces of composition are blended with pure religious feeling, and hallowed by sound doctrine and fervent devotion."

Miss Threlfall was for many years a great invalid, so prostrate indeed at times that she found it impossible even to write. Yet she was always bright and cheerful, her face seldom without a smile. Many of her hymns were dictated to a friend, who wrote them out, the manuscript being afterwards corrected by the author. A copy of "Hosanna! loud Hosanna!" sent me by Miss Eccles, Miss Threlfall's cousin, is written in a round childish hand, and was evidently copied out for the author by one of her numerous juvenile friends who visited her during her illness. Miss Eccles tells me that several of her hymns were specially composed for the Whit Monday treat which took place annually at Miss Threlfall's country home in the North of England.

During the latter years of her life Miss Threlfall resided with relatives in Dean's Yard, Westminster. Here she interested herself in parish work and won the lasting friendship of Dean Farrar and the late Dean Stanley. On her death in 1880 the former paid a glowing tribute to her memory in a special sermon which he preached on the Sunday following her decease. A hymn for children which in points of simplicity and beauty has never been surpassed is that by Mary Duncan :—

> Jesus, tender Shepherd, hear me;
> Bless Thy little lamb to-night;
> Through the darkness be Thou near me;
> Watch my sleep till morning light.
>
> All this day Thy hand has led me,
> And I thank Thee for Thy care;
> Thou hast clothed me, warmed and fed me;
> Listen to my evening prayer.
>
> Let my sins be all forgiven;
> Bless the friends I love so well;
> Take me, when I die, to heaven,
> Happy there with Thee to dwell.

This little hymn is by no means in as many collections as one would suppose seeing that to most children it is as familiar as Charles Wesley's "Gentle Jesus, meek and mild." It has been set to music by various composers, perhaps the best known tune being "Evening Prayer" by the late Sir John Stainer.

The fact that this hymn was specially written by the authoress for her own little children gives an added interest to the composition. It is generally believed to have been composed three years after her marriage, in 1839, when Mrs. Duncan was barely twenty-five years of age. Her children, of course, would at that time be scarcely old enough to comprehend the meaning of the lines, or even pronounce the words after her, but we can readily believe that she was anticipating, with all a mother's affection, the time when they would come to her knee and repeat the hymn she had written for them. That time was never to be. Only two or three months after writing the hymn Mrs. Duncan caught a severe cold which developed into pneumonia, and a few days later the young life was closed. The following year the hymn was published, and how many mothers have since taught their children the beautiful lines, who can say?

One of the most remarkable men in the Church of England to-day is the present Vicar of Lew Trenchard, the Rev. Sabine Baring-Gould. His name is appended to more works in the British Museum than that of any other living writer, and there are few subjects on which he cannot write with authority. It is, however, as a hymn writer that Mr. Baring-Gould must be considered in the present volume, and his claim to a place among writers for children lies in his being the author of—

> *Now the day is over*
> *Night is drawing nigh*
> *Shadows of the twilight*
> *Steal across the sky.*

This beautiful little hymn was written in 1865 and appeared in the *Church Times* the same year. Three years later it was included in *Hymns Ancient and Modern*, and to-day it would be difficult to find a hymnal published during the last twenty-five years, either in England or America, in which it does not appear.

The second verse of this composition, which in the original read as follows—

> Now the darkness gathers,
> Stars begin to peep,
> Birds, and beasts, and flowers
> Soon will be asleep.

has been changed by the author to—

> As the darkness deepens,
> Stars begin to peep,
> Shadows of the twilight
> Darker round us creep.

This version is, perhaps, the more poetical of the two.

Mr. Baring-Gould wrote the hymn specially for the children in his Sunday School at Horbury

Bridge. It is intended for evening singing and was founded on the text taken from Proverbs iii. 24. "When thou liest down, thou shalt not be afraid; yea, thou shalt lie down, and thy sleep shall be sweet."

One of the most prolific writers of hymns was Frances Jane Van Alstyne, more generally known as Fanny Crosby. She was born in America in 1823, and during her lifetime composed no less than two thousand hymns. Among these was "Safe in the arms of Jesus," which was written at the special request of her friend, Mr. W. H. Doane, the well-known composer. Mr. Doane had written a melody for which he had no words suitable. He therefore called on Mrs. Alstyne, played the melody over to her, and begged that she would write him some words suitable to the tune. This Mrs. Alstyne promised to do, and 'Safe in the arms of Jesus" was the result. The authoress intended it as a children's hymn, and as such it has attained great popularity.

Mrs. Alstyne in infancy lost her sight, and at the age of eleven became an inmate of the New York City Institution for the Blind. Here she remained for twenty-three years as pupil and teacher. In 1858 she married Alexander Van Alstyne, a gentleman who had set many of her hymns to music, and who was also blind. The marriage was a very happy one in spite of the terrible affliction under which both husband and wife laboured.

Most of Mrs. Alstyne's hymns appeared in Ira D. Sankey's collection and owe no inconsiderable amount of their popularity to the attractive melodies to which they are allied. It is said that the number of copies of Mrs. Alstyne's hymns sold in America and Great Britain amounts to close upon one hundred millions.

Mr. St. Hill Bourne, the Rector of Finchley, is the author of at least one composition which takes a high place among children's hymns. I refer to—

> Christ, Who once amongst us
> As a child did dwell

written in 1868 and first published seven years later. In answer to a question regarding the genesis of this hymn, the author writes—

"There is nothing very interesting to tell you about the writing of this very simple little hymn It was one of the first I composed—in fact, was written before I was ordained. I really wrote it for the children of a mission school at Clerkenwell which I superintended on Sundays. This hymn was written about the same time as my harvest hymn, 'The sower went forth sowing,' both of which have since found a place in many collections."

"Come, sing with holy gladness" was written by the late Mr. J. J. Daniell about the year 1864, and first appeared in the Appendix to *Hymns Ancient and Modern*. It is one of the most successful of all children's hymns, the third verse being especially striking—

> O boys, be strong in Jesus,
> To toil for Him is gain,
> And Jesus wrought with Joseph
> With chisel, saw, and plane;
> O maidens, live for Jesus,
> Who was a maiden's Son;
> Be patient, pure, and gentle,
> And perfect grace begun.

In a conversation with the author's brother some time ago I was informed that Mr. Daniell had not written many hymns, his time being generally occupied in writing prose works. Mr. Daniell was ordained by the Bishop of Manchester in 1848, and after filling various curacies and livings, was preferred to the Vicarage of Langley Burrell in 1879, which he occupied down to his death in 1890.

Mr. Daniell wrote many hymns for special use in his own church, but few of them have found their way into hymnals which are very widely circulated. "Come, sing with holy gladness" was written specially for children, and, his brother informed me, the author was never happier than when hearing his own "boys and maidens" singing this hymn. The tune to which it is sung is the same as that to which "Hosanna, loud Hosanna" is set, a melody which appeared in *Kocher's Zionharfe* in 1855. Mr. Daniell probably had this melody in his mind when he wrote his now famous hymn.

XII

Some General Hymns

FROM Dr. Neale's translation of the *Rhythm* of Bernard of Morlaix many hymns have been taken. Of these the most popular are, " The world is very evil," " Brief life is here our portion," " For thee, oh dear, dear country," and " Jerusalem the golden." These centos are to be found in every modern hymnal, and are favourites with all denominations. It is interesting, in connection with these hymns, to note what Dr. Neale says in the Preface to the first edition. " There would be," he modestly suggests, " no difficulty in forming several hymns, by way of cento, from the following verses, suitable to any Saint's day, to the season of Advent, or to an ordinary Sunday. If any of Bernard's verses are thus employed, I shall be thankful indeed that ' He, being dead, yet speaketh.'" Later on, when a third edition of the *Rhythm* had been called for, he wrote: " I am deeply thankful that Bernard's lines seem to have spoken to the hearts of so many. I can reckon up at least fourteen new hymnals in which more or fewer of them have found a place."

The melodies to the various hymns taken from the *Rhythm* are almost as well known, perhaps equally so, as the hymns themselves. Dr. Neale was frequently asked to what tune the words of Bernard might be sung. In the fourth edition, published in 1861, he publicly answers the query. "Of the many tunes which have been composed," he writes, "I may here mention that of Mr. Ewing, the earliest written, the best known, and with children the most popular; that of my friend, the Rev. H. L. Jenner, perhaps the most ecclesiastical; and that of another friend, Mr. Edmund Sedding, which, to my mind, best expresses the meaning of the words." The Mr. Ewing referred to was the husband of Juliana Horatia Ewing, the author of *Jackanapes* and many other well-known stories for children.

There have been many paraphrases of the 90th Psalm, some of them exceedingly fine, but they have all given place to that by Dr. Watts beginning, "Our God, our help in ages past." The exact date of this composition is uncertain, but it is generally supposed to have been written when the author was past forty. In the Methodist hymnal it is placed in that section quaintly headed "Describing Death," and is sometimes used as a funeral hymn. Though written originally "*Our* God, our help in ages past," Watts himself subsequently changed the opening line to "O God! our help in ages past," and in this form it is found in most hymnals.

It has been said that nearly all great men have been members of large families, and this is borne out by Isaac Watts, who was the eldest of nine children. The events of his life are too well known to need recalling here, but it is rather an interesting fact that, when at an advanced age he was approaching death, he found consolation, not in his own hymns, but in those of other writers. In this respect he resembled the Father of Methodism who, when he lay dying, kept repeating the opening line of Watts' fine hymn,

"I'll praise my Maker while I've breath."

Only a few moments before the end came he endeavoured to repeat the hymn, but could only gasp out " I'll praise, I'll praise," and so struggling for breath died. It was a fine tribute to the worth of Watts' hymn, and it is probable that many another Christian has also passed away with the same words on his lips. Isaac Watts died November 25, 1748, at the age of seventy-four. A monument was afterwards erected to his memory in Westminster Abbey.

One of the most successful of the many hymns which have been written during the last twenty years is George Matheson's "O Love that wilt not let me go." This hymn has been specially set to music by Dr. A. L. Peace, who wrote for it the extremely beautiful melody "St. Margaret."

Some short while since I wrote to Dr. Matheson asking him for the story of his hymn and also for a MS. in his own handwriting, in order that I might reproduce it in facsimile. It was not until after I had posted the letter that I remembered with some dismay that the great Scottish preacher was blind. By return of post, however, I received a letter written by Dr. Matheson's secretary, in which he said: "I have been blind from youth, and have long since given up all caligraphy now, attempting nothing more formidable than my signature, and this only when required for formal documents.

"The circumstances under which I wrote 'O Love that wilt not let me go,'" continues Dr. Matheson's letter, "were these. It was not composed; it came as an inspiration. I well remem-

1 MR. RICHARD BAXTER.
   From an Engraving.

2 SIR JOHN BOWRING, LL.D.
   From a Drawing.

3 MR. WILLIAM WILLIAMS.
   From an Engraving.

4 THE RIGHT REV. BISHOP RICHARD MANT, D.D.
   From an Engraving.

5 THE REV. GEORGE MATHESON, D.D.
   Photo by Horsburgh.

ber the occasion. It was at Innellan, on an evening in June, 1882. I had suffered a severe loss, and was greatly depressed. As I sat there, very sad and unhappy, the words flashed into my mind, and in a few minutes the four verses of the hymn were complete. It seemed as if they had been swiftly dictated to me by some invisible medium, complete in language and rhythm. It has been a constant source of pleasure to me that the little hymn has found so large an acceptance. Every year I receive many requests from compilers of hymnals in all parts of the world to be allowed to include 'O Love that wilt not let me go' in their collections, requests which I am only too delighted to grant."

Dr. Matheson is at present minister of St. Bernard's, Edinburgh, a position which he has now filled for many years.

There are many hymns which have associations with the Isle of Wight. We have already referred to several, and the number is increased by yet another, this being Mr. Thomas Binney's "Eternal light! eternal light!" This hymn was written at Newport, not a great way from the house in which Mrs. Luke now resides. The origin of this hymn was told me by one who in his youth was a personal friend of Thomas Binney. It appears that the late hymnist, who had a fondness for solitude, was one evening sitting at the window of his house watching the sun set. He sat there long after the light of day had disappeared, until the moon rose and the stars began to come forth. And then the thought struck him how wonderful it seemed that the sky was never free from light, it was eternal. After sitting in contemplation for some considerable time longer, the lines of the hymn began to frame themselves in his brain. Rising at last, he shut the window and retired to his own room. That night before he closed his eyes in sleep the hymn was written. It was published about the year 1820 with special music, which, however, did not commend itself very highly to the public taste. Thirty-five years later it was again set to music by a Mr. Burnett of Highgate, whose setting was very much more successful. The hymn is a favourite in America, though in American hymnals the original text is not generally adhered to. The manner in which his hymn was occasionally mutilated was sometimes a source of annoyance to Mr. Binney, who often referred rather sarcastically to the manner in which compilers unblushingly altered a hymn to suit their own particular views.

Dr. Binney, who for many years was pastor at

Newport, was the author of a great number of prose works. "Eternal light! eternal light!" is the only hymn of his which has gained any degree of popularity.

The late Thomas Toke Lynch was the author of a volume of hymns which created one of the most remarkable controversies in the history of hymnology. This book was entitled *The Rivulet*, and from it have come into common use many very beautiful compositions, the most popular, perhaps, being the one he wrote on the Holy Spirit :—

> Gracious Spirit, dwell with me,
> I myself would gracious be.

This simple little hymn has been appropriated by a great number of editors, and is the one by which Mr. Lynch's name will be longest held in remembrance.

Thomas Lynch was born in 1818 at Dunmow, in Essex, being the son of a surgeon. He was always delicate, but even as a boy bright and manly. Almost before he had reached his teens he had composed many hymns and poems, and his great ambition was to have a volume of them published. In view of this wish being fulfilled one. day he wrote a dedication, which he addressed to himself. This remarkable document began, "Dearest Myself,—As you have had some concern in writing these verses, and are besides my oldest and most intimate friend, it is but proper that I should dedicate them to you. I wish you to take this rather as a token of affection than respect. Our near relationship and close intimacy make me still retain some regard for you, although you have much injured me and thwarted many of my designs," etc. This curious address, over which Mr. Lynch had many a laugh in after years, concluded, " I remain, My dearest myself, Your affectionate though injured companion, I." Both the poems and the dedication were written before he had reached his fifteenth year.

For several years Mr. Lynch discharged the duties of a minister in London, but owing to enfeebled health he was obliged to retire. On partially recovering his strength in 1860, he took a room in Gower Street, where he continued to preach every Sunday for a couple of years, when he removed to Hampstead and became minister of Mornington Church. He died on May 9, 1871, at the age of fifty-three.

A hymn which is generally sung on Trinity Sunday, and which was written especially for that day, is Reginald Heber's magnificent paraphrase :—

*Holy, Holy, Holy, Lord God Almighty!
Early in the morning our song shall rise to Thee——
Holy, Holy, Holy, Merciful and Mighty!
God in Three Persons, Blessed Trinity!—*

This beautiful hymn is very often sung as a morning hymn, and is, in fact, a general favourite throughout the year. It was not published until after Heber's death, like very many of his hymns, but almost immediately after its appearance it began to be inserted in all the new collections. It was written on the text taken from Revelations:—
" They rest not day and night, saying, Holy, Holy, Holy, Lord God Almighty, Which was, and is, and is to come." I have in my possession a photograph of the hymn as it appears in Heber's collection of MSS. at the British Museum. It is headed " For Trinity Sunday," and is initialled with the letters " R. H." If it is not the most popular of all Heber's hymns—and many assert that it is—there is certainly no other of this author's compositions which exceeds it in beauty. It has been translated into as many foreign languages as " From Greenland's icy mountains " and is a favourite subject with students for conversion into Latin.

A word must be said regarding the triumphant and martial melody to which it is allied. This composition, known by the name of "Nicaea," was written by the late Dr. Dykes. Many attempts have been made to supplant this melody, but without success. Like that to "Eternal Father! strong to save," it stands alone and will live with the hymn. It is one of the best known of Dr. Dykes' tunes, and was very popular in Durham. On one occasion, when Dr. Dykes went into a shop in that city to make a purchase, there happened to be standing behind the counter three young assistants. A look of intelligence passed between them, and the first assistant softly uttered the word "Holy" on E, the second followed suit on G sharp and the third finished on B. Dr. Dykes looked amused and rather scandalized but said nothing. The same thing occurred on other occasions until the composer quietly asked the young men to desist, which they very politely did. This little incident was told me by a personal friend of the late composer.

"Now thank we all our God" is Miss Winkworth's translation of Martin Rinkart's great hymn. It is said to have been written in commemoration of the Peace of Westphalia, but this is a story which can hardly be relied upon. Rinkart passed through all the horrors of the Thirty Years' War, and he can certainly have seen little during that terrible time calculated to inspire in his breast a hymn of such gratitude and praise as " Now thank we all our God." One writer has suggested that he wrote the hymn in anticipation of the Peace which he knew must come some day, and this is not at all unlikely. At one time he was shut up in Eilenburg, where there were gathered together many thousands of refugees. Pestilence broke out, and they died by hundreds. All the clergymen, with the exception of Rinkart, succumbed and were buried. He was a man as brave and fearless as Martin Luther, and he continued to read the burial service over the blackened bodies of the famine's victims until he too fell exhausted. During this time he must have written the hymn which has survived for two hundred and fifty years. It has been sung on many important occasions when Peace has been declared, the last time being at St. Paul's Cathedral, when it was voiced by an immense congregation as a thanksgiving hymn for the cessation of hostilities between this country and South Africa. A curious coincidence in connexion with the singing of the hymn on that occasion was the fact that the name of the composer was closely allied to that of the man who was at one time President of the South African Republic. The name of the composer was Johann Cruger, and the hymn tune was composed about the year 1649, so that it may possibly have been written specially for Rinkart's composition. It is a fine, impressive setting, and suits the hymn admirably.

Bishop Bickersteth has written a considerable number of hymns, to some of which reference has already been made, but the one which will keep his name longest in the Church's memory is that which he wrote in 1875:—

*Peace, perfect peace, in this dark world of sin!
The blood of Jesus whispers peace within.*

This hymn steadily increases in favour year by year, and its growing popularity is in no way lessened by the very beautiful melody entitled "Pax Tecum," which was specially written for it by Mr. G. T. Coldbeck.

It has been published in the form of a card, and many hundreds of these have been distributed by the Bishop among the numerous children he has confirmed. The hymn owes its origin, the author tells me, to the impression made upon him by a sermon preached by Canon Gibbons from the text "Thou wilt keep him in perfect peace, whose mind is stayed on Thee." Canon Gibbons was a celebrated preacher, and so much was Dr. Bickersteth moved by his discourse that, on reaching home, he penned the lines almost spontaneously. They came with little effort, and the whole composition was completed in a very few minutes. He took less time over it than any other of his hymns, and yet it has become the best known. It has been favourably compared to Newman's "Lead, kindly Light," a somewhat curious coincidence when one remembers the hot water Dr. Bickersteth got into for adding an additional stanza to that favourite hymn. Mr. Richard le Gallienne, who is a critic as well as a poet, says of " Peace, perfect peace " : —" It would be difficult to name any other hymn so filled with the sense of man's security as this, which tranquillizes me at certain moments to a remarkable degree."

"O Jesu, Thou art standing" was written by the late Bishop of Wakefield in 1867. It is one of the most popular of Bishop How's compositions, and is to be found in a great number of hymnals. It has come to be associated with Holman Hunt's celebrated picture of Christ knocking at the door. Bishop How also wrote the special hymn for the late Queen Victoria's Diamond Jubilee, but though it was a fine composition and suited its purpose well, it will probably not be included in many hymnals. This, of course, is due to the fact that it was written for a unique and probably never-to-be-repeated occasion.

With regard to " O Jesu, Thou art standing," I find among my papers a letter from the author enclosing a MS. of his well-known hymn. He says:—"There is very little to be said regarding the writing of 'O Jesu, Thou art standing'; certainly nothing worth calling a story. I composed the hymn early in 1867, after I had been reading a very beautiful poem entitled 'Brothers and a Sermon.' The pathos of the verses impressed me very forcibly at the time. I read them over and over again, and finally closing the book I scribbled on an odd scrap of paper my first ideas of the verses beginning 'O Jesu, Thou art standing.' I altered them a good deal subsequently, but I am fortunate in being able to say that after the hymn left my hands it was never revised or altered in any way."

"Thy life was given for me," or, as in some collections, "I gave My life for thee," is one of Miss Havergal's most popular hymns and is an especial favourite in America. A few years before her death Miss Havergal received a letter from an unknown correspondent in Brooklyn asking for some information respecting this composition. In reply the authoress wrote :—" The hymn was the first thing I ever wrote which could be called a hymn, and it was composed when I was quite a young girl. I did not half realize what I was writing about. I scribbled it in pencil on the back of a circular, in a few minutes, and then read it over and thought, 'Well, this is not poetry. I will not go to the trouble to copy this.' And I stretched out my hand to put it in the fire: but a sudden impulse made me draw back, and I put it, crumpled and singed, into my pocket. Soon after I went out to see an old woman in an almshouse. She began to talk to me, as she always did, about her dear Saviour, and I thought I would see if the simple old woman would care for these verses which I felt sure nobody else would care to read. So I read them to her, and she was so delighted with them that, when I went back, I copied them out and kept them, and now the hymn is more widely known than any. Afterwards my father wrote for it the tune 'Baca,' to which it is now almost always sung."

As a hymnist John Wesley is principally known by his translations, and it is to him that we are indebted for having first opened up to us the beauties of German hymnody. No finer translations are to be found than his versions of Scheffler's "Thee will I love, my strength, my tower," and "Lo! God is here, let us adore."

In order to catch the thoughts that fly, John Wesley invented a system of short long-hand. When walking or riding, appropriate lines would occur to him, and these were immediately entered in a small notebook which was kept ready for the purpose. When his destination was reached the hymn was carefully gone through, the weak lines expunged and the strong ones strengthened. He was his own severest critic, and he never distributed a hymn amongst his congregation without spending an immense amount of thought upon it.

The most popular of John Wesley's translations is :—

> Now I have found the ground wherein
> Sure my soul's anchor may remain.

Mr. G. J. Stevenson in his notes says : "Some one has likened this hymn to the word of God, for in it are found no less than thirty-six separate passages of Scripture which, in language or spirit, correspond with the lines of the hymn."

When he had finished the translation John Wesley sent a copy of it to a German Moravian who was at that time in London, asking him to frankly criticise it, and if he thought it might be improved to kindly say so. It was returned with many expressions of approval and a suggestion that the third stanza should be altered. The Moravian enclosed an improved version of this verse which Wesley afterwards adopted. The "Father of Methodism" was always willing to take sound advice and ever ready to consider the opinions of his critics.

"I heard the voice of Jesus say," the finest and perhaps most loved of all Horatius Bonar's hymns, was written while the author was minister at Kelso. A reproduction of the original manuscript from Dr. Bonar's notebook, which I am able to give through the courtesy of his son, will be studied with peculiar interest by all lovers of the great Scotsman's compositions. Among hymnal manuscripts it is unique, for it gives some insight into the method the author adopted when composing his hymns. His son tells me that he would take his notebook, and while thinking out the lines of his hymn he would be busy with his pencil, making little sketches all over the margin of the page. It is evident from the MS. that Dr. Bonar, like John Wesley, made use of a kind of shorthand, though in his case the signs employed bear a strong likeness to Pitman's system. The original MS. of "I heard the voice of Jesus say" is now very much worn and faded. It is written in pencil, and the photographer who copied the original tells me that he had to give an exposure of something like three-quarters of an hour in order to get even a fairly good result.

Dr. Bonar's notebook, which is now one of the most precious relics he has left behind, contains, I believe, many other hymns, including "I was a wandering sheep," written two or three years previous to "I heard the voice of Jesus say," and that very beautiful resignation hymn, "Thy way, not mine, O Lord," written in 1855.

The name of Harriet Auber will be remembered in connection with a single hymn, though she wrote others which are to-day in use at Spurgeon's Tabernacle and elsewhere. "Our blest Redeemer, ere He breathed" was written as a Whitsuntide hymn, but has now become so great a favourite that in most collections it finds a place among those for "General Use."

Apart from the great beauty of "Our blest Redeemer, ere He breathed," the hymn is remarkable from the fact that the authoress first wrote it on a pane of glass in a window of her house at Hoddesdon, where she resided for many years. To Mr. C. W. Lock of that town I am indebted for a photograph of this interesting house, which shows the window which contained this strange manuscript, though the pane has long since been removed.

Facsimile of "I heard the voice of Jesus say."
From the original MS.

Writing from Hoddesdon, Mr. Lock says: "I remember the house well in which Miss Auber used to live, and where she died in 1862 at the patriarchal age of eighty-nine. She was buried in the churchyard immediately opposite the house. She and a Miss Mackenzie lived together, two saintly old ladies who were known and loved for many miles round. While Miss Auber wrote poetry Miss Mackenzie was the author of a considerable number of prose works of a religious nature.

" A lady resident here, whose relations lived in Miss Auber's house after the decease of the hymnist, tells me that when visiting her friends she often saw the hymn on a pane of glass in one of the bedroom windows at the back of the house, but that after her friends left Hoddesdon the pane was removed by some person and has never been recovered. No trace was ever found of it. This lady has put a cross on the window in the photograph to show which pane of glass the hymn was written on."

Though there has been a good deal of controversy as to who was the author of " All people that on earth do dwell," there can be little doubt that it was the work of William Kethe. It first appeared in 1560, and since that date few hymnals have been published in which it has not found a place. Perhaps the tune to which it is allied, and which is generally known as "the Old Hundredth," is almost as famous as the hymn itself. This tune originally appeared in the *Genevan Psalter*, which dates back to about the year 1543, and it is not therefore at all unlikely that Kethe wrote his hymn to the tune.

A somewhat interesting incident may be mentioned in connection with this composition. As sung to-day the second verse runs :—

> The Lord, ye know, is God indeed;
> Without our aid He did us make;
> We are His flock, He doth us feed,
> And for His sheep He doth us take.

In the original, however, the word in the third line was not "flock" but "folk," spelt in the Old English "folck." It is generally supposed that this alteration, which occurred not so many years after it was first published, was due to a printer's error, the "o" and the "l" being transposed. If this is correct it was rather a happy mistake, the connection between "flock" and "sheep" being decidedly appropriate. Perhaps, after all, the printer thought that he could improve on the original, and there is little doubt that he did so. It may be mentioned that Prebendary Thring in his collection has reverted to the original "folk." The exact date of William Kethe's death is not known; in all probability it occurred towards the closing years of the sixteenth century.

A hymn which is said to be a favourite with King Edward VII. is :—

*Nearer my God to Thee
Nearer to Thee!
E'en tho' it be a cross
That raiseth me!*

which was written by Sarah Flower and first published in 1841. The authoress was the daughter of a couple who first became acquainted in Newgate gaol. It was no great offence for which Benjamin Flower was imprisoned, being nothing more heinous than a spirited defence of the French Revolution. While in gaol he was visited by a Miss Eliza Gould, and the friendship thus formed ripened into love, and when his term of imprisonment was over they were married. Two children were born, Eliza and Sarah, both of whom were singularly gifted. Sarah was a poetess of great sweetness and power, while Elizabeth developed a wonderful talent for musical composition, setting nearly all her sister's poems and hymns to music.

Some time ago I had an interesting correspondence with Mrs. Bridell Fox, a near relative of the late Mrs. Sarah Adams (*née* Flower), in which she gives an attractive word portrait of the authoress of "Nearer, my God, to Thee." She says: "Sarah was tall and singularly beautiful, with noble and regular features; in manner she was gay and impulsive, her conversation full of sparkling wit and kindly humour."

"How she composed her hymns," Mrs. Fox says in another letter, "can hardly be stated. She certainly never had any idea of composing them. They were the spontaneous expression of some strong impulse or feeling at the moment; she was essentially a creature of impulse. Her translations would be, of course, to a certain extent, an exception; also, perhaps, when she was writing words for music already in use at South Place Chapel. Otherwise she wrote when she felt that the spirit moved her."

Some years ago Mrs. Fox collected Miss Flower's sacred and secular music, and this has since been lodged in the British Museum together with Mrs. Adams' religious drama *Vivia Perpetua*. Included in the former is an exquisite setting of "Nearer, my God, to Thee." It is not easy, however, and requires several good soprano voices to make it effective.

Both sisters died at comparatively early ages and within two years of each other—Eliza in December, 1846, and Sarah on August 11, 1848. At the funerals of both, the hymns and music sung were the compositions of the two gifted sisters.

Rather a remarkable story is told in connection with this hymn. When Mr. McKinley was almost *in extremis* he derived great comfort from the hymn "Nearer, my God, to Thee," and various accounts of the composition appeared in the London papers. One correspondent told a story which was related to him by the Rev. Dr. Moulton, who was for over thirty years a missionary in the Tongan Islands. "On one of his periodical visits to the smaller islands," wrote the anonymous correspondent, "he landed at one rarely even visited by missionaries, and there heard that an old Tongan, who had some years before been converted to Christianity, was dying. The doctor hastened to the hut of the sufferer, and there a curious sight met his view. The old man had been propped up by his friends so that he clung by his two arms to a beam stretching across the room; there he half hung with closed eyes and a face drawn with agony constantly murmuring some words. The doctor drew silently near to him, thinking that the dying man was making some last request. 'Judge of my astonishment,' he said in relating the incident, 'when I heard these words uttered over and over again—in Tongan of course—"Nearer, O God, to Thee! Nearer to Thee,"' In those days—almost forty years ago—the hymn of the cultured, saintly Englishwoman had not reached the Tongan natives, but the same spirit that inspired the thought in her doubtless inspired it in the heart of the poor, untutored Christian Tongan."

In the Olney Hymn Book there is no more beautiful composition than John Newton's "How sweet the name of Jesus sounds." It was written when the author had passed his fiftieth year. It is probably the best known of all Newton's hymns with the exception of his "Glorious things of Thee are spoken."

Newton was the son of a sailor and spent many years at sea, where, on his own confession, he spent the life of a reckless and profligate sailor. He was at one time in the service of an African slave-dealer, and for some years commanded a slave ship. There was no kind of wickedness which he did not commit or boast of having committed. But when he reached his thirtieth year he came under the influence of the Nonconformists, and forsaking his seafaring life he gave himself up to

1 THE REV. W. BULLOCK, D.D.
Photo by Notman, Nova Scotia.
2 THE REV. W. ST. HILL BOURNE.   3 MR. WILLIAM WHITING.
Photo by H. Edmonds, Hull.   *From a Photo.*
4 MR. W. CHATTERTON DIX.   5 THE REV. HENRY FRANCIS LYTE, M.A.
Photo by Lindon Hatt, Clifton.   *From an Engraving.*
6 THE REV. HENRY ALFORD, D.D.
Photo by Maull & Co.

preaching and became curate of Olney, where most of his hymns were written. He has published his own life, wherein he gives a faithful and frank account of himself, dwelling with considerable emphasis on those years of wickedness which preceded his conversion. He lived to the good old age of eighty-two and continued preaching almost to the last. The following little story is related regarding the hymnist which is not without interest :—

"When he had passed his fourscore years he continued to preach. As it was with difficulty that he could see to read his manuscript he took a servant with him into the pulpit, who stood behind him and with a wooden pointer would trace out the lines. One Sunday morning Newton came to the words in his sermon, 'Jesus Christ is precious,' and wishing to emphasize them he repeated : 'Jesus Christ is precious.' His servant thinking he was getting confused whispered, 'Go on, go on, you said that before'; when Newton looking round replied, 'John, I said that twice, and I am going to say it again'; then with redoubled force he sounded out the words, 'JESUS CHRIST IS PRECIOUS.'"

To the late Rev. William Bullock, D.D., Dean of Halifax, Nova Scotia, the Church is indebted for "We love the place, O God," one of the best known of modern hymns. It has been my privilege to see the composition in the original, and it is remarkable to note the amount of alteration it has undergone at the hands of compilers. The last two verses as written by Dean Bullock have appeared in few hymnals, and as they may be new to many of my readers their reproduction here may not be without interest:—

> We love Thy saints who come
> Thy mercy to proclaim,
> To call the wanderers home,
> And magnify Thy name.
>
> Our first and latest love
> To Zion shall be given—
> The house of God above,
> On earth the gate of heaven.

With regard to this hymn Mr. R. H. Bullock, the Dean's son, who still lives in Halifax, says: "We have a strong impression that the hymn was composed in 1827 for the consecration of the church in Trinity Bay, Newfoundland, of which he was the Rector. Two or three years ago this hymn was sung at the consecration of a new church erected on the same site, when the sermon which was preached there in 1827 was preached again from the original manuscript. My father more often than not read his sermons, and we have therefore been able to preserve very many of them.

"All through his life my father cultivated poetry, and I note in his journal (which we still possess) the account of a voyage round the island of Newfoundland, with the Governor, in his yacht in 1828, when several pieces of domestic and religious poetry were composed at sea. My father was passionately fond of the water, and was never so happy as when at the helm running before a stiff breeze."

Dean Bullock was born at Prettiwell, Apex, and after receiving his education at the "Blue Coat School" passed into the Royal Navy, where he remained for some years and attained the rank of lieutenant. He was employed with his brother, the late Admiral Frederick Bullock, in surveying the coast of Newfoundland, and it was while thus engaged that he resolved to take Holy Orders and become a missionary in that colony.

Should you ever visit Trinity Bay you will find that the name and memory of the late Dean (he died in 1874) are both held dear, and that his influence is still widely felt. This little settlement has furnished no less than five ministers of the Gospel, and with expressions of gratitude and affection the settlers will tell you to whose influence this is due.

It has been said of the hymns by the late Dr. J. S. B. Monsell that few of them, if any, will celebrate their second century. This is perhaps too severe a judgment, for a couple at least will live—"O worship the Lord in the beauty of holiness," and "Fight the good fight with all thy might." Either of these hymns is worthy the pen of any hymnist. The latter came prominently before the public during the years of the South African war, and was heartily sung by congregations of all denominations. In America, too, it is a great favourite, and played a prominent part in religious services during the war between that country and the Philippines. It has been set to music several times by different composers, the most popular tunes being "St. Crispin," by Sir G. J. Elvey, and "Pentecost," by William Boyd.

"O worship the Lord in the beauty of holiness" is perhaps Monsell's best known hymn. It was written about the year 1860, after the author had passed his fiftieth year, so that it was not, as I have seen it stated, one of Dr. Monsell's earliest compositions. After its publication Dr. Monsell wrote a second version, but it is the origina which

SOME GENERAL HYMNS

is usually found in our hymnals. In many cases, however, the hymn begins, " Worship the Lord in the beauty of holiness," the interjection being omitted. This hymn was very successfully set to music by Sir Henry Smart, his tune, " Meredun," being the one to which it is generally now sung.

John Samuel Bewley Monsell was born in Londonderry in 1811. In 1834 he took orders, and subsequently became Vicar of Egham and afterwards of Guildford in Surrey. He published several volumes of poems and hymns, as well as some prose works. His death was due to an accident. His church at Guildford was undergoing repairs, and Dr. Monsell, who took considerable interest in the renovation, used often to watch the men at work. One day he stood in the aisle, and was looking up at some alterations which were being made in the roof when a large piece of masonry fell. To the consternation of the workmen, the stone struck Dr. Monsell on the head, felling him to the ground, where he lay unconscious. He was tenderly carried to the rectory, where everything was done to save his life, but without success. He died on April 9, 1875, at the age of sixty-four.

Nearly all the hymnal manuscripts of the late

hymnist are in the possession of his son, Colonel Monsell, who resides in London.

Between the two hymns, "Come unto Me, ye weary" and "I heard the voice of Jesus say," there is a good deal of similarity. Both are what one might call Invitation hymns; both are written in verses of eight lines each; and though the metre is not the same, the difference is so slight as almost to be unnoticeable; while the author of the remarkably beautiful tunes to both, known as "Come unto Me" and "Vox Dilecti," is the same, namely John Bacchus Dykes. Curiously enough, too, in many hymnals they follow each other. There was, however, a difference of over twenty years between the time of their composition, "Come unto Me, ye weary" not having been written until 1867, while, as already mentioned, "I heard the voice of Jesus say" was written in 1845.

Not a great while before his death Mr. Chatterton Dix sent me a manuscript copy of his hymn:—

> "Come unto Me, ye weary,
> And I will give you rest."
>
> O blessed voice of Jesus,
> Which comes to hearts oppressed!

together with a few remarks as to the circumstances under which he composed it. " I was ill and depressed at the time," he says, "and it was almost to idle away the hours that I wrote the hymn. I had been ill for many weeks, and felt weary and faint, and the hymn really expresses the languidness of body from which I was suffering at the time. Soon after its composition—and it took me some time to write out, for my hand trembled, and I could with difficulty hold the pen—I recovered, and I always look back to that hymn as the turning-point in my illness. It is a somewhat curious fact that most of my best known hymns were written when I was suffering from some bodily ailment. Dr. Dykes' setting I consider one of the most beautiful in the hymnal." Like the late Cardinal Newman, Mr. Dix was almost tempted to say that it had much to do with the success of the hymn. "Come unto Me, ye weary" was suggested to the author by the words, "Come unto Me, all ye that labour and are heavy laden, and I will give you rest." Mr. Dix died at Clifton in 1898 at the age of sixty.

Considering the very large number of hymns which were written by the late Sir John Bowring it is rather remarkable that so few should have found their way into collections used by the

Church of England. One, however, which appears to be sung by all denominations is that very fine hymn beginning " In the Cross of Christ I glory." This was one of Sir John Bowring's earliest hymns, being written when he was between twenty and thirty years of age. It is popular in all countries where the English tongue is spoken and has been translated into several languages.

To Charles II's chaplain, Richard Baxter, we owe one of the most beautiful resignation hymns ever penned. " Lord, it belongs not to my care" must have been composed when the author was quite an old man, and was not published until shortly before his death. It is said to have been a great favourite with his wife and was sung by her during her last illness. Richard Baxter wrote many other hymns not one of which, however, has so staunchly stood the test of time as " Lord, it belongs not to my care." It was Baxter, it will be remembered, who, when greeted by the terrible Judge Jeffreys with the remark, " Richard, I see the rogue in thy face," replied " I had not known before that my face was a mirror." Baxter died in 1691 at the age of seventy-six.

A hymn which, though originally written for Trinity Sunday, has become a general favourite for any season of the year is Bishop Mant's "Bright the vision that delighted," sometimes commencing "Round the Lord in glory seated," as in Thring's collection. It was written in the early thirties and published in the author's *Ancient Hymns*. From thence it passed into numerous collections and ultimately took its place as the most popular of all this writer's hymns. Bishop Mant died in 1848.

The number of hymns which have come to us from the Welsh is not great, but among them is one which I cannot refrain mentioning before bringing this little volume to a close. It is William Williams' "Guide me, O Thou great Jehovah." The hymn was first published in the original Welsh in 1745, but it was not until some twenty-five years later that the first English translation appeared. This translation, which is partially the one now in general use, was made by Peter Williams, a clergyman of Carmarthen. Whether any relationship existed between author and translator I do not know, but as Williams is by no means an uncommon name in the Principality it is not improbable that the similarity in names is merely a coincidence. After Peter Williams' version had appeared, the author of the original, not being altogether satisfied with it, determined to make a translation of his own. This he did so successfully that in nearly all modern hymnals the last two verses are generally those taken from his own translation.

William Williams was born near Llandovery in 1717, and after ordination became curate in various Welsh parishes. He was a fine preacher, and had a large following. His hymns, most of which were written at odd moments, were collected and published in book form. They became immensely popular in Wales, and are to-day sung in the original tongue in many of the villages and towns. The only composition, however, which may be said to be universally known in England is "Guide me, O Thou great Jehovah." For this hymn Sir George Elvey wrote the very fine melody "Pilgrimage," to which it is usually sung. Mr. Williams died in 1791 at the age of seventy-four.

Index of Authors' Names

Adams (*née* Flower), Sarah, 313
Alderson, Eliza Sibbald, 90
Alexander, Cecil Frances, 280
Alford, Henry, 141, 241
Allen, James, 107
Ambrose, St., 1, 30, 37, 49
Anatolius, St., 22, 220
Armes, Dr. Philip, 208
Auber, Harriet, 43, 310

Bach, J. Sebastian, 135
Baker, Sir Henry, 7, 17, 19, 28, 34, 70, 109, 110, 159, 169, 173, 178, 197, 251
Baring-Gould, S., 145, 147, 287
Barnby, Sir Joseph, 129, 195, 234
Baxter, Richard, 324
Benedictus, Jacobus de, 104
Bennett, G. J., 225
Bernard of Morlaix, 293
Bickersteth, Bishop, 20, 25, 227, 304
Biggs, L. C., 26
Binney, Thomas, 267, 297
Bliss, Mrs. Worthington, 142, 233

Blunt, A. Gerald, 245
Bode, J. E., 185
Bonar, Horatius, 80, 121, 309
Bourne, W. St. Hill, 240, 290
Bowring, Sir John, 323
Boyd, William, 320
Brady, Nicholas, 64
Bridge, Sir Frederick, 241
Bridges, Matthew, 129
Bright, William, 33, 169
Bullock, William, 317
Byrom, John, 58

Campbell, Jane M., 244
Campbell, Robert, 134
Cassander, Georgius, 38
Caswall, Edward, 26, 49, 82, 104, 106, 176
Celano, Thomas of, 40
Cennick, John, 50
Chandler, John, 37, 279
Clark, Jeremiah, 139
Claudius, Matthias, 243
Clephane, Elizabeth C., 218
Coldbeck, G. T., 304
Coles, V. S. S., 164
Collins, Henry, 176
Collyer, William B., 46
Conder, Josiah, 166

INDEX OF AUTHORS' NAMES

Cotterill, Thomas, 46
Cowper, William, 94, 180
Cox, Frances E., 124, 247
Coxe, Arthur C., 211
Crosby, Fanny, 289
Crossby, John, 60
Cruger, Johann, 303

Daniell, J. J., 291
Dix, William Chatterton, 63, 140, 244, 322
Doane, W. H., 257, 289
Doddridge, Phillip, 44, 162, 187
Doudney, Sarah, 232
Downton, Henry, 204, 244
Dykes, Dr. J. B., 8, 19, 25, 55, 83, 88, 90, 105, 148, 166, 178, 182, 195, 226, 249, 251, 265, 302, 322

Edmeston, James, 20
Ellerton, John, 15, 17, 109, 179, 195, 198, 205, 231, 248
Elliott, Charlotte, 215, 239
Elvey, Sir G. J., 243, 320, 326
Ewing, Alexander, 253, 294

Faber, Father, 9, 177
Farmer, John, 75
Farrar, Frederick W., 74, 286
Ferrars, Earl, 107
Fortunatus, Venantius, 143
Foster, Myles, 210
Fuller-Maitland, Frances, 183

Fuller-Maitland, Mrs., 183

Gadsby, Henry, 142
Gauntlet, Dr., 145, 196
Gellert, C. F., 124
Gladstone, W. E., 181
Gounod, Charles, 281
Gurney, Archer T., 122, 284
Gurney (*née* Blomfield), Dorothy, 124, 193
Gurney, John Hampden, 283

Hamerton, Samuel C., 73
Hampton, John, 144
Handel, George Frederick, 131
Hankey, Katherine, 219, 256
Harris, F. W., 116
Havergal, Frances Ridley, 46, 84, 273, 306
Haydn, Joseph, 105
Hayne, L. G., 81
Heathcote, Sir William, 6
Heber, Reginald, 70, 111, 200, 229, 252, 300
Hensley, Lewis, 48
Hodges, George S., 265
Hopkins, E. J., 279
How, Bishop W. W., 275, 305
Hulton, Everard, 232
Huntingdon, Lady, 108
Hyne, L. G., 49

Ingemann, B. S., 149
Innocent III, Pope, 104
Irons, Joseph, 43

INDEX OF AUTHORS' NAMES 329

Irons, Wm. Josiah, 43, 176
Jenner, H. L., 294
John, St., of Damascus, 125, 136
Joseph, St., of the Studium, 154

Keble, John, 4, 150, 195
Kelly, Thomas, 57, 113, 114,
Ken, Bishop, 1, 22, 38 [139
Kethe, William, 311
King, Joseph, 46
Knecht, J. H., 154
Kocher, Conrad, 64

Lowry, Robert, 269
Lynch, Thomas Toke, 299
Lyte, Henry Francis, 11, 212, 238
Luke (née Thompson), Jemima, 260
Lundie, R. H., 81
Luther, Martin, 46, 68

Maclagan, Archbishop, 115, 169, 250
Madan, Martin, 50, 52
Mant, Bishop Richard, 325
Marriott, John, 209
Mason, Lowell, 100
Matheson, George, 296
Maude, Mary F., 170
Maurice, P., 246
Mercer, William, 67
Midlane, Albert, 267
Miller, E., 164

Miller, Emily Huntington, 277
Milman, Arthur, 112
Milman, Dean H. H., 71, 110, 237
Monk, W. H., 16, 47, 109, 160, 180, 251
Monsell, John S. B., 320
Montgomery, James, 8, 72, 174, 209
Moultrie, Gerald, 150, 231

Neale, John Mason, 22, 27, 30, 70, 121, 125, 129, 133, 136, 143, 149, 154, 176, 220, 293
Newman, John Henry, 23, 49, 87, 89
Newton, John, 316

Oakeley, Frederick, 65
Oakeley, Sir Herbert, 6
Ouseley, Sir F. A. G., 109

Palestrina, 105, 122
Palmer, Ray, 99, 174
Peace, A. L., 296
Perronet, Edward, 132
Plumtre, Dean E. H., 159
Pollock, Thomas B., 117
Pott, Francis, 85, 121
Potter, Thomas J., 155
Prudentius, 69.
Prynne, George Rundle, 270

Redhead, Richard, 237
Ringwaldt, Bartholomaus, 45

INDEX OF AUTHORS' NAMES

Rinkart, Martin, 302
Ritter, Paul, 7
Rossini, 105
Sankey, Ira D., 219, 233, 256, 290
Schaff, Phillip, 138
Scheffler, Johann, 307
Schenk, H. T., 247
Schulz, J. A. P., 244
Scott, Sir Walter, 40
Sears, Edmund Hamilton, 68
Sedding, Edmund, 294
Selbourne, Lord, 63
Shirley, Walter, 107
Shrubsole, William, 132
Smart, Sir Henry, 134, 143, 321
Stainer, Sir John, 241, 251
Stanley, Arthur Penrhyn, 138, 286
Stone, Samuel John, 151, 187, 202
Sullivan, Sir Arthur, 124, 128, 145, 234, 245
Summers, T. O., 266

Tate, Nahum, 64
Temple, Dr., 151
Tennyson, Lord, 188, 238
Threlfall, Jeannette, 284
Thring, Godfrey, 21, 30, 35, 46, 116, 130, 157, 199, 226, 244
Toplady, Augustus Montagu, 92
Tuttiett, Lawrence, 54, 79
Twells, Henry, 27, 35

Wainwright, John, 60
Warner, Anna B., 263
Watts, Isaac, 77, 100, 189, 205, 210, 294
Weisse, Michael, 127
Wesley, Charles, 8, 50, 56, 57, 83, 101, 120, 130, 156, 191, 251, 255
Wesley, John, 51, 57, 84, 307
Wesley, Samuel, 57
White, Henry Kirke, 182
Whitefield, George, 51
Whiting, William, 223
Williams, Isaac, 176
Williams, William, 325
Winkworth, Catherine, 68, 127, 247, 302
Wordsworth, Bishop Christopher, 31, 128, 248, 284

Xavier, Francis, 106

Young, Andrew, 259

Index of First Lines of Hymns

A few more years shall roll, 80
Abide with me, 11, 14, 17
Above the clear blue sky, 279
According to Thy gracious word, 174
Adeste Fideles, 65
All hail the power of Jesus' Name, 131
All people that on earth do dwell, 311
All praise to Thee, my God, this night, 1
All things bright and beautiful, 282
Alleluia, alleluia, Hearts to heaven and voices raise, 128
Alleluia ! Sing to Jesus, 139
Almighty Father, hear our cry, 227
And now, belovèd, Lord, Thy Soul resigning, 90
And now, O Father, mindful of the love, 169
And now the wants are told, 34

Angel voices ever singing, 122
Angels from the realms of glory, 72
As pants the hart for cooling streams, 65
As with gladness men of old, 63 [28
At even ere the sun was set,
At the Cross her station keeping, 104, 110
At the Lamb's high feast we sing, 135
Awake, my soul, and with the sun, 1, 38

Be present at our table, Lord, 52
Behold the glories of the Lamb, 78
Bound upon the accursèd Tree, 111
Brief life is here our portion, 196, 293
Bright the vision that delighted, 325
Brightest and best of the sons of the morning, 70

INDEX OF FIRST LINES OF HYMNS

Brightly gleams our banner, 155
Brother, thou art gone before us, 234

Calm on the listening ear of night, 68
Christ is risen! Christ is risen! 122, 284
Christ the Lord is risen again, 127
Christ, who once amongst us, 241, 290
Christ, Whose glory fills the skies, 8
Christians, awake, salute the happy morn, 58
Come, praise your Lord and Saviour, 275
Come, sing with holy gladness, 291
Come unto Me, ye weary, 140, 244, 322
Come, ye faithful, raise the strain, 125
Come, ye thankful people, come, 241
Crown Him with many crowns, 129

Day by day the little daisy, 283
Day of wrath, O day of mourning, 42
Days and moments quickly flying, 81
Dies Irae, 40

Eternal Father, strong to save, 223
Eternal Light, Eternal Light, 267, 297

Fair waved the golden corn, 284
Far off our brethren's cry, 204
Father, before Thy throne of light, 75
Father, let me dedicate, 79
Father, while the shadows fall, 278
Fierce raged the tempest on the deep, 226
Fierce was the wild billow, 220
Fight the good fight with all thy might, 320
Flow fast, my tears, the cause is great, 108
For thee, Oh dear, dear country, 293
Forth in Thy name, O Lord, I go, 101
Forward! be our watchword, 141, 241
From Greenland's icy mountains, 200
From heaven above to earth I come, 68

Gentle Jesus, meek and mild, 255
Give us the wings of faith to rise, 210

INDEX OF FIRST LINES OF HYMNS 333

Glorious things of Thee are spoken, 316
Go to dark Gethsemane, 109
God moves in a mysterious way, 98
God of mercy, God of Grace, 212
God of the living, in whose eyes, 231
God the Father, great and holy, 75
Golden harps are sounding, 274
Gracious Spirit, dwell with me, 299
Great God, what do I see and hear? 45
Guide me, O Thou great Jehovah, 325

Hail the day that sees Him rise, 156
Hark! a joyful voice is thrilling, 49
Hark! a thrilling voice is sounding, 49
Hark! in the presence of our God, 269
Hark! my soul, it is the Lord, 180
Hark, the glad sound, the Saviour comes, 44
Hark! the herald angels sing, 56, 58, 131
Hark! the sound of holy voices, 247

He is gone — beyond the skies, 138
Here, Lord, we offer Thee all that is fairest, 245
Holy, holy, holy, Lord God Almighty, 71, 301
Hosanna! loud hosanna, 284, 292
Hosanna, we sing like the children dear, 264
How sweet the Name of Jesus sounds, 316
How welcome was the call, 197
Hues of the rich unfolding morn, 4

I heard the Voice of Jesus say, 309, 322
I love to hear the story, 277
I think when I read that sweet story of old, 260
I was a wandering sheep, 310
I'll praise my Maker while I've breath, 295
In the Cross of Christ I glory, 324
In the field with their flocks abiding, 74
In the name of our Salvation, 149
It came upon the midnight clear, 68
It is a thing most wonderful, 276
"It is finished!" 116

Jerusalem the Golden, 253, 293
Jesu, Lover of my soul, 58, 92, 131, 191
Jesu, meek and gentle, 271
Jesu, meek and lowly, 177
Jesu, my Lord, my God, my all, 176
Jesus, blessèd Saviour, 84
Jesus Christ is risen to-day, 119
Jesus lives! no longer now, 124, 225
Jesus loves me, this I know, 263
Jesus shall reign where'er the sun, 205
Just as I am, without one plea, 215, 239

King of Saints, to whom the number, 205

Lead, kindly light, 23, 88
Lead us, Heavenly Father, lead us, 22
Lift up your heads, eternal gates, 122
Lift up your heads, ye gates of brass, 209
Light's abode, celestial Salem, 133
Lo! God is here, let us adore, 307
Lo! He comes with clouds descending, 45, 50, 54, 132
Lo, to us a child is born, 286

Lord! her watch Thy Church is keeping, 204
Lord, it belongs not to my care, 324
Lord of Glory, who has bought us, 91
Lord of the harvest, it is right and meet, 204

March, march onward, soldiers true, 160
My faith looks up to Thee, 98, 174
My God and Father! while I stray, 239
My God, and is Thy Table spread, 162
My God, I love Thee, not because, 106
My spirit longeth for Thee, 62

Nearer, my God, to Thee, 313
New every morning is the love, 4
Not all the blood of beasts, 100
Now I have found the ground wherein, 308
Now, my tongue, the mystery telling, 176
Now thank we all our God, 302
Now that the daylight fills the skies, 30
Now the day is over, 288
Now the labourer's task is o'er, 230

INDEX OF FIRST LINES OF HYMNS 335

O Bread to Pilgrims given, 174
O come, all ye faithful, 65, 67
O Deus ego amo Te, 106
O Father, all-creating, 198
O food that weary pilgrims love, 173
O God, our help in ages past, 77, 79, 294
O God, whose all-creating might, 168
O happy band of pilgrims, 154
O happy day that fixed my choice, 187
O Jesu, I have promised, 185
O Jesu, Lord of Heavenly grace, 37
O Jesu, Thou art standing, 305
O little child, lie still and sleep, 264
O Love that wilt not let me go, 296
O perfect Love, all human thought transcending, 124, 193.
O praise the Lord our God, 161
O quickly come, dread Judge of all, 54
O Strength and Stay, 194
O Thou before whose Presence, 187
O timely happy, timely wise, 4
O worship the Lord in the beauty of holiness, 320
Of the Father's love begotten, 69
Oft in sorrow, oft in woe, 182
Oh, what the joy, 225
Once in royal David's city, 282
Once, only once, and once for all, 169
Onward, Christian soldiers, 144
Our blest Redeemer, ere He breathed, 43, 310

Pange, lingua gloriosi Corporis mysterium, 176
Peace, perfect peace, in this dark world of sin, 304
Pleasant are Thy courts above, 213
Praise to the Holiest in the height, 87

Rejoice, the Lord is King, 130
Rejoice, ye pure in heart, 159
Rock of ages, cleft for me, 89, 91, 237
Round the Lord in glory seated, 325

Safe in the arms of Jesus, 289
Saviour, again to Thy dear Name we raise, 17, 109
Saviour, blessèd Saviour, 157

INDEX OF FIRST LINES OF HYMNS

Saviour, breathe an evening blessing, 20
Saviour, sprinkle many nations, 211
Shall we gather at the river? 269
Sleep on, belovèd, 232
Sol praeceps rapitur, proxima nox adest, 26
Son of God, Eternal Word, 31
Splendor paternae gloriae, 37
Stabat Mater Dolorosa, 103
Sun of my soul, 4, 6, 17
Sunset and evening star, 238
Sweet Saviour, bless us ere we go, 9
Sweet the moments, rich in blessing, 107

Tell me the old, old story, 219, 256
Ten thousand times ten thousand, 249
That day of wrath, that dreadful day, 41
The Church's one Foundation, 151
The day is gently sinking to a close, 31
The day is past and over, 22
The Day of Resurrection, 136
The daylight fades, 266
The foe behind, the deep before, 129
The Head that once was crowned with thorns, 139
The King of Love my Shepherd is, 178
The morning bright, 266
The radiant morn hath pass'd away, 35
The Royal Banners forward go, 143
The saints of God! their conflicts past, 250
The Son of God goes forth to war, 252
The Sower went forth sowing, 241
The strife is o'er, the battle done, 121
The sun is sinking fast, 26
The Voice that breathed o'er Eden, 195
The world is very evil, 293
The year is gone beyond recall, 85
Thee will I love, my Strength, my Tower, 307
There is a fountain filled with blood, 94
There's a Friend for little children, 267
There is a green hill far away, 280
There is a happy land, far, far away, 259
There is a land of pure delight, 102
There were ninety and nine that safely lay, 218
Thine for ever! God of love, 156, 170

INDEX OF FIRST LINES OF HYMNS 337

Thou art coming, O my Saviour, 46
Thou hidden love of God, 251
Thou Judge of quick and dead, 83
Thou, Whose Almighty Word, 208
Through all the changing scenes of life, 65
Through midnight gloom from Macedon, 203
Through the night of doubt and sorrow, 147
Thy kingdom come, O God, 48
Thy life was given for me, 306
Thy way, not mine, O Lord, 310
'Tis gone, that bright and orbèd blaze, 5
To Thee, O Lord, our hearts we raise, 244

Vox clara ecce intonat, 49

Waken, Christian children, 73
We are but little children weak, 284
We are soldiers of Christ, Who is mighty to save, 117
We have not known Thee as we ought, 117
We love the place, O God, 318
We plough the fields and scatter, 243
We pray Thee, Heavenly Father, 164
We sing the praise of Him Who died, 114
We thank Thee, Lord, for this our food, 52
When I survey the wondrous Cross, 77, 164, 189
When our heads are bow'd with woe, 237
When through the torn sail the wild tempest is streaming, 229
While shepherds watched their flocks by night, 64, 210
Who are these like stars appearing? 246

Ye faithful, approach ye, 66
Ye servants of the Lord, 44